Networks on Networks
(Second Edition)

Role of connectivity in physics of geobiology and geochemistry

Online at: https://doi.org/10.1088/978-0-7503-5698-5

Networks on Networks (Second Edition)

Role of connectivity in physics of geobiology and geochemistry

Allen G Hunt

Department of Physics, Wright State University, Dayton, OH, USA

Muhammad Sahimi

*Department of Chemical Engineering and Materials Science,
University of Southern California, Los Angeles, CA, USA*

IOP Publishing, Bristol, UK

ISBN   978-0-7503-5698-5 (ebook)
ISBN   978-0-7503-5696-1 (print)
ISBN   978-0-7503-5699-2 (myPrint)
ISBN   978-0-7503-5695-4 (mobi)

DOI   10.1088/978-0-7503-5698-5

Version: 20241201

IOP ebooks

British Library Cataloguing-in-Publication Data: A catalogue record for this book is available from the British Library.

Published by IOP Publishing, wholly owned by The Institute of Physics, London

IOP Publishing, No.2 The Distillery, Glassfields, Avon Street, Bristol, BS2 0GR, UK

US Office: IOP Publishing, Inc., 190 North Independence Mall West, Suite 601, Philadelphia, PA 19106, USA

To those who fight against tyranny, colonialism, and apartheid in the Middle East.

Contents

Preface

Foreword to the first edition by Bruce Milne

Ecology, in the broadest sense that examines relationships within the environment, is necessarily an eclectic domain of interest spanning from quantum phenomena at the reaction centers of photosynthesis, upward to earth orbital fluctuations that create glacial cycles, and further beyond as rare asteroid impacts cause great extinction events. Ecology is of both physics and biology and of both coupled together. Academic ecology is concerned primarily with two things: (1) regulation of energy capture and flow through ecosystems as constitutes biological organization, and (2) regulation, maintenance, and functionality of biotic diversity. The two are connected thermodynamically through biophysical realities and framed by evolutionary theory. A landscape perspective embraces complexities introduced by temporal and spatial factors that affect storage and transport of organisms, matter, information, and energy. Ecological theory seeks to explain more patterns from fewer particulars by building upon axiomatic constructs that reduce uncertainty. Accordingly, the last few decades have seen great advances in biology, ecology, hydrology, soil physics and related disciplines due to applications of several closely related technical tools and theories. These include allometric scaling, dimensional analysis, fractals, and percolation theory. At root, these pivot on a common axiomatic notion, which is the similitude $y = cx\alpha$, where by virtue of operational precedence the exponent reconciles one side of the equation with the other. We could say this construct affords a scientific conceit whereby everything we cannot say about the system is forced to live in the exponent whose value can be estimated from data, or better, obtained from theory. Closely allied is the Buckingham Pi theorem which promises simplification of technical description to a relatively small number of independent dimensions, thereby reducing the clutter of redundant variables. The trove of ever-more readily accessible scientific literature is a kind of time machine, with the present volume by Allen Hunt and Stefano Manzoni a milestone worthy of careful study. From an arbitrary starting date a student can venture back and forth in the literature to reconstruct trajectories of questions, theoretical approaches, facts, tests of theory, and the opening of new routes of enquiry. Yet, standing at the event horizon we occupy at this and every moment, with its infinite set of possible discoveries, tribulations, and dead ends ahead, the community struggles to sniff out the most fertile routes while facing the nagging trade-offs of human capital, technology, disciplinary boundaries, and time itself. At this same moment we can marvel at where we have arrived, collectively and solitarily, to reap intellectual satisfaction. We can feel scientific awe, thanks to the cogent curiosity and relentless pursuit of technical mastery, ultimately to find robust and even universal patterns that bridge between disparate physical phenomena, biological species, and environmental conditions. Indeed, to those familiar with the culture of physics, demonstrations of universal behavior as shown in this volume is deeply rewarding evidence of profound consistency between theory and the world.

The present work is as much a treatise on percolation theory that characterizes myriad phenomena in soils, vegetation, and crops as it is a demonstration of how to define, think about, and then solve problems. The authors' approach is informed by an aesthetic implicit in percolation theory itself, namely that something very relevant happens at a unit scale, perhaps a point on a lattice, a root tip, soil pore, or bacterium. They demonstrate that biologically relevant transport in complex media is largely one of advection through tortuous networks of accessible conduits, rather than a story of diffusion, which is too slow to satisfy biological demand for energy and nutrients. The narrative arc of the book sustains a somewhat tangential curiosity about a plausible basis for the Gaia hypothesis, which, to be supported, needs a still incomplete biophysical basis and an as-yet articulated evolutionary understanding, especially as concerns the question of natural selection among indiscrete and symbiotic taxonomic entities. Along these lines this volume echoes the scope and imagination of Vladimir I Vernadsky, whose essays in the 1920s on geochemistry and the biosphere introduced concepts of geological constraint on biota, geochemical potential of life, life as means of solar energy dissipation, geochemical invariance in the face of biological innovation, and transport speeds of bacteria due to population expansion. Percolation theory as applied within addresses what is essentially vertical transport from the soil pore upward through stomata across that thin layer of the terrestrial biosphere between bedrock and tree top, the so-called critical zone. The authors' foundational contributions will find interest among climate change modelers, landscape ecologists, and ecohydrologists who may reasonably entertain percolation theory for networks in two-dimensional real space, as Hunt shows elsewhere for Hack's law of stream length, and as my colleagues and I show with spatial phase transitions of piñon–juniper woodland ecotones in New Mexico and again as a Horton law of riparian vegetation diversity in a network of 13 000 streams in Kansas. To follow Hunt and Manzoni's protocol means making an inventory of knowledge gaps that exist in the context of spatially extensive landscapes on terrain as it modulates atmospheric flows. There is a need to list constraints that pertain to landscapes, analogous to those in the vadose zone such as hydraulic conductivity, critical values for porosity, and moisture content at which percolation dominates. The authors' protocol could serve the interests of many natural scientists. In summary, the striking relevance of percolation theory to seemingly disparate phenomena testifies to its power to capture deeply salient behaviors that govern the persistence of living systems. The current work is both an outline of critical thought about extremely challenging problems in the physical and natural sciences and simultaneously an inspiration to current and future investigators from a variety of disciplines that actually share much in common.

Bruce T Milne,
University of New Mexico, September 2015

Preface to the second edition

Our guides to writing the first edition of *Networks on Networks* were that the biological networks that describe the system behavior of the terrestrial biosphere actually function over a physical network, and that the optimal means of treating the physical properties of this network relevant for biological and chemical processes would employ percolation theory, since this physical network consists of physical pathways that connect distinct physical locations. To lowest order, the soils and rocks can be divided into two parts: one, the pore space, which allows communication, and the second one, the grains, which do not. The physical properties of this network, including their scale dependence, determine not only what entities can communicate with each other, but also the speed by which they do. If we wish to know how fast soil forms, or vegetation grows, this must be connected to how fast fluids flow and solutes are transported over the network. Thus, some new fundamental guides for reasoning in this context were formulated, such as the best foundation for addressing the interaction between plants (directed networks) and soils (random networks) should surely be to treat the soil as the random network that it is. 'The physics of geobiology and geochemistry'—the subtitle of the first edition of this book—then, applied the geological maxim that the present is the key to the past to address what features of the biosphere and shallow geochemistry could, with this foundation, be seen in a new light.

This basis for investigation turned out to be far more productive than the authors suspected. With hindsight, it can be seen that the single most important advance noted in the first edition was the use of percolation scaling power-law relationships to describe the height growth of plants (and the equivalent increase in root radial extent), as well as the depth of chemical weathering and associated soil horizons. However, without a second edition, the promise of the first edition would not be fully realized, or addressed in a unified manner.

The second edition not only represents an updated version of the first edition, but also contains several new chapters that represent the research progress made since 2016, when the first edition was published. Dr Stefano Manzoni, who was the co-author of the first edition, wished, however, to take his research in a different direction and, therefore, did not participate in the preparation of the present edition. The loss of his expertise in intra-plant processes is missed in the present edition. Thus, one of us (AGH) invited the other (MS) to participate in preparing the second edition of the book. The two of us have had a long-term collaboration, going back 15 years, and have published many papers together, including several that will be described throughout this edition. It should also be noted here, as well, that the use of spatio-temporal scaling relationships for vegetation growth and soil formation formulated by AGH, the heart of the first edition, was based on the reading of publications of MS and his concept that flow over all disordered networks is dominated by the critical paths that are defined by percolation. As a consequence, the spatio-temporal scaling of solute transport and its relevance to soil formation should be described by (mostly) universal power laws and exponents of percolation. This is the point that will be emphasized throughout this edition.

But what is new goes far beyond refining existing calculations. In short, combining the scaling relationships in a way to maximize plant productivity allowed solution of the 'central problem of hydrology' (Horton's 1931 address to the American Geophysical Union regarding the establishment of the hydrology section).

The maximization of plant productivity had already in 1959 been introduced by Odum in *Fundamentals of Ecology* as a potential guide for plant evolution, but he had only considered the competing necessities within the plants for growth, as well as protection against water shortage. Our substitution of the limiting effects due to the water needs of the living entity of the soil itself reproduced the known properties of the water balance and its derivatives, the net primary productivity of plant ecosystems, and the elasticity of streamflow. The solution united once again time and space scales that span from the pore to the continental. And, it is easy to show that the entire method of solution was, in its broad outline, foreseen by the authors of the Eagleson report, 'Opportunities in the Hydrologic Sciences' (National Academy of Sciences, 1991), as can be seen by reading its pages 65–66, and confirmed also by one of the writers (Ed Waymire).

The second edition of our book documents how—quoting from 'Challenges and Opportunities in the Hydrologic Sciences' (National Academy of Sciences, 2012), the 20-year follow-up to the Eagleson report—'Water affects life, which influences water' is given physical form. The scaling relationships document how water affects life and the optimization then directs how life affects water in the determination of the evapotranspiration across climates. The substitution of the derived evapotranspiration back into the plant growth rates generates the observed net primary productivity. As stated in the Eagleson report (1991), the hydrologic cycle has become a 'template' for the carbon cycle.

Importantly, a similar loop is closed in ecology by the same research. The maximum in productivity is enhanced by plant diversity, whereas the result for the net primary productivity can now be used to predict plant species richness across climate zones, addressing a problem in ecology that goes back to Alexander von Humboldt, the so-called 'latitudinal gradient of species richness.' Further, in line with maximum entropy theory of ecology, which assigns an entropy to the species distribution and interprets the maximum entropy as a minimum in (Earth history-related) information, or equilibrium in speciation, it becomes possible to detect where, geographically, such information is depressing the species numbers.

Finally, we can now also report that the predicted time for co-evolution of vascular plants and their direct symbionts and consumers (bacteria and fungi) to spread across a continent, derived from the same plant growth scaling relationships, and given in the first edition as 80 Myr, is, in fact, very nearly the time it took for the planet to emerge from Paleozoic ice ages triggered by the plants' stepwise invasion of the land, each time 60 Myr. Thus, in fact, predictions of percolation theory in the soil do, at any rate, give the correct time scale for biosphere evolution as documented from climate.

The unification provided by these fundamental new results is the driving force for preparing a second edition. But there are many 'smaller-scale' interesting results that followed, too, such as the potential to predict stability against shallow landslides. We hope that the reader can enjoy the arc of the narrative in the new edition and use it to trigger additional advances in geochemistry and geobiology.

We should thank our collaborators that, over the years, have helped us gain better understanding of the problems that we discuss in this book, a list of whom is too long to be given here. But, we would like to particularly thank Dr Behzad Ghanbarian of

Kansas State University whose helping hands and intellect have always been there for us to rely on, and Stefano Manzoni, whose contributions to the chapters on water transport in plants and allometry still stand, providing connections with the later chapters as well as potential further basis for unification of concepts.

Allen G Hunt, Dayton, Ohio
Muhammad Sahimi, Los Angeles, California
May 2024

Preface to the first edition

Order from chaos is simultaneously a mantra of physics and a reality in biology. Physicist Norman Packard suggested that life developed and thrives at the edge of chaos. Questions remain, however, as to how much practical knowledge of biology can be traced to existing physical principles, and how much physics has to change in order to address the complexity of biology. Phil Anderson, a physics Nobel laureate, contributed to popularizing a new notion of the end of 'reductionism'. In this view, it is necessary to abandon the quest of reducing complex behavior to known physical results, and to identify emergent behaviors and principles. In the present book, however, we have sought physical rules that can underlie the behavior of biota as well as the geochemistry of soil development. We looked for fundamental principles, such as the dominance of water flow paths with the least cumulative resistance, that could maintain their relevance across a wide range of spatial and temporal scales, together with the appropriate description of solute transport associated with such flow paths. Thus, ultimately, we address both nutrient and water transport limitations of processes from chemical weathering to vascular plant growth. The physical principles guiding our effort are established in different, but related concepts and fields of research, so that in fact our book applies reductionist techniques guided by analogy. The fact that fundamental traits extend across biotic and abiotic processes, i.e. the same fluid flow rate is relevant to both, but that distinctions in topology of the connected paths lead to dramatic differences in growth rates, helps unite the study of these nominally different disciplines of geochemistry and geobiology within the same framework. It has been our goal in writing this book to share the excitement of learning, and one of the most exciting portions to us has been the ability to bring some order to the question of the extent to which soils can facilitate plant growth, and what limitations on plant sizes, metabolism, occurrence and correlations can be formulated thereby. While we bring order to the soil constraints on growth, we also generate some uncertainties in the scaling relationships of plant growth and metabolism. Although we have made a first attempt to incorporate edaphic constraints into allometric scaling, this is but an initial foray into the forest.

Allen G Hunt, Dayton, Ohio
Stefano Manzoni, Stockholm, Sweden

Author biographies

Allen G Hunt

Allen Hunt is University Professor in the Physics Department of Wright State University. He has published over 150 refereed papers and 5 books on subjects across ecology, geology/geophysics, chemistry, physics, and hydrology. He has been a Fulbright scholar and appeared in Who's Who Among American Teachers.

Muhammad Sahimi

Muhammad Sahimi is Professor of chemical engineering and materials science, and the NIOC Chair in petroleum engineering at the University of Southern California in Los Angeles. His research interests include flow, transport, reaction, sorption, and deformation in porous media, percolation theory, fracture and failure of heterogeneous materials and rock, and application of artificial intelligence to such problems. He has published over 400 papers and three books, is a Fellow of the American Institute of Chemical Engineers and the American Physical Society, and has received numerous research and teaching awards, including, among others, Humboldt Foundation Research Fellowship Award, the Khwarizmi International Award for Distinguished Achievements in Science, Life-time Achievements Award and Honorary Membership of the International Society for Porous Media, and the Kimberly-Clark Distinguished Lectureship Award.

IOP Publishing

Networks on Networks (Second Edition)
Role of connectivity in physics of geobiology and geochemistry
Allen G Hunt and Muhammad Sahimi

Chapter 1

Introduction

1.1 Background

Biogeochemistry is present to a great extent in the top few meters of Earth's terrestrial surface, which is what we refer to as soil. Throughout Earth's history, this layer has been eroded and deposited, or simply buried, hence creating a record of the history of life and of climates, of weathering of rocks and atmospheric chemistry, and erosion and the water cycle, which has been preserved. From the perspective of the recorded evidence of the history of life and its adaptation to, and alteration of, the planetary surface, the study of such paleosols and sedimentary deposits belongs to geobiology. In other aspects, the study of soils, both past and present, teaches us most profoundly the chemical, but also physical, alteration of whatever was laid in place before, either in bedrock or some sedimentary deposit, such as dunes, alluvium, landslide materials, or volcanic deposits. The chemical alteration belongs to the field of low-temperature geochemistry; to the extent that it is inextricably linked with biology, we use the term biogeochemistry to describe the science behind it. We will, however, see that it is really the flux of liquid water into and through soils that sets the rates of organic, as well as inorganic, processes, making hydrology a central part of the study of soils. Since it is within the soil and its plants that the fate of precipitation falling on Earth's surface is decided—whether to be evaporated back to the atmosphere, or to run-off back to the oceans—the soil has become a central topic of the study of hydrology as well. Professional organizations have certainly begun to recognize this fact: The soil physics division of the Soil Science Society of America has now added to its name 'and Hydrology,' while the National Science Foundation (NSF) of the United States supports 'Critical Zone Research' centered within, but not restricted to, soil, and the American Geophysical Union offers a session on 'Soil Swirl' topic at its annual meeting.

Soil is, therefore, a topic that attracts interest across many fields of science, including geochemistry, biology, ecology, hydrology, geomorphology, and geophysics, as well as chemical and civil engineering. Somewhat surprisingly, however,

doi:10.1088/978-0-7503-5698-5ch1

soil scientists have long maintained the integrity of separate disciplines, such as soil physics, soil chemistry, soil biology, etc. The historical soil science structure gives a nod to the traditional ranking in various branches of sciences, with physics at its foundation. This way of thinking was challenged in the latter part of the 20th Century with threats of a complete overthrow in the current century. Such challenges tend to originate in biology where emphasis is placed on individual distinctions at all levels of taxonomy, though countered by a rapidly growing interest in a systems approach. In spite of such currents and countercurrents, the fundamental thesis in this book is that understanding of the entire soil system must be based on a correct understanding of the soil physics, such as water flow and solute transport. Even here, however, the impact of the study of complexity is key (Sahimi and Hunt 2021).

Beginning with the pioneering work of Fatt (1956a, 1956b, 1956c), and continued throught the 1980s and 1990s, it became clear that the morphology of any porous medium, and in particular soil, can be mapped onto a network of interconnected pore throats—narrow channels through which fluids flow—and pore bodies—chambers in which most of the porosity of the pore space resides. In the modern theory of networks, the analogs of pore throats and pore bodies in a pore network are referred to as the bonds (or links) and sites (or nodes). Once this mapping is implemented, one can employ the vast arsenal of statistical physics of disordered media to describe various properties of soil. In particular, one can invoke the concepts of percolation theory (Stauffer and Aharony 1994, Sahimi 2023) in order to study various phenomena in soil, particularly fluid flow and solute transport. As we will explain later in this book, percolation theory predicts that there is a a minimum, or a critical, connectivity of the pore throats (bonds) or pore bodies (sites) above which a sample-spanning cluster of connected pore bodies and pore throats exists through which fluid flow and solute transport occur, whereas if the connectivity falls below the critical value, no macroscopic fluid flow and transport can take place. In addition, percolation theory predicts that most macroscopic properties of pore networks, representing soil, follow universal power laws near the connectivity threshold, with exponents that are largely independent of the details of soil morphology, and depend only on whether we view the soil under study as a two- or three-dimensional pore network.

Thus, for example, invoking modern statistical physics of disordered media, and in particular percolation theory, we will show that one can use the power laws that percolation theory provides us with to predict soil depth based on scale-up of solute transport past a single grain over durations of minutes to distances (and depths) of over 100 m, to time scales exceeding 100 million years, which represents a concrete example of the geological tenet of uniformitarianism that 'the present is a key to the past.' Such predictions, though they support the traditional hierarchy of soil science, do not uphold the traditional curriculum of soil physics. Ironically, it is the shortcomings of soil physics that prevent it from achieving foundational importance, which perhaps offers a reason as to why its role in soil science has diminished in recent decades.

At first, it may seem strange that physical laws that govern fluid flow and transport could be the basis for understanding soil, since soil is vastly different in many characteristics from a simple abiotic porous medium. It is dirty, it sticks together, and it has smells and colors. It changes with the weather and conditions. It is a hugely complex system, yet teeming with life, and evolving and adapting over time, physically, chemically, biologically. As such, the soil has been defined as the 'most complicated biomaterial on the planet' (Young and Crawford 2004, p 1634). The biological feedbacks in soil formation and the links between processes at various distinct spatial and time scales make the study of soil a daunting task. The differences between soils and engineered (nearly homogeneous) media are enormous, including organic content, chemical composition, porosity, flow and transport properties, conduction, and other phenomena. Situated in between are natural heterogeneous media (regolith), such as those freshly exposed at Earth's surface by such processes as landslides (Trustrum and de Rose 1988), glacial retreat (Mavris *et al* 2010), mining (Frouz *et al* 2008), or even treethrow (Borman *et al* 1995). In view of their heterogeneity, such media have more in common with soils, yet are not classified as such, owing to the lack of development of soil horizons, their minimal weathering characteristics, and low organic content. Nonetheless, the typical simplification to a study of abiotic media may not be as big a hindrance to understanding the real nature of soil, as is the common restriction to studying nearly uniform systems (or starting with a non-viable theory of soil physics adapted to homogeneous systems).

The health of the soil *in situ* is a foundation for civilization (Hillel 2004, 2005, Montgomery 2007). But studies of soil *in situ* are hampered by their complexity and variability. Studying soil in the laboratory has been nearly impossible: simply maintaining the porosity of a sample taken from a field and brought to a laboratory is difficult, and preserving the biota intact is an even much greater challenge. Nevertheless, real progress has been made. Exploiting the chemical warfare between various species of bacteria and against other microorganisms, the recent ability to culture 'living' interacting soils (Ling *et al* 2015), is expected to trigger a manifold increase in the rate at which antibiotics are discovered, converting soils, in the near future at least, to a biomedical research tool for preventing infection (Ling *et al* 2015), rather than something merely needing to be cleansed from a wound.

In recognition of its importance to the future of human life on Earth, the United Nations chose 2015 to be the Year of Soil. Soil is also a significant fraction of what has come to be known as the 'Critical Zone,' an outgrowth of the United States NSF-funded research. The Critical Zone includes the high permeability layer between the tops of the trees and bedrock, which is key to its relevance to human life, as well as to its classification, as we will see later in this book. In many ways, particularly insofar as the physical medium and the biological constituents interact, soil is probably the least understood of the components of the Critical Zone. But a clear understanding of the physics of soil is a necessary starting point for understanding its relevance to the Critical Zone.

Looking at the problem from a societal angle, soil, and indeed the entire Critical Zone, has long deserved attention due to their importance to civilization, which prospers and fails on the ability of the soil in its domain to sustain agriculture (Montgomery 2007). Beyond its direct relevance to human society due to its role in agriculture, storage of water, repository for wastes, and foundation for our buildings, soil is 'the crucible of life, a self-regulating, biophysical factory, acting like a composite living entity' (Hillel 2005). In fact, such a somewhat subdued description recalls Lovelock and Margulis' (1974) Gaia hypothesis that argued that Earth's surface does not merely act like a living entity, but *is* one. This conclusion developed based on the geological evidence that Earth's skin managed to modulate the atmospheric composition to its own advantage, and to quite extraordinary disequilibria. In particular, the concentrations of O_2 and CH_4 constitute a violation of the rules of equilibrium chemistry by *thirty* orders of magnitude, while gaseous nitrogen should not be present at all. As stated by the author, 'so great is the disequilibrium among the gases of Earth's atmosphere that it tends to a combustible mixture, whereas the gases of Mars and Venus are close to equilibrium and more like combustion products.' While the Gaia hypothesis has generated more controversy than agreement, the line of reasoning developed in this book bears further on that discussion. We will return to the hypothesis in the last chapter of this book.

Starting in the 1980s, but accelerating recently, scientists have devoted more general attention to the biogeochemical processes in soil. The additional attention is a consequence of the relevance of soil to changes in global nutrient cycling, especially carbon, nitrogen, and phosphorus. Weathering of silicate minerals, a process essential to the formation of soil, accounts for the greatest sink for atmospheric carbon over time scales of millennia and longer, while the activity of organisms in the soil plays a similar role over shorter time scales. We will see that the distinction can be understood if one understands, (1) the relative magnitudes of carbon storage repositories, and (2) the fundamental transport properties of soil. The relevance of soil formation and its evolution to the global carbon cycle binds the issues of climate change and secondary effects and feedbacks on the biosphere, to the fate of Earth's soils. The sources, sinks, and pathways of carbon and nutrients are mapped out in biological networks that incorporate nodes and links, as well as transport across the boundaries, to the hydrosphere and atmosphere. Many of the important properties of such networks are purely biochemical in nature, such as which organisms communicate with which other organisms; which are at war with each other; which act symbiotically, or do so parasitically. The argument advanced in this book is, however, that the communications between the biological entities, as well as their growth, are realized over physical networks underlying the biota. A similar connection between ecological processes and physical structure has been recognized and studied in the case of rivers, along which strong ecological patterns develop and self-organize (Bertuzzo *et al* 2007, Muneepeerakul *et al* 2008). Similar developments at smaller scales relevant to soil science are missing. While a complete understanding of soil will necessitate full understanding of the couplings and feedbacks (which change even the physical characteristics of the medium) and beyond, to address the influence of various spatial and temporal scales on each other, and so forth, our perspective is that, unless the fundamental physics of the transport processes is understood, the basis for understanding the remaining complexity would be missing.

1.2 Fundamental scaling relationships: advection versus diffusion

Soil, the porous medium that is the skin of Earth, is highly permeable. What does this mean in practice? It implies that flow of water through this medium can be relatively rapid. Why is this relevant? Because, for example, speed is an important component of nutrient and water transport within soils and from soil to plants. In turn, nutrient and water supply to plants mediates the growth of the canopy towards sunlight, or of the root tips and fungal hyphae, thereby affecting species composition and ecosystem development. The entire Critical Zone features constraints on flow due to soil heterogeneity and structure. Such constraints operate on the flow of water and the solutes that it contains. Within the vegetation itself, the flow is less randomly heterogeneous than it is structured. But within soil, the flow is heavily influenced by the small-scale disorder of the medium, such as the variations in the sizes of the pores. The functions of vascular plants are closely tied to their ability to extract not merely water, but also nutrients, from the soil, a process that is driven by evapotranspiration at the top, and involves flow of water through the roots and xylem to the stomata of the leaves (Cruiziat *et al* 2002, Manzoni *et al* 2013, Larcher 2003). Nutrient-carrying water must, however, flow through the soil to reach the roots, or the roots must grow to reach the source of water and nutrients, or some combination of the two. Thus, since the soil and the organism are in series, the properties of soil impose constraints on, as well as provide opportunities for, plants to access water and nutrients. Such possibilities include growth towards nutrient sources, as well as the determination of the internal architecture governing flow.

The rate at which water flows in the Critical Zone is specified by the permeability and a pressure gradient generated by the organism (osmotic pressure is a well-known example). Alternatively, gravity or evapo(transpi)ration can generate pressure gradients that control vertical flow. The cohesion-tension theory explains water transport in plant tissues (xylem) as its flow under tension, following the water potential gradient from the soil to the leaves and atmosphere (Cruiziat *et al* 2002, Tyree 1997). Solutes so transported by moving water are said to be 'advected,' and their transport is 'advective.' Nutrient transport is predominantly passive, carried along by the flowing water, but, particularly at smaller scales, is partly diffusive as well and is driven by (for example, concentration) gradients actively established and maintained by plants (Larcher 2003). Nevertheless, owing to the necessity for plants to access nutrients within the soil, its solute transport properties are relevant to the transport of nutrients within plants as well.

Therefore, processes for locating, accessing, taking up, and transporting nutrients involve the fundamental physical processes of advection (by flowing water) and diffusion. How is diffusion distinct from advection? Consider, first, advection. When, for example, the flow direction is from an area with greater concentration of any particular solute to an area with lower concentration, the result is net transport of solute in the direction of the flow. Diffusion, based on stochastic molecular motion, tends to equalize solute concentrations by more often transporting the solute from a region of high concentration to low concentration than the reverse (Fick's first law of diffusion), and proceeds whether the fluid is flowing or not. How these processes are modeled is ultimately the key to our understanding of the roles they play and, thereby, also to our understanding of the constraints that soil imposes on the growth of vascular plants.

In order to understand the limitations on plant growth that stem from the behavior of the soil, one must understand properly solute transport within the soil. Solute transport by water flowing through a porous medium is typically modeled using the advection–diffusion equation (ADE) (Bear 1972, Sahimi 2011):

$$\frac{\partial C}{\partial t} + v \cdot \nabla C = D_h \nabla^2 C. \tag{1.1}$$

Here, C is the concentration of a solute at a point in space and time, so that the left side of the equation is the rate of its change at an arbitrary point in space, while v is the velocity of the fluid, and D_h is a 'hydrodynamic dispersion' coefficient, which represents an *effective* diffusion coefficient that combines both advection and molecular diffusion. Thus, the term on the right side is used to model a diffusion-like process that ultimately produces Gaussian spreading, while the second term on the left side represents the advective transport. The conceptual basis for using this equation in disordered porous media, such as soils, is flawed, since it is the variable fluid advection rates in space that produce the majority of the solute spreading, and this flow variability has no connection with microscopic molecular diffusion, nor is it described by an equation of the form of equation (1.1).

There are also a number of practical problems with equation (1.1), as documented by various authors over the last half a century (Aronofsky and Heller 1957, Scheidegger 1959, Lallemand-Barres and Peaudecerf 1978, Silliman and Simpson 1987, Sahimi 1987, Sahimi and Imdakm 1988, Arya *et al* 1988, Neuman 1990, Gelhar *et al* 1992, Berkowitz and Scher 1995, Xu and Eckstein 1995, Margolin and Berkowitz 2000, Cortis and Berkowitz 2004). Specifically, such problems include the fact that equation (1.1) does not predict the increase in the dispersivity, i.e., the ratio D_h/v, with experiments' scale that is observed on scales from microns (Baumann *et al* 2010) to kilometers (Gelhar *et al* 1992); it does not predict the long-tailed solute distributions that are the rule rather than the exception (Coats and Smith 1964, Berkowitz and Scher 1995, Margolin and Berkowitz 2000, Cortis and Berkowitz 2004, Cushman and O'Malley 2015); it actually underpredicts solute arrivals at both short and long times (Cortis and Berkowitz 2004), and it does not allow for the possibility that the solute velocity can be less than that of the fluid (Sahimi 1987), a point which has been verified specifically for porous media (Roberts *et al* 1986, Berkowitz *et al* 2000, Bromley and Hinz 2004, Koestel *et al* 2012). Such time-dependent velocities are also known to be required by many of the heavy-tailed distributions (Scher *et al* 1991, Sahimi 2012) detected in porous media, and which are now recognized as being the norm (Cortis and Berkowitz 2004), rather than an anomaly (Cushman and O'Malley 2015).

The practical problems with the ADE also includes the hydrodynamic dispersion term. The dispersion coefficient D_h does not describe diffusion, as realistic choices for its value depend on velocity and length scale (Sahimi 2011). Furthermore, it cannot be predicted (Gelhar *et al* 1992) based on the ADE and, at large scales, the molecular diffusion coefficient D is not even important. For the present purposes, however, the chief defect of the ADE is that it predicts a solute velocity equal to that of the fluid, whereas it will turn out that it is precisely the distinction between the solute and fluid velocities that makes it possible to determine the effects of solute transport limitations on soil production and vascular plant growth, as well as what the chief advantage of root topology is. Nevertheless, the theoretical and experimentally observed proportionality of observed solute velocities to the flow velocity (Maher 2010, Molins *et al* 2012, Salehikhoo *et al* 2013, Sahimi 2012, Hunt *et al* 2015) reinforces the competitive advantage of high flow velocities, implying that, (1) strategies to enhance water fluxes tend also to optimize the access to nutrients, and (2) the theoretical predictions that make the solute fluxes proportional to the water flux allows, in principle, simultaneous optimization of both.

One consequence of the rapid movement of water in the Critical Zone is, therefore, that a wide range of solutes and single-celled living organisms can be transported rapidly through the medium. Transport of solutes into and through living organisms is necessary for life: nutrients and carbon sources must be delivered to cells, and waste products removed. In order to promote vigorous growth, or multiplication, of organisms, transport should be relatively rapid. But solute transport through complex, natural porous media tends to be slow. Solutes, which are carried along passively by flowing fluids, travel distances proportional to the fluid flow rate (Molins *et al* 2012, Maher 2010, Salehikhoo *et al* 2013, Hunt *et al* 2015) although their actual movement is typically slower than the fluid (Hunt *et al* 2015). The fluid flows a distance that is linear in time. Understanding the properties of solute transport within the soil is relevant for understanding the structure of vascular plant roots, whose architecture serves not merely to maximize surface area for adsorption, but also to optimize the rate at which the nutrients are accessed.

An understanding of the relevance of such fluid flow to life is further facilitated by considering the general structure of transport in living organisms, both plant and animal. Single-celled organisms (as well as some small or non-vascular plants) survive primarily by utilizing the process of diffusion (Koch 1990). Diffusion is a remarkably efficient process at sufficiently small length scales. Simply put, one can scale diffusion lengths, x, and times, t, as $D \sim x^2/t$, where D is the diffusion coefficient of the molecular species in the medium of interest. This means that the diffusion distance x increases as the square root of the time, whereas fluid advection distances increase linearly with the time. The advection process in plants is regulated chiefly, though not exclusively, by transpiration (although this is a unidirectional process), a process that requires no direct energetic input from the organism. At short enough time or length scales, diffusion is faster than a linear process—note the infinite slope of the square root at $t = 0$—but with increasing time scales, it must, at some crossover distance, become slower than linear; see figure 1.1.

Thus, depending on achievable flow rates, advection must become a more efficient means of transporting solutes beyond some distance. Against the backdrop of the vast range of length scales relevent to living organisms, ranging from nanometers to kilometers, one can approximate the crossover distance as being

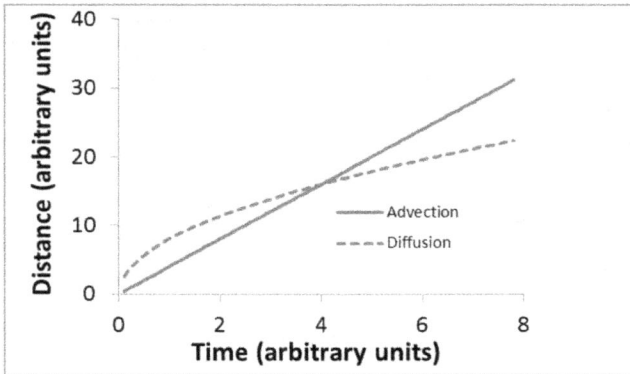

Figure 1.1. (Homogeneous) advection as compared with (homogeneous) diffusive transport. Such a comparison is characterized by the Péclet number (Saffman 1959, Pfannkuch 1963), as long as the relevant length scale is restricted to a single, or few, pores. Upscaled solute transport by advection does not follow a simple linear dependence in time.

roughly the size of a single cell, a distance measured in microns. It turns out that there is a similar crossover distance in soils, on the order of several pores, a distance that is typically about tens of microns. Dendritic circulation systems, unusual in the abiotic subsurface, but typical of vascular plants, provide the architecture for optimal transport from large to small scales and vice versa, while the crossover scale, somewhat larger than a single cell, distinguishes between the dominant processes.

Animals also have a circulatory system. In lung alveoli, diffusive exchange of carbon dioxide and oxygen with the atmosphere occurs at small scales. The blood removes the carbon dioxide and supplies the oxygen to the individual cells through transport, which is mediated by the pressure differences generated by the beating heart, whereas the advective exchange of the same gases with the atmosphere is done by the contraction of the diaphragm. The dendritic structure of both respiratory and circulatory systems has evolved to optimize the exchange, but the crossover scale can be calculated accurately using the constructal theory (Bejan 1997), and the contrast in the permeabilities of the membranes and the capillaries. The comparison between plant and animal strategies may help to inform an analogous discussion of the soil itself. According to the constructal theory, (1) the generation of images of patterns and rhythms in nature is a phenomenon of physics, and (ii) the phenomenon follows that principle—the constructal law—according to which 'for a finite-size flow system to persist in time—to live—it must evolve such that it provides greater and greater access to the currents that flow through it.'

A result of fundamental importance is that there exists in soils a similar crossover in efficiencies, but apparently without such large-scale organized hierarchical flow networks, aside from the individual multi-celled organisms. Recall that typical diffusion coefficients of CO_2 in water and in HCO_3^- at room temperature, which are the constituents in silicate weathering reactions, are, respectively, 1.8×10^{-9} m^2 s^{-1} (Lu et al 2013) and 1.19×10^{-9} m^2 s^{-1} (Lide 1994). In soils, with porosity of 0.3 or

0.4 (and a diffusion coefficient proportional to approximately the square of the porosity), this translates to effective diffusion coefficients closer to 10^{-10} m^2 s^{-1}. In fact, many relevant substances typically have diffusion coefficients closer to 10^{-11} m^2 s^{-1} in porous media (Dane and Topp 2002), and such solutes include nitrate and glucose (Watt *et al* 2006).

A measure of the competition between advection versus diffusion for transporting materials through a single pore is the Péclet number, which is the ratio of the diffusion and the advection time scales. When the diffusion time is large compared to the advection time, advection is faster enough to dominate the transport process. For typical pore-scale vertical flow velocities (Blöschl and Sivapalan 1995), the flow velocity v in unconsolidated porous media is about 10^{-6} m s^{-1}, while the typical pore sizes are roughly 10 microns. Thus, the ratio of the two time scales—the Péclet number Pe—is Pe $= xv/D$, which is very nearly one for many of the relevant solutes. With increase in length scale to a few pore lengths, Pe exceeds one and advection becomes dominant. In contrast, under unsaturated conditions predominant in the rooting zone of most ecosystems, the hydraulically active pores are smaller than those under saturated conditions, implying that the contribution by diffusion is significant. It is at least an interesting coincidence that the pore size required for the crossover from diffusion-dominated to advection-dominated transport over a length scale of a few pores is compatible with the typical range of pore sizes present, and needed by vascular plants as a source of water (0.2–30 μm) (Watt *et al* 2006). Moreover, the fact that the observed pore sizes produce the crossover from the diffusive to the advective transport is ultimately a consequence of the strength of Earth's gravitational field, which produces the requisite pressure gradient. This discussion reveals the need to have a theoretical analysis that can accurately predict both advection and diffusion processes.

Let us dig deeper into the competition between diffusion and advection processes and their roles. As pointed out earlier, it is important to note that the usually utilized advection–dispersion equation, equation (1.1), which is supposed to account simultaneously for the effects of both advection and diffusion processes, is in fact incapable of predicting observed solute transport data, even in the absence of complexities due to organic molecules and living organisms. In fact, Scheidegger (1959) had already noted the shortcoming of the ADE. Of course, use of continuum flow dynamics for macro-organisms, such as the human body, would never be considered, although it can be applied within a single blood vessel. However, the mismatch between the predictions of of the ADE and the data for solute transport in soils is different from its troubles within animals, for example.

The problems in soils stem from the inability of the ADE for addressing the complications of heterogeneity that soils contain, rather than its ignorance of the hierarchical structures. In each case, however, the problems relate to the connectivity of more highly permeable regions. In geological media, including soils, such preferential flow paths are associated with non-Gaussian solute distributions (Sahimi 1987, National Research Council 1996, Moreno and Tsang 1994, Bruderer-Weng *et al* 2004). Non-Gaussian distributions have been discussed in the magazine *Physics Today* as being consistent with nonlinear scaling of solute transport time with transport distance in amorphous semiconductors and polymers (Scher *et al* 1991). There, the results reported by Pfister (1974), Pfister and Griffiths (1977), Pfister and Scher (1978), Tiedje (1984) and Bos *et al* (1989) predicted typical transport distances of mobile charges proportional to a power of the time that is less than one. Such a power-law relationship between transport time and transport distance is critical to the purpose and content of this book. It is worthwhile to note that the same theory and the same parameters apply to the length-time scaling relationships in both electronic transport (Ghanbarian-Alavijeh *et al* 2012) and the solute transport in porous media (Hunt *et al* 2014).

The failure of the ADE can be avoided by developing pore network models that exploit the discrete connections of the pore space. Such models for solute transport were first developed in a series of papers by Sahimi *et al* (1982, 1983, 1986a, 1986b), and Sahimi and Imdakm (1988), and elaborated on later on by Sahimi (1987, 2012). The classical continuum models for diffusion in soils, water flow, and various forms of conduction are also unsuccessful, when the connectivity characteristics of the porous medium is ignored. Given that all the organisms and microorganisms, which utilize or inhabit the soil, must also operate within the same framework or space, and that the preferred means to address the connections between such organisms is by considering biological networks, it seems obvious that the entire framework of traditional soil physics, chemistry, and biology, predicated on continuum mechanics and its consequences, requires revision. The implications will extend through geochemistry to geobiology as well. The grounds for this conclusion are therefore: (1) a network model of the physical medium can be made compatible with network models of the interactions between the organisms, micro or macro, which inhabit the space, communicate, and grow and move within the medium; (2) the physical properties of the medium are better characterized utilizing the network models; (3) the transport of solutes, which may limit growth, or communication, cannot be attempted at large space and time scales without accounting for spatially-variable advection, whereas any analysis of advection using the traditional continuum models, based on the classical continuum hydrodynamics, is fatally flawed; (4) the advantages of root architecture do not even show up properly without the relevant solute transport theory, and (5) time scales of transport cannot, in general, be properly calculated otherwise.

The research necessary to construct such a framework has been in progress for the past two decades, and is summarized in this book.

1.3 Summary

The traditional means of studying soils is as a continuum and is, therefore, based on continuum-scale differential equations, i.e., the classical theories of hydrodynamics and transport processes. The approach produces, however, a fundamental mismatch in the study of the biological processes and organisms in physical media, such as soils, whose relationships are best characterized using networks. Flow and transport between and even within the biological entities proceed over physical networks too. The rates associated with biological processes are dependent on flow, diffusion, and transport rates through the pore space. Physical networks that represent the pore space are best studied using discritized equations, rather than continuum differential equations. It has already been demonstrated that, so long as the discretization is chosen appropriately (obviously on the scale of the pores), the experimental properties of the medium are reproduced much more accurately by the network models. Such difference equations, when applied to unsteady-state flow calculations, are equivalent to solving random impedance networks. In 2013, an article entitled, *What is Wrong with Soil Physics?* (Hunt *et al* 2013) described the consequences for the study of soil physics of using the capillary bundle models typically invoked with continuum methods, and the consequent failure to address the connectivity issues in networks. Additionally, the parameters describing the physical networks, such as connectivity (and its scale dependence), will have relevance to the biological networks. Finally, the conceptualization of the calculation of flow through soil in terms of a minimum in energy dissipation was developed in the context of random impedance networks, and is based on percolation theory.

Thus, this book is devoted to developing an understanding of porous media, processes and properties, in terms of networks. The applications are mainly to soil formation and vegetation growth limitations, the water cycle and drainage basin evolution. Such growth limitations bear on the traditionally poorly understood, but active, areas of research, such as the decline of forest productivity with age. The theoretical approach that is utilized, namely, the aforementioned percolation theory, provides predictions for flow, diffusion, and advective solute transport (Hunt *et al* 2014) that are more accurate than the continuum models. Percolation theory is well-suited for application to random networks, for which the original predictions of electrical conductivity were derived (Ambegaokar *et al* 1971, Pollak 1972). The two most important aspects of percolation theory exploited in this book are, (1) its ability for identifying the characteristic rates and optimal paths, and (2) its prediction for solute transit times along such paths. The first application uses the critical percolation probability to define the most permeable portion of the network; the second is based on percolation scaling laws.

Naturally, if soil physics by itself cannot be understood within the confines of the traditional continuum formulation, one can only expect greater problems to be encountered when the complexity of the process under investigation is increased. Thus, when it comes to the scaling of chemical reactions with time in porous media, there is no continuum theory that can simultaneously describe the observed proportionality of reaction rates to the flow speeds, as well as to a negative power of the time (White and Brantley 2003, Maher 2010). More complicated scenarios and justifications must be developed. But network approaches based on percolation theory do predict this, and agree with the data over time scales that range from seconds to over one hundred million years. The proportionality of reaction rates to flow speeds, as shown in figure 1.2, indicates the relevance of advective, rather than diffusive, solute transport in the reactions. Thus, chemical processes are dependent on the water flow through the medium, and indeed on that flow through the optimal paths. Such arguments do not, however, change the fact that the criterion for selecting a dominant transport mechanism at the scale of a single pore can be based

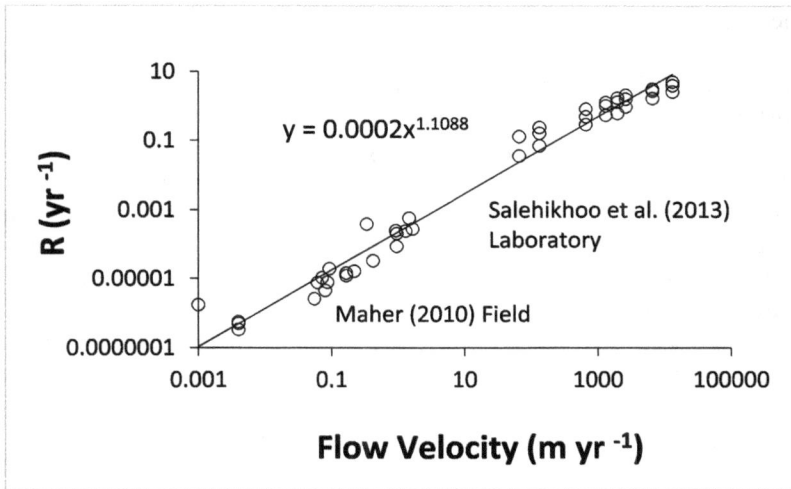

Figure 1.2. Proportionality of chemical weathering reactions to flow velocity in porous media. The results from carbonate rocks in laboratory samples extend to much larger velocities (1300 m yr^{-1}) than any that are found in the field, whereas pore-scale flow rates cannot exceed maximum precipitation rates (10 m yr^{-1}) by very much; such large reaction rates as associated with the high flow rates are possible because carbonates weather orders of magnitude more rapidly than do silicates. Hunt *et al* (2021) John Wiley & Sons. Copyright 2021 American Geophysical Union.

on the continuum mechanics, and the result implicates diffusion at significantly smaller scales, advection at larger scales.

The crossover in the relevance of diffusion relative to advection at a fundamental network length scale turns out to be a feature that soils have in common with multicellular organisms: while the most important transport of dissolved compounds at length scales less than a single cell (roughly a micron) is diffusion, at larger scales organisms must rely on advection, or the time scales for chemical transport would be much too long to sustain life, at least as we know it. Furthermore, for the growth of plants, even purely physical advective processes in a porous medium take far too long, if they are constrained only by the choice of pathways optimized in three dimensions. The pathways chosen must be organized hierarchically as well, i.e., dendritically. In this vein, pun intended, the development of soils to a depth of a meter can take a half a million years, while root growth to a length of one meter can proceed in weeks. The difference in these two time scales can be understood by appealing to the physical processes of transport through distinct media, providing that the media are treated as the networks that they actually are.

In fact, the analogy to multi-celled organisms focuses our attention on any mathematical structure, which can distinguish between multi-cell and single-cell organisms, as well as the porous medium from a single pore. The crossover in relevance from diffusion at the level of a single cell to advection at larger scales is mirrored by a similar crossover in relevance in the soil. That this length scale in soils should coincide (more or less) with that of the pore scale is at least a remarkable coincidence, since the relevant Péclet number depends on the vertical flow velocity, and that value depends on the acceleration of gravity and the pore sizes. In other words, one could add a second remarkable coincidence to the chemical arguments for the existence of a global organism, namely, that the pore sizes in soils are aligned with the gravitational potential in such a way that fluid flow rates are fast enough to overcome the limitations of diffusion when progressing beyond the scale of single pores. In more practical terms, the discussions of this book provide, for example, a quantitative scaling argument for nutrient transport limitations to growth, and why it is advantageous for roots to thus grow towards nutrients, rather than waiting for the nutrients to come to them.

Currently the pace of research into the physical properties of porous media is greatly outstripping the rate at which the implications of this research can be incorporated into the biological component. Such a mismatch is confusing, but it also presents an opportunity to forge links and, thus, a motivation for this work. This research is highly relevant in the study of geochemistry and geobiology, meaning that the stakes of getting it right are high.

The geochemical importance to the study of soil is tied strongly to its rate of formation through the chemical weathering of silicate minerals, with consequences for terrestrial-oceanic carbon fluxes and oceanic carbon burial on time scales of millions of years and longer (Berner 1992, Algeo and Scheckler 1998). Weathering of one mole of silicate minerals, such as that found in basalt, draws down one mole of CO_2 from the atmosphere, which is then typically stored in carbonate sediments. Changes in the rate of chemical weathering of silicate rocks affect climate and, possibly, extinctions (Raymo 1994, Vance *et al* 2009), as well as landscape evolution (Anderson and Anderson 2010).

We will show that the rate of soil formation is governed by the physics of non-reactive solute transport in a typical heterogeneous near-surface medium. The scaling function is predictive and links time scales of seconds with 100 million years (Hunt 2015, 2016, Yu and Hunt 2017). The geobiological component centers around the solute transport scaling law that defines the influence of the soil on the growth of vascular plants and fungi. This particular theoretical development helps under-standing quantitatively of the decline in forest productivity with stand age and is, thus, again relevant to the global carbon budget, though on shorter time scales. The biological scaling law has been verified over time scales, from hours to 10^5 years. Both the scaling laws, as well as the scaling function that describes the effects of human intervention on biota, namely, the growth of crops, originate in the physics of flow in a heterogeneous porous medium, as seen from their common origin at a single point on a space–time diagram. The significance of the three scaling relationships, taken together, is that, in addition to helping define natural inputs to the global carbon cycle, influential anthropogenic factors are also better under-stood. The point of divergence for all three relationships, roughly thirty microns and one minute, defines a typical subsurface fluid velocity of about a micron per second (Blöschl and Sivapalan 1995, Clapp and Hornberger 1978, Freeze and Cherry 1979) and, thus, ties the physics of flow and transport to the geochemistry of soil formation, and the geobiological questions regarding the influence of soil on the growth of land plant. Whether or not the aforementioned typical flow velocity is common to soils and abiotic media is an important question that still needs to be answered, since if biological characteristics tend either to increase or decrease typical flow velocities, then an important feedback can be discovered.

In the following chapters we will describe, first, the theoretical methods on which our general approach is based. Then, we discuss the theoretical basis for understanding the most important aspects of the soil physics, after which we will give a summary of the state of knowledge of soil physics and its impact on geochemistry, including the process of soil formation itself. Then, we will present a summary of knowledge of the biological realm of the Critical Zone, with a focus on the physical properties. Finally, potential implications for the Gaia hypothesis are discussed in the concluding chapter.

References

Algeo T J and Scheckler S E 1998 Terrestrial-marine teleconnections in the Devonian: links between the evolution of landplants, weathering processes, and marine anoxic events *Philos. Trans. Roy. Soc. Lond. Ser.* B **353** 113

Ambegaokar V N, Halperin B I and Langer J S 1971 Hopping conductivity in disordered systems *Phys. Rev.* B **4** 2612

Anderson R S and Anderson S P 2010 *Geomorphology: The Mechanics and Chemistry of Landscapes* (Cambridge: Cambridge University Press)

Aronofsky J S and Heller J P 1957 A diffusion model to explain mixing of flowing miscible fluids in porous media *Trans. AIME* **210** 345

Arya A, Hewett T A, Larson R G and Lake L W 1988 Dispersion and reservoir heterogeneity *SPE Reservoir Eng.* **3** 139

Baumann T, Toops L and Niessner R 2010 Colloid dispersion on the pore scale *Water Res.* **44** 1246

Bear J 1972 *Dynamics of Fluids in Porous Media* (Amsterdam: Elsevier)

Bejan A 1997 *Advanced Engineering Thermodynamics* (New York: Wiley)

Berkowitz B and Scher H 1995 On characterization of anomalous dispersion in porous and fractured media *Water Resour. Res.* **31** 1461

Berkowitz B, Scher H and Silliman S E 2000 Anomalous transport in laboratory-scale, heterogeneous porous media *Water Resour. Res.* **36** 149

Berner R A 1992 Weathering, plants, and the long-term carbon-cycle *Geochim. Cosmochim. Acta* **56** 3225

Bertuzzo B E, Maritan A, Gatto M, Rodriguez-Iturbe I and Rinaldo A 2007 River networks and ecological corridors: Reactive transport on fractals, migration fronts, hydrochory *Water Resour. Res.* **43** W04419

Blöschl G and Sivapalan M 1995 Scale issues in hydrological modelling: A review *Hydrol. Process.* **9** 251

Borman P T, Spaltenstein H, McClellan M H, Ugolini F C, Cromack K and Nay S M 1995 Rapid soil development after windthrow disturbance in pristine forests *J. Ecol.* **83** 747

Bos F C, Guion T and Burland D M 1989 Dispersive nature of hole transport in polyvinylcarbazole *Phys. Rev.* B **39** 12633

Bromley M and Hinz C 2004 Non-Fickian transport in homogeneous unsaturated repacked sand *Water Resour. Res.* **40** W07402

Bruderer-Weng C, Cowie P, Bernabé Y *et al* 2004 Relating flow channeling to tracer dispersion in heterogeneous networks *Adv. Water Resour.* **27** 843

Clapp R B and Hornberger G M 1978 Empirical equations for some soil hydraulic properties *Water Resour. Res.* **14** 601

Coats K H and Smith B D 1964 Dead-end pore volume and dispersion in porous media *Soc. Pet. Eng. J.* **4** 73

Cortis A and Berkowitz B 2004 Anomalous transport in 'classical' soil and sand columns *Soil Sci. Soc. Am. J.* **68** 1539

Cruiziat P, Cochard H and Ameglio T 2002 Hydraulic architecture of trees: main concepts and results *Ann. For. Sci.* **59** 723

Cushman J H and O'Malley D 2015 Fickian dispersion is anomalous *J. Hydrol.* **531** 2015

Dane J H and Topp C G (ed) 2002 *Methods of Soil Analysis* (Madison, WI: Soil Science Society of America) ch 6.2

Fatt I 1956a The network model of porous media: I. Capillary pressure characteristics *Trans. AIME* **207** 155

Fatt I 1956b The network model of porous media: II. Dynamic properties of a single size tube network *Trans. AIME* **207** 160

Fatt I 1956c The network model of porous media: III. Dynamic properties of networks with tube radius distribution *Trans. AIME* **207** 164

Freeze R A and Cherry J A 1979 *Groundwater* (Englewood Cliffs, N J: Prentice-Hall)

Frouz J, Prach K, Pižla V, Hánel L, Starý J, Tajovsky K, Maternad J, Balik V, Kalcik J and Rehounkova K 2008 Interactions between soil development, vegetation and soil fauna during spontaneous succession in post mining sites *Eur. J. Soil Biol.* **44** 109

Gelhar L W, Welty C and Rehfeldt K R 1992 A critical review of data on field-scale dispersion in aquifers *Water Resour. Res.* **28** 1955

Ghanbarian-Alavijeh B, Skinner T E and Hunt A G 2012 Saturation dependence of dispersion in porous media *Phys. Rev. E* **86** 066316

Hillel D 2004 *Introduction to Environmental Soil Physics* (Amsterdam: Elsevier)

Hillel D 2005 Soil: crucible of life *J. Nat. Resour. Life Sci. Educ.* **34** 60

Hunt A G 2015a Predicting rates of weathering rind formation *Vadose Zone J.* **14** 1

Hunt A G 2016 Soil depth and soil production *Complexity* **21** 42

Hunt A G, Ewing R P and Horton R 2013 What's wrong with soil physics? *Soil Sci. Soc. Am. J.* **77** 1877

Hunt A G, Ewing R P and Ghanbarian B 2014 *Percolation Theory for Flow in Porous Media* 3rd edn (Berlin: Springer)

Hunt A G, Ghanbarian-Alavijeh B, Skinner T E and Ewing R P 2015 Scaling of geochemical reaction rates via advective solute transport *Chaos* **25** 075403

Hunt A G, Skinner T E, Ewing R P and Ghanbarian-Alavijeh B 2011 Dispersion of solutes in porous media *Eur. Phys. J. B* **80** 411

Hunt A, Egli M and Faybishenko B 2021 *Hydrogeology, Chemical Weathering, and Soil Formation* (Wiley)

Koch A L 1990 Diffusion–the crucial process in many aspects of the biology of bacteria *Advances in Microbial Ecology* **vol 11** (Springer) pp 37–70

Koestel J K, Moeys J and Jarvis N J 2012 Meta-analysis of the effects of soil properties, site factors and experimental conditions on solute transport *Hydrol. Earth Syst. Sci.* **16** 1647

Larcher W 2003 *Physiological Plant Ecology* 4th edn (Berlin: Springer)

Lallemand-Barres A and Peaudecerf P 1978 Recherche des relations entre la valeur de la dispersivité macroscopique d'un milieu aquifere, ses autres charactéristiques et les conditions de mésure (Research for relations between the macroscopic dispersivity value of an aquifer, its other characteristics and measurement conditions) *Bulletin du B. R. G. M. (deuxiéme série)* **section III** 227

Lide D R 1994 *CRC Handbook of Chemistry and Physics* 75th edn (Boca Raton, FL: CRC Press)

Ling L L *et al* 2015 A new antibiotic kills pathogens without detectable resistance *Nature* **517** 455

Lovelock J E and Margulis L 1974 Atmospheric homeostasis by and for the biosphere: the Gaia hypothesis *Tellus* **26** 1

Lu W, Guo H, Chou I M, Burruss R C and Lanlan L 2013 Determination of diffusion coefficients of carbon dioxide in water between 268 and 473 K in a high pressure capillary optical cell with *in situ* Raman Spectroscopy *Geochim. Cosmochim. Acta* **115** 183

Maher K 2010 The dependence of chemical weathering rates on fluid residence time *Earth Plan. Sci. Lett.* **294** 101

Manzoni S, Vico G, Katul G and Porporato A 2013 Biological constraints on water transport in the soil-plant-atmosphere system *Adv. Water Resour.* **51** 292

Margolin G and Berkowitz B 2000 Application of continuous time random walks to transport in porous media *J. Phys. Chem.* B **104** 3942

Mavris C, Egli M, Pltze M, Blum J D, Mirabella A, Giacci D and Haeberli W 2010 Initial stages of weathering and soil formation in the Morteratsch proglacial area (Upper Engadine Switzerland) *Geoderma* **155** 359

Molins S, Trebotich D, Steefel C I and Shen C 2012 An investigation of the effect of pore scale flow on average geochemical reaction rates using direct numerical simulation *Water Resour. Res.* **48** W03527

Montgomery D 2007 *Dirt, The Erosion of Civilization* (Berkeley, CA: University of California Press)

Moreno L and Tsang C F 1994 Flow channeling in strongly heterogeneous porous media: A numerical study *Water Resour. Res.* **30** 1421

Muneepeerakul R, Bertuzzo E, Lynch H J, Fagan W F, Rinaldo A and Rodriguez-Iturbe I 2008 Neutral metacommunity models predict fish diversity patterns in Mississippi-Missouri basin *Nature* **453** 220

National Research Council 1996 *Rock Fractures and Fluid Flow* (Washington, DC: National Academy Press)

Neuman S P 1990 Universal scaling of hydraulic conductivities and dispersivities in geologic media *Water Resour. Res.* **26** 1749

Pfannkuch H 1963 *Contribution à L'étude des Déplacements de Fluides Miscibles dans un Milieu Poreux (Contribution to the Study of the Displacement of Miscible Fluids in a Porous Medium)* (Paris: Revue de l'Institut français du pétrole et Annales des combustibles liquides)

Pfister G 1974 Pressure-dependent electronic transport in amorphous As_2Se_3 *Phys. Rev. Lett.* **35** 1474

Pfister G and Griffiths C H 1978 Temperature-dependence of transient hole hopping transport in disordered organic solids–carbazole polymers *Phys. Rev. Lett.* **40** 659

Pfister G and Scher H 1977 Time-dependent electronic transport in amorphous solids—As_2Se_3 *Phys. Rev.* B **15** 2062

Pollak M 1972 A percolation treatment of DC hopping conduction *J. Non Cryst. Solids* **11** 1

Raymo M E 1994 The Himalayas, organic-carbon burial, and climate in the Miocene *Paleoceanography* **9** 399

Roberts P V and Mackay D M 1986 A natural gradient experiment on solute transport in a sand aquifer Technical report, Stanford University, Department of Civil Engineering

Saffman P G 1959 A theory of dispersion in a porous medium *J. Fluid Mech.* **6** 321–49

Sahimi M and Hunt A G (ed) 2021 *Complex Media and Percolation Theory* (Berlin: Springer)

Sahimi M 1987 Hydrodynamic dispersion near the percolation threshold: scaling and probability densities *J. Phys.* A **20** L1293

Sahimi M 2011 *Flow and Transport in Porous Media and Fractured Rock* 2nd edn (New York: Wiley-VCH)

Sahimi M 2012 Dispersion in porous media, continuous-time random walks, and percolation *Phys. Rev.* E **85** 016316

Sahimi M 2023 *Applications of Percolation Theory* 2nd edn (Berlin: Springer)

Sahimi M, Davis H T and Scriven L E 1983 Dispersion in disordered porous media *Chem. Eng. Commun.* **23** 329

Sahimi M, Heiba A A, Hughes B D, Scriven L E and Davis H T 1982 Dispersion in flow through porous media SPE Paper 10 969

Sahimi M, Heiba A A, Davis H T and Scriven L E 1986a Dispersion in flow through porous media: II. Two-phase flow *Chem. Eng. Sci.* **41** 2123

Sahimi M, Hughes B D, Scriven L E and Davis H T 1986b Dispersion in flow through porous media: I. One-phase Flow *Chem. Eng. Sci.* **41** 2103

Sahimi M and Imdakm A O 1988 The effect of morphological disorder on hydrodynamic dispersion in flow through porous media *J. Phys.* A **21** 3833

Salehikhoo F, Li L and Brantley S L 2013 Magnesite dissolution rates at different spatial scales: the role of mineral spatial distribution and flow velocity *Geochim. Cosmochim. Acta* **108** 91

Scheidegger A E 1959 An evaluation of the accuracy of the diffusivity equation for describing miscible displacement in porous media *Proc. of Theory of Fluid Flow in Porous Media Conf. (University of Oklahoma)*

Scher H, Shlesinger M and Bendler J 1991 Time-scale invariance in transport and relaxation *Phys. Today* **44** 26

Silliman S E and Simpson E S 1987 Laboratory evidence of the scale effect in dispersion of solutes in porous media *Water Resour. Res.* **23** 1667

Stauffer D and Aharony A 1994 *Introduction to Percolation Theory* 2nd edn (London: Taylor & Francis)

Tiedje T 1984 Information about band-tail states from time-of-flight experiments *Semiconductors and Semimetals* **vol 21C** ed J Pankov (New York: Academic) p 207

Trustrum N A and de Rose R C 1988 Soil depth-age relationships of landslides on deforested hillsides, Taranaki, New Zealand *Geomorphology* **1** 143

Tyree M T 1997 The Cohesion-Tension theory of sap ascent: current controversies *J. Exp. Bot.* **48** 1753

Vance D, Teagle D A H and Foster G L 2009 Variable quaternary chemical weathering fluxes and imbalances in marine geochemical budgets *Nature* **458** 493

Watt M, Silk W K and Passioura J B 2006 Rates of root and organism growth, soil conditions and temporal and spatial development of the rhizosphere *Ann. Bot.* **97** 839

White A F and Brantley S L 2003 The effect of time on the weathering rates of silicate minerals: why do weathering rates differ in the lab and in the field? *Chem. Geol.* **202** 479

Xu M and Eckstein Y 1995 Use of weighted least-squares method in evaluation of the relationship between dispersivity and field scale *Ground Water* **33** 905

Young I M and Crawford J W 2004 Interactions and self-organization in the soil-microbe complex *Science* **304** 1634

Yu F and Hunt A G 2017 An examination of the steady-state assumption in soil development models with application to landscape evolution *Earth Surf. Process. Landforms* **42** 2599

IOP Publishing

Networks on Networks (Second Edition)
Role of connectivity in physics of geobiology and geochemistry
Allen G Hunt and Muhammad Sahimi

Chapter 2

Networks in ecological systems

2.1 Background: ecological networks

Nature contains all types of complex networks that play a critical role in how Earth survives and thrives. Understanding the structure of such networks, as well as the role that they play in various processes, is critical to modeling various phenomena that occur in Nature. Ecological networks can take a variety of forms, but are all based on the concept of edges linking nodes. Independent of the particular phenomenon and ecologcal system that one may study, the construction of the edges in the network is perhaps best accomplished by direct observation of, for example, pollinators associated with flowers, or predators consuming prey, which means that individuals of both species must, at least, be in the same place at the same time (Guseva *et al* 2022). But this particular kind of interaction is only one of several that are addressed in network theory.

In general, nodes represent 'individuals' with the links describing their interactions in a form typically associated with social behavior of animals. The most common application is when the nodes and links represent, respectively, species and their dominant interaction, such as predator–prey, competition, and synergism. But a third application (Poisot *et al* 2016), in which the nodes represent communities or ecosystems and the links represent the fluxes between them, corresponds to the fundamental goal of Ecosystem Science at the United States National Science Foundation. As can be seen, the scale of the links of such a network can vary from the micrometer to the kilometer, as the scale of the organisms increases from bacteria to ecosystems.

Network complexity may be represented simply in the number of links (see, e.g., Karimi *et al* 2019). As might be expected, naturally developing forest networks manifest evidence of greater complexity (link number) than vineyards. On the other hand, species richness is higher in vineyards. Another measure of complexity is in the range of the number of links connected to individual nodes, i.e., the heterogeneity of the network. Nodes occupied by dominating genera, called hubs, can be identified by

2-1

their large number of links. A change in hubs may reflect biome distinctions, or human interference. Moreover, positive links describe facilitative interactions, and negative links describe inhibitive interactions (Ma 2018), with the ratio of the number of two types of links suggested to represent a measure of health in a human microbiome. Certainly a ratio greater than 1 implies a mostly facilitative network. In a study of developing biological soil crusts (BSC), the ratio was found to always be greater than 1, indicating that such structures, essential for soil fixing and plant growth in many arid regions, require a symbiotic relationship among the micro-organisms making up the BSC (Zhou *et al* 2020). Interestingly, however, this ratio was largest in the earliest stages of development, decreasing towards 1 over time. In the process, however, even in the increasingly competitive environment of maturing crusts, it was the increasing role of fungi in decomposition-produced byproducts that helped to stabilize the soil.

While any network is characterized by several important properties, we focus in this book on the crucial role of *connectivity*. Any network with a complex structure consists of branches or links of various sizes and other characteristics that are connected together at certain points or nodes. Without enough connectivity, the network collapses and cannot operate even inefficiently, let alone efficiently. Thus, understanding the role of connectivity is crucial to gaining deep insights into how natural networks, and in particular those that are of interest to us in this book, operate.

2.2 Soil networks

Soil networks are typically represented at the level of the interactions between microbes, with the nature of the interactions having been traditionally assumed to be more competitive in Nature. This traditional assumption has more recently been challenged in various ways. One recent network-based study indicates that the decomposition of soil organic matter requires cooperation between multiple species (Liu *et al* 2024), including across trophic levels, from bacteria to invertebrates. Moreover, this complex degree of cooperation is argued to help sustain diverse soil networks with greater stability against disturbances (Liu *et al* 2024). Another recent network-based approach (Verdú *et al* 2023) demonstrates a different kind of exception to the simple (transitive) dominance traditionally assumed among competing species, i.e., $A > B$ and $B > C$ imply $A > C$. If, rather, C can outcompete A, there is an identified mechanism by which a population of microbes competing in the same environment for the same resources does not evolve towards a monoculture of A. Such a loop, or cycle, can also lead to greater resilience to changes in external conditions, if only due to increased genetic diversity.

At the same time though, soil is a porous medium with all the complexities that are due to its heterogeneity. The heterogeneity manifests itself in the soil's morphology; that is, its porosity, pore-size distribution, pore interconnectivity, and pore surface roughness. Soil, similar to almost all other types of porous media, was modeled for decades by a bundle of parallel capillary tubes. Although by using several adjustable parameters, the bundle of capillary tube model could provide reasonable fits to data for some properties of soil, it gradually became clear that the model fails if soil and other type of porous media are highly heterogeneous, or one is interested in studying two- and three-phase fluid flow and transport in such porous media, because under such conditions the effect of the connectivity becomes paramount, and cannot be ignored, whereas a bundle of capillary tubes model is always connected, and cannot take into account the experimental observations that if a finite fraction of the tubes is disconnected from the rest of the network, the network as a whole cannot operate at macroscale.

This motivated development of network models of porous media. Such models are also intuitively appealing, because it is clear that a fluid's path in a porous medium branches out and, later on, joins one another. It was, however, Mohanty (1981) who first developed a rigorous procedure for mapping any porous medium onto an equivalent network. As he showed, such networks consist of pore throats— the narrow passages through which flow and transport of fluids occur—which are connected by pore bodies (for simplicity, usually referred to as pores), the chambers in which most of the porosity (volume fraction of the pore space) resides. The connectivity or coordination number is the number of pore throats connected to the same pore body. The pioneering work of Mohanty was followed up by others in order to obtain a more accurate and representative pore network of a porous medium; see Dong and Blunt (2009) and references therein, with much more details given by Sahimi (2011). The network that results from the mapping usually has a random topology, with its coordination number or connectivity varying from pore body to pore body. Thus, random networks, such as the Voronoi network have also been used in the literature as models of porous media. Computer simulations of Jerauld *et al* (1984a, 1984b) showed, however, that as long as the average coordination number of a topologically-random network is very close to, or identical with, the coordination number of a regular network, the effective flow and transport properties of the two networks are virtually equal.

Although the procedure for mapping of porous media, including soil, onto an equivalent network was first developed by Mohanty (1981), the intuitve idea has a much older history, which is almost a century old. Bjerrum and Manegold (1927) developed a random network made of randomly distributed points in space, connected to one another by cylindrical tubes, to study transport in porous media. But, the computational limitations of that era time severely limited their ability for carrying out any extensive computations. The first application of network models to modelling two-phase flow in porous media was pioneered by Fatt (1956a, 1956b, 1956c), who used various two-dimensional (2D) networks of bonds representing the pore throats, and assumed that pore bodies do not affect fluid flow and, thus, attributed no volume to them. The radii of the bonds were selected from a probability density function, representing the pore-size distribution of the medium. The length of each bond was assumed to be proportional to the inverse of its radius. Rose (1957) and Dodd and Kiel (1959) used pore network models to study immiscible displacement processes in porous media. Ksenzhek (1963) used a pore network model to predict the capillary pressure curves for porous media. Thus, although in the physics literature two seminal papers of Kirkpatrick (1971, 1973) are generally credited for popularizing the use of networks of interconnected bonds for modeling transport in disordered media, the earlier pioneering works had already used such models to study transport processes in disordered porous media. Note that, although not of prime interest in this book, fractures in porous formations form interconnected networks with well-defined morphological properties, including connectivity; see Sahimi (2011, 2023) for comprehenseve discussions.

2.3 Root networks

Any plant has three organs, with roots being one of them. The most important function of the roots is taking in water and nutrients, so that plants can grow. Plants are also anchored by roots, which help them survive when there is too much water, such as when there is flooding, or too little water and nutrients in severe droughts.

How does the root network develop? Roots grow by elongation of their tip. In a region of tissue called meristem, cells in the growing root divide at the interior of the root. The growth of roots begins when the seedling germinates. The root that results from the seedling or embryonic plant is what is usually referred to as the *primary* root. However, as we all have observed, after a few weeks of growth there is more than one vertical root; that is, the plants develop other roots that are called *secondary* roots that branch out from the primary one. Since roots grown in soil, their growth is affected by the soil's morphology, which is highly heterogeneous and, therefore, roots do not all grow straight in a vertical direction, but rather many of their parts grow sidewise and are known as *lateral* roots. Thus, the resistance to growth of roots due to the morphology of the soil is an important factor in the emerging structure of root network. For example, compactness of the soil alters root growth since in compact soils the pores between soil particles are reduced in size and even number. To counter the resistance to their growth by soil, roots thicken; their rate of growth-elongation decreases, and branching patterns are modified, often resulting in an increase in lateral growth of roots.

Root networks are complex, baffling scientists for a long time. The difficulty in their characterization is that roots lack clearly defined branching points, equivalent to axial buds. In addition, the spatial heterogeneity of the rhizosphere affects development of the root network. Thus, efforts have focused on developing a practical descriptor that represents the branching attributes of root networks accurately. Fractal geometry offers one promising descriptor for characterizing the structure of root networks. The idea is appealing because the repetitive branching of roots generates self-similarity, at least in an averaged way, which is a fundamental characteristic of fractal objects, although objections to the idea were also raised (Berntson and Stoll 1996). A simple way of determining the fractal dimension of root networks is by the *box-counting method*. One covers the system by non-overlapping $d-$ dimensional boxes of linear size r, and determines the number $N(r)$ of the roots that intercept each box. The fractal dimension D_f is then defined by

$$D_f = \lim_{r \to 0} \frac{\ln N}{\ln(1/r)}.$$

(2.1)

One may also define the fractal dimension through the relation between the system's mass M and its characteristic length scale L. In that case, self-similarity requires,

$$M \sim L^{D_f}. \tag{2.2}$$

The two methods produce identical values of D_f.

The idea of fractal geometry has been applied to a variety of root networks, and the estimated D_f were used to demonstrate its variation with genotype (Tatsumi *et al* 1989, Fitter and Stickland 1992, Lynch and van Beem 1994), with plant age (Fitter and Stickland 1992), and with growth conditions (Eghball *et al* 1993, Berntson 1994, Lynch and van Beem 1994). The analyses were carried out for roots that had been extracted from the soil and spread on a flat surface and, therefore, represented 2D networks. Eshel (1998) was probably the first who determined the fractal dimension of a 3D root network, using the box-counting method. For more recent references see Yang *et al* (2022). We will return to this problem later in this book where we describe the percolation model of water balance.

2.4 Vegetation networks

Ecosystems exhibit remarkable resilience to environmental and human-induced changes, and one primary issue has always been identifying factors that contribute to the resilience. One important contributing factor is the way an ecosystem's plants are spatially arranged. As we learn throughout this book, there are stark differences between those in which the plants form isolated clusters or patchy clumps, and those that are arranged as a network that extends over the entire terrain. It has been hypothesized that at least some biological systems operate in the vicinity of a phase transition, such as one in which the patchy clumps join together and form an extended network. Such a critical state fosters numerous functional advantages and allows the system to optimize its ability to react collectively. A good review of the subject is given by Muñoz (2018). The hypothesis has led to better understanding of many natural phenomena, such as sandpiles (Bak *et al* 1987). The spatial distribution of earthquakes epi- or hypocenters has also been suggested to form a connected percolation cluster (Sahimi *et al* 1993, Tafti *et al* 2013), which has received experimental verification (see Sahimi 2023, for detailed discussions). Such systems exhibit power-law distributed events, which is the signature of a system that operates at or near a second-order (i.e., continuous and gradual) phase transition.

Of particular interest to us in this book is the vegetation patterns in ecosystems that are arid or semi-arid. One example of a large-scale vegetation pattern is the Namibian fairy circles (Tarnita *et al* 2017). Fairy circles are are circular patches of land barren of plants that vary between 2 and 12 m in diameter, and are often encircled by a ring of stimulated growth of grass. They form in the arid grasslands of the Namib desert in western parts of Southern Africa, as well as in Western Australia. These patterns exhibit clear spatial scales, and although several research groups have proposed various hypotheses about the process that leads to their formation, none have conclusively proven how they are formed. One hypothesis is that they are a consequence of vegetation patterns that arise naturally from competition between grasses. But, generally speaking, understanding collective phenomena in large-scale complex ecological systems, such as rainforests, and the patterns that they exhibit, represents a fundamental open problem in theoretical ecology.

The concepts of percolation theory have been used (Martín *et al* 2020) to demonstrate that, in semi-arid environments, dynamic environmental variability promotes the emergence of vegetation patches that have broadly distributed cluster sizes (a cluster is a set of connected microscopic elements of a system; see chapter 3). Moreover, use of cluster-based approaches have led to identification of the scales of spatial aggregation and the corresponding tree clusters in Malaysian tree species (Plotkin *et al* 2002). Extensive data for various regions, ranging from dry deciduous forests to evergreen wet forests, indicate empirically that most species are more aggregated, rather than being random (Condit *et al* 2000). In addition, anomalous density fluctuations, an important fingerprint of long-range correlations that are developed in connected networks near their connectivity threshold, have been identified in the spatial distribution of various tropical species (Villegas *et al* 2021a). Species richness, which will be described later in this book, depends on the sampling area, and this fact has been exploited to argue that specific ecosystems are in a state of incipient criticality.

Analyzing empirical data for Barro Colorado Island (in the man-made Gatun Lake in the middle of the Panama Canal), Villegas *et al* (2024) showed that the statistics of vegetation clusters follow those of percolation theory. The database contains sufficiently high-resolution data with eight censuses, collected every five years from the 1980s, of more than 4^5 trees and shrubs with diameter at breast height greater than 0.01 m, which belong to about 300 species in 50 ha (1000×500 m^2) and provide position and species for each plant (see Condit *et al* 2000). The data make it possible to resolve vegetation clusters of conspecific and heterospecific plants, which were utilized by Villegas *et al* (2024). To analyze the data, Villegas *et al* (2024) used the concepts of continuum percolation theory. These will be described in the next chapter but, for now, it suffices to say that, based on some predefined distance r, two individual points will belong to the same cluster if their Euclidean distance is less than or equal to r, and a percolating cluster exists if a path can be drawn connecting all points with edges of length smaller than r. Using the mean nearest-neighbor distance (MNN) r_{mnn}, defined as the nearest-neighbor mean point-to-point Euclidean distance, one can study a point process independently of the area in which it takes place. Thus, the control parameter r is normalized by the interparticle distance of the point process, $\hat{r} = r/r_{mnn}$, allowing one to interpret the results of cluster analysis in terms of the concepts of percolation phase transition. The function $n_s(s, \hat{r})$ in the database represents the cluster-size probability distribution, and is used to calculate a quantity called susceptibility, representing mean cluster size, $S(\hat{r}) = \sum_s s^2 n_s(s, \hat{r})/s n_s(s, \hat{r})$, where s is the size of a cluster of a given $\hat{r}$. At the critical distance $\hat{r}_c$, one expects to have

$$n_s \sim s^{-\tau}. \tag{2.3}$$

in 2D percolation, one has, $\tau = 187/91 \simeq 2.05$

Villegas *et al* (2024) analyzed the cluster properties of two important cases, namely, the community (all the plants together) and single-species levels or shrubs. For small r, the network consisted of small clusters, each with a few closely-packed plants, but for larger r, the cluster size expands until—at some threshold value r_c— one single 'giant cluster' includes nearly all the plant population. The sudden switch from isolated clusters to a system-spanning network represents a percolation transition. The analysis of Villegas *et al* (2024) indicated that, at the community level, one has a percolation phase transition at about, $\hat{r}_c \simeq 2.4 \pm 0.1$. At the critical point $\hat{r}_c$, various species aggregate, exhibiting a power-law distribution of cluster sizes, equation (2.3), with an exponent, $\tau \approx 2.0 \pm 0.1$. But, when Villegas *et al* (2024) analyzed the most abundant species—the shrub *Hybanthus prunifolius*— which has been shown to quantify the collective behavior of the entire system in agent-based models (Villegas *et al* 2021b), a distinct pattern emerged. For this shrub, there was no abrupt transition, but rather there was a range of r values that produce a percolating cluster that spans the sample area and includes a large fraction of the population, as well as an absolute lack of characteristic scales. This implies that there is a broad region where multiple resolution scales hierarchically percolate and, consequently, the distribution of clusters in the system is expected to follow an intrinsic power law for a wide range of values $\hat{r}$. In other words, the single-species distribution has no 'special' length scale, but exhibits system-wide clustering over many different scales, which is the hallmark of a critical state.

Thus, vegetation patterns (clusters) form connected networks and resemble percolation systems at or near the connectivity threshold.

2.5 River networks

River networks are complex social and ecological systems that are hierarchically organized, and represent important components of watershed landscapes, as they act as a hydrological continuum for species dynamics and ecological processes, in addition to being a major contributor to the watershed's hydrological processes. Thus, characterization of the architecture of a river network has been a long-standing problem, and has been studied since at least the 1940s. In particular, connectivity of a river network is critical to the way it affects the environment in its area. If a river network is well-connected, it promotes biological exchanges, energy and nutrient cycling, and maintains relative stability in a changing environment due to urbanization and other factors. Hydrological connectivity is affected by the spatial organization of river networks. In turn, the network's dynamic behavior is indicative of potential spatial and temporal heterogeneity of hydrological processes, while the biological complexity is a result of interactions between flow and the physical river network at various scales (Zhang *et al* 2023).

Horton (1945) and Strahler (1957) were presumably the first who developed a systematic methodology for characterizing the connectivity of river networks. In their methodology, one defines a source stream as the section that runs from a channel head to the first intersection with another stream. Such source streams are then classified as the network's first-order stream segments. One then removes all source streams and identifies the new source streams of the remaining network, which represent the network's second-order stream segments. If the process is repeated until one stream segment is left, its order Ω is also the river network's order. A river network also contains side streams and absorbing streams. The former is any stream that joins into a stream of higher order, with the latter being the absorbing stream. Side streams and absorbing streams have their own orders. A river network may also have an isolated stream or streams whose relative rank can be ambiguous.

In a series of important paper over a period of nearly two decades, Tokunaga (1966, 1978, 1974) introduced the idea of measuring side stream statistics, which constitute the main part of a platform from which all other river network scaling laws follow. Tokunaga's law also has close connections with drainage density, is a measure of how a river network fills space. In Tokunaga's methodology for a river network, on counts $\langle T_{\mu,\nu} \rangle$, it is the average number of side streams of order ν that enter an sbsorbing stream of order μ. For $\Omega \geqslant \mu \geqslant \nu \geqslant 1$, $\langle T_{\mu,\nu} \rangle$ represents the entries of a lower-diagonal matrix **T**. For example, for Mississippi river, if $\mu = 4$, then, $\langle T_{4,\nu} \rangle = 12$, 3.3, and 1.1, for $\nu = 1$, 2, and 3.

River networks represent self-similar fractal structures, and are characterized by a fractal dimension D_f (see above). This will be described in chapter 10. Given the self-similarity, Tokunaga made two key observations. One is that $\langle T_{\mu,\nu} \rangle$ should not depend on either μ or ν, but only on their difference, $k = \mu - \nu$. The second observation was that if we change k, $\langle T_{\mu,\nu} \rangle$ must themselves change by a systematic ratio. Together, the two statements lead to the Tokunaga's law:

$$\langle T_{\mu,\nu} \rangle = \langle T_k \rangle = \langle T_1 \rangle (R_T)^{k-1}, \tag{2.4}$$

Thus, only two parameters, $\langle T_1 \rangle$ and the ratio R_T are needed to characterize the matrix **T** and its entries $\langle T_{\mu,\nu} \rangle$. $\langle T_1 \rangle$ represents the average number of side streams of one order lower than the absorbing stream, and is typically on the order of 1.0–1.5. Since such side streams are the dominant in river networks, their number estimates the network's breadth. Thus, larger values of $\langle T_1 \rangle$ correspond to wider networks, whereas smaller $\langle T_1 \rangle$ implies networks with relatively thinner profiles. R_T, on the other hand, is a measure of how the density of side streams of decreasing order increases and, thus, represents a measure of changing length scales. Dodds and Rothman (2001) analyze the consequences of Tokunaga's law, present several generalizations, and establish a link with other well-known laws regarding river networks.

Summarizing, river networks represent complex self-similar structure whose connectivity can be characterized accurately. In chapter 10, we will use percolation theory to characterize the self-similarity of river networks.

References

Bak P, Tang C and Wiesenfeld K 1987 Self-organized criticality: an explanation of the $1/f$ noise *Phys. Rev. Lett.* **59** 381

Berntson G M 1994 Modeling root architecture: are there tradeoffs between efficiency and potential of resource capture? *New Phytol.* **127** 483

Berntson G M and Stoll P 1996 Finite spatial scales of self-similarity in real-world branching structures http://plantecohostharvard.edu/gmbWWW/OnlinePubl/FracMeth/

Bjerrum N and Manegold E 1927 Ueber kollodium-membranen, II *Kolloid Z. USSR* **43** 5–14

Condit R *et al* 2000 Spatial patterns in the distribution of tropical tree species *Science* **288** 1414

Dodd C G and Keil O G 1959 Evaluation of Monte-Carlo methods in studying fluid-fluid displacement and wettability in porous rocks *J. Phys. Chem.* **63** 1646

Dodds P S and Rothman D H 2001 Geometry of river networks III. Characterization of component connectivity *Phys. Rev.* E **63** 016117

Dong H and Blunt M J 2009 Pore-network extraction from micro-computerized-tomography images *Phys. Rev.* E **80** 036307

Eghball B, Settimi J R, Maranville J W and Parkhurst A M 1993 Fractal analysis for morphological description of corn roots under nitrogen stress *Agron. J.* **85** 287

Eshel A 1998 On the fractal dimensions of a root system *Plant Cell Environ* **21** 247

Fatt I 1956a The network model of porous media: I. Capillary pressure characteristics *Trans. AIME* **207** 155

Fatt I 1956b The network model of porous media: II. Dynamic properties of a single size tube network *Trans. AIME* **207** 160

Fatt I 1956c The network model of porous media: III. Dynamic properties of networks with tube radius distribution *Trans. AIME* **207** 164

Fitter A H and Stickland T R 1992 Fractal characterization of root system architecture *Funct. Ecol.* **6** 632

Guseva K, Darcy S, Simon E, Alteio L V, Montesinos-Navarro A and Kaiser C 2022 From diversity to complexity: microbial networks in soils *Soil Biol. Biochem.* **169** 108604

Horton R E 1945 Erosional development of streams and their drainage basins; hydrophysical approach to quantitative morphology *Bull. Geol. Soc. Am.* **56** 275

Jerauld G R, Hatfield J C, Scriven L E and Davis H T 1984a Percolation and conduction on Voronoi and triangular networks: a case study in topological disorder *J. Phys.* C **17** 1519

Jerauld G R, Scriven L E and Davis H T 1984b Percolation and conduction on the 3D Voronoi and regular networks: a second case study in topological disorder *J. Phys.* C **17** 3429

Karimi B, Dequiedt S, Terrat S, Jolivet C, Arrouays D, Wincker P *et al* 2019 Biogeography of soil bacterial networks along a gradient of cropping intensity *Sci. Rep.* **9** 3812

Kirkpatrick S 1971 Classical transport in disordered media: scaling and effective-medium theories *Phys. Rev. Lett.* **27** 1722

Kirkpatrick S 1973 Percolation and conduction *Rev. Mod. Phys.* **45** 574

Ksenzhek O S 1963 Capillary equilibrium in porous media with intersecting pores *Russ. J. Phys. Chem.* **37** 691

Liu X, Chu H, Godoy O, Fan K, Gao G F, Yang T *et al* 2024 Positive associations fuel soil biodiversity and ecological networks worldwide *Proc. Natl Acad. Sci. USA* **121** e2308769121

Lynch J P and van Beem J J 1994 Growth and architecture of seedling roots of common bean genotypes *Crop Sci.* **33** 1253

Ma Z 2018 The P/N (positive-to-negative links) ratio in complex networks–a promising in-silico biomarker for detecting changes occurring in the human microbiome *Microb. Ecol.* **75** 1063

Martín P V, Domínguez-García V and Muñoz M A 2020 Intermittent percolation and the scale-free distribution of vegetation clusters *New J. Phys.* **22** 083014

Mohanty K K 1981 Fluids in porous media: two-phase distribution and flow *PhD Thesis* University of Minnesota

Muñoz M A 2018 Colloquium: Criticality and dynamical scaling in living systems *Rev. Mod. Phys.* **90** 031001

Plotkin J B, Chave J and Ashton P S 2002 Cluster analysis of spatial patterns in Malaysian tree species *Am. Nat.* **160** 629

Poisot T, Stouffer D B and Kéfi S 2016 Describe, understand and predict: why do we need networks in ecology? *Funct. Ecol.* **30** 1878

Rose D W 1957 Studies of waterflood performance. III. Use of network models *Illnois State Geol. Surc. Circ.* **p 237**

Sahimi M 2011 *Flow and Transport in Porous Media and Fractured Rock* 2nd edn (New York: Wiley-VCH)

Sahimi M 2023 *Applications of Percolation Theory* 2nd edn (Berlin: Springer)

Sahimi M, Robertson M C and Sammis C G 1993 Fractal distribution of earthquake hypocenters and its relation with fault patterns and percolation *Phys. Rev. Lett.* **70** 2186

Strahler A 1957 Quantitative analysis of watershed geomorphology *Trans. Am. Geophys. Union* **38** 913

Tafti T A, Sahimi M, Aminzadeh F and Sammis C G 2013 Use of microseismicity for determining the structure of the fracture network of large-scale porous media *Phys. Rev. E* **87** 032152

Tarnita C E, Bonachela J A, Sheffer E, Guyton J A, Coverdale T C, Long R A and Pringle R M 2017 A theoretical foundation for multi-scale regular vegetation patterns *Nature* **541** 398

Tatsumi J, Yamauchi A and Kono Y 1989 Fractal analysis of plant root systems *Ann. Bot.* **64** 499

Tokunaga E 1966 The composition of drainage network in Toyohira-River basin and valuation of Horton's first law *Geophys. Bull. Hokkaido Univ.* **15** 1

Tokunaga E 1978 Consideration on the composition of drainage networks and their evolution *Geogr. Rep., Tokyo Metrop. Univ.* **13** 1

Tokunaga E 1974 Ordering of divide segments and law of divide segment numbers *Trans. Jpn. Geomorphol. Union* **5** 71

Verdú M, Alcántara J M, Navarro-Cano J A and Goberna M 2023 Transitivity and intransitivity in soil bacterial networks *ISME J.* **17** 2135

Villegas P, Cavagna A, Cencini M, Fort H and Grigera T S 2021 Joint assessment of density correlations and fluctuations for analysing spatial tree patterns *R. Soc. Open Sci.* **8** 202200

Villegas P, Gili T and Caldarelli G 2021 Emergent spatial patterns of coexistence in species-rich plant communities *Phys. Rev. E* **104** 034305

Villegas P, Gili T, Caldarelli G and Gabrielli A 2024 Evidence of scale-free clusters of vegetation in tropical rainforests *Phys. Rev. E* **109** L042402

Yang Q, Cheng W, Hao Z, Zhang Q, Yang D, Teng D, Zhang Y, Wang X, Shen H and Lei S 2022 Study on the fractal characteristics of the plant root system and its relationship with soil strength in tailing ponds *Wirel. Commun. Mob. Comput.* **2022** 9499465

Zhang X, Li F and Zhao Y 2023 Impact of changes in river network structure on hydrological connectivity of watersheds *Ecol. Indic.* **146** 109848

Zhou H, Gao Y, Jia X, Wang M, Ding J, Cheng L *et al* 2020 Network analysis reveals the strengthening of microbial interaction in biological soil crust development in the Mu Us Sandy Land, northwestern China *Soil Biol. Biochem.* **144** 107782

IOP Publishing

Networks on Networks (Second Edition)
Role of connectivity in physics of geobiology and geochemistry
Allen G Hunt and Muhammad Sahimi

Chapter 3

Percolation theory, effective-medium approximation, and upscaling

3.1 Background

This chapter describes basic concepts of percolation theory, as well as a theoretical approach, usually referred to as the critical-path analysis, to link the theory to macroscopic quantities of porous media and materials, such as conductivity and diffusivity. The concepts, methods, and predictions described in this chapter will be used in chapters 3 and 4 to derive, respectively, closed-form equations for the effective conductivity and diffusivity of porous media as a function of fluid saturation, as well those that are relevant for describing solute transport.

In general, percolation theory describes emergent properties related to the connectivity of a large number of objects that typically have some spatial extent, and their spatial relationships are statistically prescribed. The chief relevance of percolation theory to complex interconnected systems in general, and soils in particular, is its ability for predicting global properties given local information (Sahimi 2011, 2023, Hunt *et al* 2014). Here, we describe flow, conduction, diffusion, and other transport processes in porous media. The relationships between local and global properties are not trivial: sometimes the global properties relate to universal topological (connectivity) properties described specifically by percolation theory, and in other cases to system-dependent characteristics of percolation. This important distinction is a product of the particular relevance of percolation in the problems of interest in this book.

A system is said to be percolating when a sufficient fraction p of the entities in question in the networks that were described in chapters 1 and 2, such as sites, bonds, etc, is connected locally, so that a global connection emerges (Stauffer and Aharony 1994). The global connection is a cluster of locally-connected entities that is unbounded in size, except any limitation that is imposed on it due to the finite size of the system. Percolation theory predicts that when the fraction p exceeds a critical

doi:10.1088/978-0-7503-5698-5ch3

value p_c, called the *percolation threshold*, the size of the largest cluster of interconnected objects is infinite or, in a finite system, is sample-spanning. p_c depends on the microscopic details of the system or networks. Those with higher local connectivities have lower p_c, and vice versa.

The basis of percolation in local connections can also be seen as a condition for allowing passage, or communication throughout a system. A pathogen can pass from one organism to another; charge can be transported from one entity to another, fire can propagate from one tree to another, and so forth. Percolation theory then predicts specific expressions that describe, for example, the probability that a cluster of infected organisms reaches a given size, or the probability that a series of connected conducting entities spans a finite system. For the conduction problem, if the probability of a system-spanning connection is unity, which is the case for a system above its percolation threshold, percolation theory also predicts the general behavior of the conductivity. For an ensemble of finite systems below the percolation threshold, it predicts the probability of measuring a given conductivity.

Upscaling is the process of generating verifiable predictions of properties of a porous medium at large spatial scales based on an understanding of their variability at smaller or local scales. Upscaling has been a focus of soil physics and hydrogeology for the half quarter century (Freeze 1975, Gelhar 1986, Deutsch 1989, Desbarats 1992, Kolterman and Gorelick 1996, Scheibe and Yabusaki 1998, Cushman *et al* 2002, Ebrahimi and Sahimi 2002, 2004, Knudby *et al* 2006, Vereecken *et al* 2007, Neuweiler and Vogel 2007, Rasaei and Sahimi 2008, 2009, Moslehi *et al* 2016). How to upscale flow and transport of solutes in a solution is a question of understanding the effect on the macroscopic properties of the heterogeneity of the pore space. This problem is linked with the characterization of porous media and understanding the effect of the relevant spatial scales. In fact, the fields of hydrogeology and soil physics have become preoccupied with identifying differences in parameters and processes at various space and time scales (Blöschl and Sivapalan 1995, Loague and Corwin 2006). Focusing on individual scales, rather than cross-scale information flow, have, however, sometimes complicated problems, giving rise to new ones, rather than leading to general solutions (Rasaei and Sahimi 2008, 2009, Hunt and Ewing 2014). Emphasizing the differences between various scales also tends to isolate investigators of soil physics from those in hydrogeology. In contrast, the concepts of percolation theory allow, by effectively bridging scales, identification of strategies for unification across scales (Sahimi 1993, 2023, Hunt *et al* 2014), hence allowing better communication between distinct disciplines, as well as distillation of principles. Although typical discussions of upscaling of flow and transport properies neglect the percolation approach, its use in this regard goes back to the early 1970s (Ambekaokar *et al* 1971, Pollak 1972), and in comparison with other methods was already demonstrated (Kirkpatrick 1971, 1973, Seager and Pike 1974) to be the superior way to predict transport properties in strongly heterogeneous media. Such demonstrations continued through the 1990s (Bernabe and Bruderer 1998) and beyond (Ghanbarian and Hunt 2012, Hunt *et al* 2014, and references therein).

3.2 Percolation theory and scaling properties

The concepts and ideas of percolation theory may be applied to porous media in two distinct ways. The first type of application defines a porous medium as a binary material: a given location represents either a pore or the solid matrix, and flow through it is either possible or not. In this particular application, how close the porous medium is to its percolation threshold affects its behavior strongly, and can even determine whether the appropriate mathematical tools to utilize are percolation-based, or are based on the effective-medium approximation (EMA). The EMA —sometimes called *poor man's perrcolation* (Sahimi 2023)—prescribes a qualitatively similar, but quantitatively different kind of behavior from what is expected in percolation theory. We will return to the EMA later in this book. The second kind of application considers a porous medium as having a wide range of local pore conductances and, hence, flow rates. In this case, percolation theory identifies the most permeable portions of the porous medium. The mathematical basis utilizes the value of the percolation threshold p_c in the cumulative distribution of local pore conductances in such a way as to identify the largest possible value of the smallest conductance (the bottleneck) that allows a continuously-connected path (cluster) through the medium. In either case, it is possible to transform the system's variables into percolation variables, p, p_c, etc, and to utilize the machinery of percolation theory.

There are three classical percolation problems in three basic varieties, namely, site, bond, and continuum percolation problems; see figure 3.1. Site and bond percolation problems may be defined on either regular networks, such as a square network, or on irregular networks, such as a Voronoi network (Winterfeld *et al* 1981, Jerauld *et al* 1984). A Voronoi network is constructed as follows. Consider n Poisson points, randomly distributed on a plane (or in 3D space), each of which is the basis for a Voronoi polygon (polyhedron in 3D), which is that part of the space that is closer to its Poisson point than to any other Poisson point. A Voronoi network is constructed by connecting to each other the Poisson points in the neighboring polygons or polyhedra. Site and bond percolation can also be defined on tree structures with constant branching ratios, which are known as Bethe lattices (or Cayley trees). Although interesting in their own right, Bethe lattices are not considered further here, but fractal tree-like networks will be discussed later in this book in the context of water transport and allometric relations in plants.

In order to simplify the discussion, first consider a regular network. The sites are then the nodes in the network, while the bonds are the links between the nodes. If the discussion is based on site percolation, imagine swelling the sites so that nearest-neighbor sites are in contact, while the bond lengths shrink to zero, as shown in figure 3.1, bottom left. In site percolation, the relevant variable is the probability p that a site is 'occupied;' the links (bonds) then automatically exist between two neighboring occupied sites. For the conduction problem, occupation of a site corresponds to the entity at that site to be conducting. A simple physical site percolation problem is constructed by the random emplacement of equal-sized metallic (conducting) and non-metallic (non-conducting) spheres in a large

Generic regular network
(bond and site features vary)

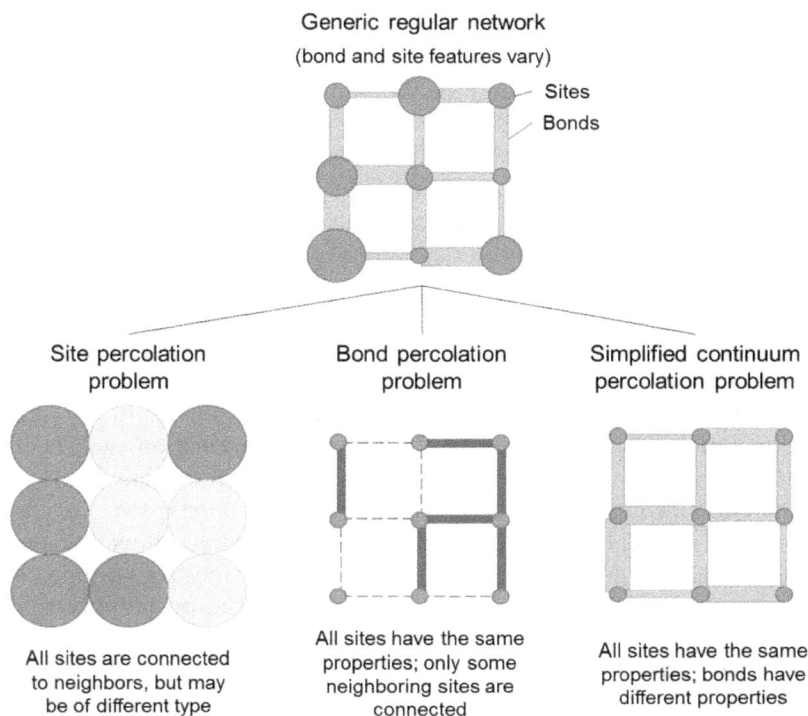

Figure 3.1. Schematic of various types of percolation problems.

container. If two metal spheres touch each other, a (thermal, electric, magnetic, etc) current can flow from one to the other. If the fraction p of the metallic spheres exceeds a critical value—the percolation threshold—a continuous conducting path will be formed across the system. The larger the fraction of the metallic spheres, the better connected the path, and the higher the conductivity of the system will be. This was first demonstrated by simple and elegant experiments of Malliaris and Turner (1971) and by experiments for a freshman physics course at Harvard University (Fitzpatrick *et al* 1974). Percolation theory predicts that the electrical conductivity is a function of the fraction of the metallic spheres, and that for $p \leqslant p_c$, the macroscopic conductivity vanishes.

In the simplest bond percolation, the sites are all identical, while the stochastic structure of the connected cluster stems from the connections between neighboring sites; see figure 3.1, bottom middle. Imagine the sites shrinking to points, while the links are possible between them. The links, or bonds, are then inserted at random with probability p. The bonds could consist of identical wires or resistors, or their conductances can be distributed according to a probability density function. Then, the purpose of the problem is again to estimate the macroscopic conductivity of the system, which is a function of the fraction p of the conducting bonds that are connected, as long as $p > p_c$. In a more complex bond percolation problem, the sites can be distributed randomly and have variable energies, while the bonds can have variable resistances and all of them be connected. Another simple bond percolation problem is one in which one considers a window screen in the form of a square grid. One then cuts at random a fraction p of the elements of the lattice. At some critical fraction $p < p_c$, which is $p_c = 0.5$ for the square grid, the window screen will lose its connectedness and fall apart. Percolation theory addresses directly the question, *at what fraction of the cut bonds does the screen fall apart?* That is, what is the percolation threshold p_c? Other quantities of interest, such as, the size of the largest hole (cut bonds) in the screen, if $p < p_c$, and the structure of such holes, are also addressed by percolation theory.

Site and bond percolation problems correspond roughly to finite-element and finite-difference approximations to differential equations. For a porous medium, sites and bonds correspond to pore bodies and pore throats. Of course, if the structure at the pore scale is properly taken into account, then solution of the set of difference equations for, for example, fluid flow or conduction, will be far more accurate than the solution of the original differential equations that cannot take into account the effect of the disordered morphology of the pore space.

In continuum percolation, arbitrarily shaped objects either allow passage of flow, current, etc, or they do not, which correspond to conductances of varying values; see figure 3.1, bottom right. In this context, p represents the probability that a pore throat takes on a given conductance. At the pore scale, again, such regions may be distinguished as individual pores bodies (or pore throats), while at larger spatial scales—such as geologic media—they might be regions with given permeabilities, elastic moduli, etc. Once again, the most important value of p is p_c, the critical value corresponding to a series of local conductances that allow percolation to occur. In an infinitely large system, p_c is precisely defined: values of $p > p_c$ guarantee 'percolation,' i.e., the existence of an infinitely large cluster of interconnected sites (bonds or objects), while $p < p_c$ guarantees that percolation does not occur. For a given finite-size system, this transition may occur at a value of p somewhat greater than or less than p_c of an infinite system

Continuum percolation problems first received attention in the 1970s, when experiments were carried out by a network of sintered glass and metallic particles. The glass particles may have different sizes and shapes from the metallic particles (which are typically smaller), while the sintering process tends to change the shapes of the particles, producing a net flow of material into the pore space. The irregularity of the particles' shapes can, therefore, be contrasted with the regular geometry of the site percolation problem. In the following chapters, water and solute transport in soils will be formulated as continuum percolation problems in two- (saturated conditions) or three-phase systems (unsaturated conditions). When one fluid displaces another in a three-phase system, it can be advantageous to use *invasion percolation* (Wilkinson and Willemson 1983), but in many cases this particular distinction is unnecessary (Sahimi and Yortsos 1990). We will discuss this subject later in this book.

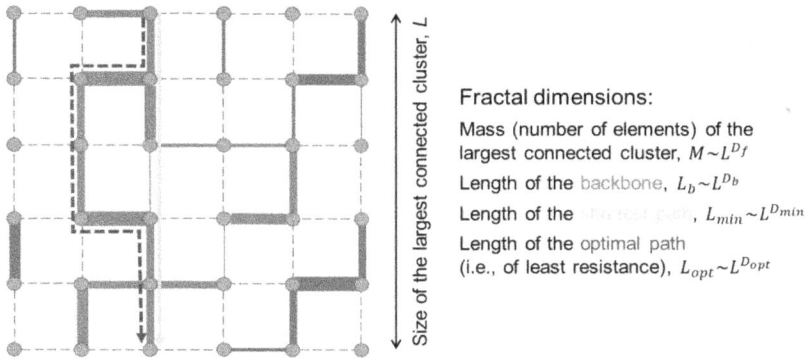

Fractal dimensions:

Mass (number of elements) of the largest connected cluster, $M \sim L^{D_f}$

Length of the backbone, $L_b \sim L^{D_b}$

Length of the shortest path, $L_{min} \sim L^{D_{min}}$

Length of the optimal path (i.e., of least resistance), $L_{opt} \sim L^{D_{opt}}$

Figure 3.2. Schematic representation of backbone (red), shortest path (yellow arrow), and optimal path (violet dashed arrow), and definitions of the fractal dimensions and exponents, linking cluster properties to its Euclidean size. In the schematic network, thin dashed bonds are not conductive, while solid bonds have nonzero conductance values increasing with line thickness. Note that $D_b = D_{bb}$ and $D_{opt} = D_{op}$ throughout this book.

3.3 Structure of percolation clusters

Consider, again, the bond percolation problem on a square lattice in which a fraction p of bonds connect the sites, as shown in figure 3.2. For very small values of p, the bonds only connect pairs of nearest-neighbor sites. As p increases, more pairs of sites become connected, and clusters of interconnected sites emerge gradually. As p approaches p_c, many of the clusters merge and form larger clusters, with complex internal structure. Our goal is to quantify the number and structure of such clusters as a function of p. The structure has been described using various quantities, such as their perimeter, density, mass—by which we mean the number of sites, M—the shortest path length, and the ramification of the clusters. Two factors contribute to the perimeter of the clusters, the number of sites in the cluster with neighboring sites not in the cluster. One is proportional to the volume of the system (Kunz and Souillard 1978), while the second, similar to surface area, is proportional to the volume raised to the power $(1 - 1/d)$, where d is the Euclidean dimension (Stauffer and Aharony 1994). The radius of a large cluster is not given in terms of its volume by the usual relationships valid for Euclidean objects. In fact, large clusters at, or near, the percolation threshold are *fractal* and statistically self-similar objects (see chapter 2) without scale reference, except at the smallest scale at which the scale of the lattice becomes visible. Therefore, it is useful to define the fractal dimensions of various features of the clusters as a function of its Euclidean size L. The fractal dimension D_f of the entire cluster is defined by, $D_f = d \ln(M)/d \ln(L)$, leading to the scaling relation for the mass, $M \sim L^{D_f}$. Other quantities can also be considered, such the shortest path, which is also a fractal object with a fractal dimension D_{min}, and the optimal or the least resistive path with a fractal dimension D_{op}; see figure 3.2. Both D_{min} and D_{op} are defined in a manner similar to the way we defined D_f.

Although an infinite (sample-spanning) connected cluster exists above the percolation threshold, many finite clusters exist as well, and in fact, depending on

p, only a fraction of the connected bonds or sites are actually connected to the infinite cluster. This fraction is critical to the study of hysteresis in wetting and drying. For p greater than, but still close to, p_c, a large portion of the sites on the infinite cluster are located on what are called *dead-end bonds*, which are those that are connected to the rest of the infinite cluster by only one bond. If current were to flow across the system through the infinite cluster, the dead-end bonds would carry no current. If the dead-end bonds or sites are removed from the cluster, what remains is called the *backbone*, which is that part of the infinite cluster that carries current, with every bond or site in it multiply connected, as it has a large number of loops. The backbone also contains the *red* bonds, those for which no alternate path exists. Thus, if a red bond is cut, the flow of current is interrupted. Red bonds are associated with the largest drops in the potential field (pressure, temperature, etc), which is why they are dubbed red or 'hot.' If the length scale viewed is not too large, then large finite clusters just below the percolation threshold have the same appearance as the infinite cluster just above percolation.

Next, we introduce some quantities that characterize the shape of the percolation clusters and their properties. These quantities and their dependence on the probability p will be used in the subsequent chapters. Near the percolation threshold p_c, the correlation length, which describes the Euclidean radius of the largest connected cluster for $p < p_c$, is given by

$$\xi_p = \xi_0(p_c - p)^{-\nu}. \tag{3.1}$$

See table 3.1 for a summary of percolation exponents. As mentioned above, percolation clusters are statistically self-similar fractal structures over any length scale smaller than ξ_p, and Euclidean at length scales larger than ξ_p (Sahimi 1993). Since ξ_p diverges at p_c, then the percolation clusters are fractal objects at p_c when viewed over *any* length scale. For $p > p_c$, ξ_p represents the separation of current-carrying paths, and is a measure of the radius of the largest holes (of unoccupied sites or bonds) in the network. The fraction of sites connected to the infinite cluster is P_∞ and, near p_c, is proportional to $(p - p_c)^\beta$. The mass fractal dimensionality D_f of the infinite cluster is given by, $D_f = d - \nu/\beta$. Thus, $D_f \simeq 2.53$ in 3D systems, and $D_f = 91/48 \simeq 1.9$ in 2D (Stauffer and Aharony 1994). These values have been confirmed for residual fluids—those that remain behind after the fluids that initially saturate a porous medium are displaced by a second fluid and lose their macroscopic

Table 3.1. Values of the the critical exponents according to percolation theory and the EMA.

Exponent	Model	$d=2$	$d=3$
β	Percolation	5/36	0.41
ν	Percolation	4/3	0.89
μ	Percolation	1.3	2.0
μ	EMA	1.0	1.0

connection at p_c—in both 2D (Lenormand and Zarcone 1985) and in 3D (Clement *et al* 1987), leaving no doubt about the relevance of percolation and its associated exponents defined here to problems in porous media. All the exponents and fractal dimensionalities described so far are universal, implying that they depend only on the Euclidean dimensionality of the system, and not on such properties as the connectivity, pore-size distribution, etc.

The mass fractal dimensionality D_{bb} of the backbone depends on the system and the type of percolation problem. Thus, the backbone of invasion percolation problems (relevant to unsaturated media, when one fluid invades a porous medium to expel a second fluid in the pore space) and random percolation (perfectly dry or perfectly saturated media) can have distinct fractal dimensions. In the case of invasion percolation, the fractal dimension may also depend on whether the fluids involved are compressible or incompressible, and on whether the percolation problem is site percolation (displacement of a non-wetting fluid by a wetting one, referred to as imbibition), or bond percolation (drainage, the opposite of imbibition). Moreover, whether the defending (in-place) fluid during invasion remains trapped (i.e., it is incompressible) or not also affects the values of the fractal dimensions, especially in 2D problems. Of course, these values also depend on the dimensionality of the system. An excellent compilation of relevant values is given by Sheppard *et al* (1999), who developed the most efficient algorithm to compute the most accurate estimates of the aforementioned fractal dimensions. The distinction between these values will have relevance to solute transport. In general, when invasion percolation is relevant, solute path lengths are shorter with lower tortuosity, and the solute velocities differ less from fluid velocities than under saturated conditions. Of course, the fluid flow rates are also smaller, since the hydraulic conductivity diminishes with diminishing water content. Table 3.2 also distinguishes between trapping and non-trapping percolation problems. The latter involve fluids that can be approximated as incompressible and, therefore, can remain 'trapped' during both imbibition and drainage, once a cluster of them is surrounded by the displacing (invading) fluid, hence impeding the formation of a connected

Table 3.2. Values of the fractal dimension of the largest percolation cluster (D_f), its backbone (D_{bb}), the shortest path length (D_{min}), and the optimal path length (D_{op}). RP, NTSIP, TSIP, and TBIP refer, respectively, to random percolation, non-trapping site invasion percolation, trapping site invasion percolation (imbibition), and trapping bond invasion percolation (drainage). Estimates are from Sheppard *et al* (1999).

Model	D_f		D_{bb}		D_{min}		D_{op}	
	$d=2$	$d=3$	$d=2$	$d=3$	$d=2$	$d=3$	$d=2$	$d=3$
RP & NTSIP	1.896	2.523	1.643	1.870	1.131	1.347		
TSIP	1.825	2.528	1.217	1.861	1.214	1.370	1.210	1.43
TBIP	1.825	2.528	1.217	1.458	1.217	1.458		

cluster spanning the entire sample. Because in the practical cases dealt with here, both air and water are essentially incompressible and can, therefore, be trapped, only the last two rows of table 3.2 are really relevant to typical unsaturated porous media problems.

The shortest pathway, also called the *chemical path*, between two points that are connected on the infinite cluster, scales as the $D_{\min}$ power of the Euclidean separation L between the two points, i.e., $L^{D_{\min}}$; see figure 3.2. Flow through the shortest path is not generally faster, because flow rates in the connected cluster depend on the series of local conductances—specifically the least conductive bond. Therefore, as in the example illustrated in figure 3.2, the fastest path follows the least resistive path, but may be longer then the shortest one. The length of the optimal path through a rough energy landscape scales as the D_{op} power of L, as shown by Sheppard *et al* (1999).

3.4 Scaling of the macroscopic conductivity

Near the percolation threshold p_c, in both bond and site percolation problems, in which the bonds have equal strength, or they are distributed according to a relatively narrow distribution, the effective conductivity g_e of the medium is given by (Kirkpatrick 1973, Sahimi *et al* 1983b),

$$g_e = g_0(p - p_c)^\mu, \tag{3.2}$$

where g_0 is a constant, with its value being related to the resistances of the bonds, and the morphology of the network. Its prediction is not the focus here. The exponent μ takes on the value 2 in three dimensions (Clerc *et al* 2000, Gingold and Lobb 1990) and 1.3 in two dimensions (Normand and Herrmann 1990). Two-dimensional values are appropriate when the conduction path is forced to remain on a surface, such as along fractures, or along the surfaces of minerals. In such situations, however, p_c tends to be larger, so that in fact the conductivity at smaller values of p is not larger than in three dimensions. The power law (3.2) is also applicable, in many cases, to continuum percolation. If the bonds do not have equal conductance, and the conductances are broadly distributed, a more suitable strategy for estimating the conductivity may be the critical-path analysis, as described below.

It is often pointed out that continuum percolation can produce non-universal values of the exponent μ (Kogut and Straley 1979, Sahimi *et al* 1983a, Halperin *et al* 1985). The non-universality arises when the bonds' conductances are not equal, and in fact their statistical distribution satisfies certain constraints. The same non-universality is also obtained using the critical-path analysis (Hunt 2004a), which, moreover, provides verified predictions even when power laws of the form of equation (3.2) are not generally valid (Hunt and Gee 2002, Ghanbarian and Hunt 2012).

3.5 Water partitioning in the pore space

Although water partitioning in a pore space is not exactly a percolation problem, some understanding of the phenomenon is necessary before addressing saturation-dependent properties. Thus, a minor digression is required at this point. A curved interface between two immiscible fluids indicates a pressure difference between the two phases. For example, when water is at a slightly higher pressure than the air, the pressure difference, usually referred to as the capillary pressure P_c, is related to the meniscus curvature by the Young–Laplace equation:

$$P_c = P_w - P_a = \gamma \left(\frac{1}{r_1} + \frac{1}{r_2} \right), \tag{3.3}$$

where the relevant interfacial energy γ—the surface tension—is between the two fluids, and r denotes the radius of meniscus curvature, with subscripts 1 and 2 denoting (any) two orthogonal directions (assumed equal in the following). For the case of capillary rise, with liquid at equilibrium in a vertical tube of radius r with circular cross-section, the downward force is $2\pi r h (\rho_w - \rho_a) g$, balanced by the upward capillary force, $2\pi r \gamma \cos(\theta)$, where ρ_w and ρ_a are, respectively, the densities of water and air, g is the gravitational constant, and θ is the contact angle. Therefore,

$$h = \frac{2\gamma \cos\theta}{gr(\rho_w - \rho_a)} \equiv \frac{A}{r} \tag{3.4}$$

Here, A is the combination of all the constants in equation (3.4), and is equal to 1.48×10^{-5} m^2 under standard conditions (20 °C), Note that all the water in the tube, which is above the free water surface, is at a negative pressure. Because pressure is exerted by a standing column of water, earth scientists often express pressure in terms of the height of the water column required to produce it.

The strong wetting condition is the assertion that all pores with radii $r > A/h$, where h is the height above a water table, are filled with air, while pores with smaller radii are filled with water. For our purpose, this assumption is sufficient, except for the question of accessibility, which is a topic of percolation theory, and treated next. Note that, in a medium with largest pore, r_m, it is also possible to define an air-entry pressure, $h_A = A/r_m$. This is not actually the pressure at which water starts to drain, however, since air cannot find a connected path through the medium until sufficient air-filled pores become allowable to form an interconnected path.

3.6 Accessibility

For $p > p_c$, a fraction $X_a(p)$ of the connected bonds (or sites) is actually connected to the infinite cluster. All other sites and bonds will be on finite clusters. Near p_c, $X_a(p)$ follows the power law, $X_a(p) \sim (p - p_c)^\beta$, where β is the exponent defined earlier. The importance of this physical fact is felt especially in flow of two immiscible fluids in a porous medium, such as wetting and drying and the associated hysteresis. An important contributor to the hysteresis, the so-called 'link-bottle effect,' is well-known to the porous media community (Lenhard *et al* 1991, Lenhard 1992). According to the capillary relationship, equation (3.4), when an air–water–solid system is at equilibrium, a given pore can be filled with water only, if its radius is less than some value: $r < A/h$. Water is 'allowed' in such pores, i.e., it can invade and fill it up. But, according to the pore-body and pore-throat picture of porous media, the tension required to remove water from such a pore is higher than what allows water in: in the removal process (drainage) the meniscus must 'fit' through a pore throat, whereas in the filling process (imbibition) the meniscus must span the pore body. This fundamental asymmetry in the wetting and drying processes means that, at a given water content, the pressure in a drying curve should be higher than in a wetting curve. This factor is due to the geometry of an individual pore, but in a system with self-similar properties, it relates to every pore. According to Lenhard *et al* (1991) and Lenhard (1992), the ratio of the two pressures is typically about 2.

Consider imbibition under a high tension (negative water pressure) into an initially dry medium. Water cannot access most pores that are allowed, because the paths to such pores pass through other pores that are not small enough to allow water (Heiba *et al* 1982, 1992). This problem is obviously related to percolation theory, but it could be argued that wetting should be discussed only in the context of invasion percolation (Wilkinson and Willemson 1983) and wetting fronts. Nevertheless, Hunt (2004b) showed that an expression of the actual moisture content in terms of the allowable moisture content, and the percolation accessibility function X_a defined above, both written in terms of the capillary rise h, when combined with the cited ratio, 2, of pore-body to pore-throat radii, produces excellent agreement with experiment, with no adjustable parameters whatsoever. For our purpose, we will largely neglect imbibition and drainage dynamics and focus on longer time scales at which the details of drying-wetting events can be neglected. In the case of diffusion or solute transport, which depends on the topology of the percolation backbone, the distinction between random and invasion percolation is critical.

3.7 Finite-size scaling

The subject of finite-size scaling was first clearly elucidated by Fisher (1971). It is an essential aspect of percolation theory, but cannot be considered in detail here. The most important aspect of finite-size scaling is that, if a property y depends on the percolation variables as, $y \sim (p - p_c)^q$, then the dependence of y on the system size L for $p = p_c$ is, $y \sim L^{-q/\nu}$, where ν is the exponent of the percolation correlation length defined above. Thus, a quantity that diverges (vanishes) in the approach to the percolation threshold ($q < 0; q > 0$), also diverges (vanishes) at the percolation threshold when L is unbounded, but with an exponent modified by the exponent of the correlation length ν. This is highly useful, particularly in numerical calculations, where it may be otherwise impossible to isolate effects of finite-size systems.

3.8 Critical-path analysis

To understand and quantify the contributions of a pore-size (or pore-conductance) distribution to the hydraulic conductivity, one invokes the critical-path analysis (CPA). It can be utilized in studying of unsaturated porous media, modeled by pore networks, characterized by wide ranges of pore conductances. The conceptual basis for the CPA is the application of simplified descriptions of two idealized bond configurations: (i) The equivalent resistance of two bonds in series, whose resistances are greatly different, is essentially the same as the larger of the two; (ii) the equivalent resistance of the same two resistances, but in parallel configuration, is the smaller one. In a complex porous medium that contains series and parallel configurations of pores, we seek the interconnected path through local conductances that yields the largest possible value of its smallest conductance, which we call the *critical conductance*. The smallest conductance on the path defines the conductance of the path because, according to (i), the smallest conductance in a series is the most limiting one. When a multitude of paths (each controlled by their smallest conductance) occurs in parallel, condition (ii) stipulates that the largest possible value of the smallest conductance on each path contributes most to the medium conductance. The tool to define the critical conductance is provided by percolation theory.

The CPA applies percolation theory to select the critical conductance from a distribution of local preconductances; see figure 3.3. We consider the case of conduction in a fixed network with randomly distributed local conductances, i.e., the conductances linking each pair of neighboring sites that are identically and independently distributed. Defining such a conductance distribution on a symmetric lattice does not give rise to any conceptual difficulties, though, as will be seen later, the definition in continuum percolation problems is somewhat more subtle. Thus, for simplicity, we begin with bond percolation. Then, one introduces the normalized local conductance distribution, $w(g)$, by

$$\int_0^\infty w(g)dg = 1. \tag{3.5}$$

The cumulative distribution, $W(g < g')$ is then given by,

$$W(g < g') = \int_0^{g'} w(g)dg. \tag{3.6}$$

Network of random local conductances:
what is the local conductance controlling the macroscopic conductance?

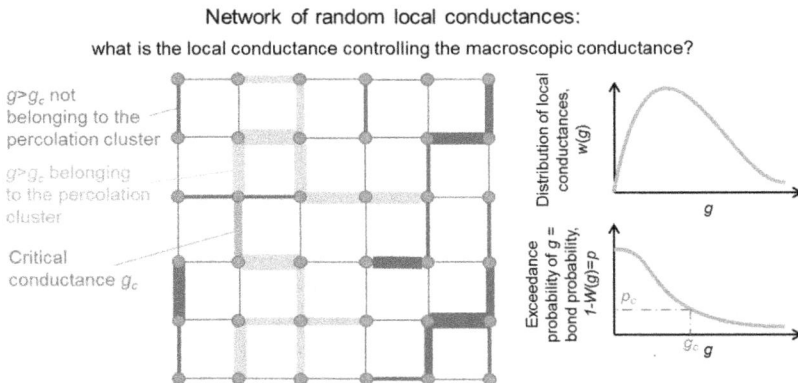

Figure 3.3. Schematic representation of critical path and critical conductance.

We are interested in the complementary function, the exceedance probability of g', defined by

$$1 - W(g < g') = \int_{g'}^{\infty} w(g)dg. \tag{3.7}$$

Because the individual conductances are independently and identically distributed, $1 - W(g < g')$ may refer to either the probability that a given bond has a conductance $g < g'$, or the fraction of such bonds with $g < g'$. The latter definition underlies the application of percolation theory. Since $1 - W(g < g')$ equals a fraction between 0 and 1, it may, therefore, be identified with a bond probability p, a definition that allows selection of a subset of conductances that includes all those with conductances larger than a specified minimum conductance g'. If bond percolation threshold p_c is known for the network, we can determine the value of $g' \equiv g_c$ for which the subset of all conductances larger than or equal to g' percolates, by setting $1 - W(g' = g_c) = p_c$. The choice of p_c guarantees that an infinitely large cluster of interconnected conductances with no conductance smaller than g_c exists. g_c is then what we called the *critical conductance*. It has been shown that the placement of the critical conductances controls the entire field of pressure drops along the pores (Bernabé and Bruderer 1998), if the original distribution $w(g)$ is wide enough. This is the mathematical reason for preferring a deterministic framework for the calculation of the conductivity of a porous medium to a stochastic one.

In order to apply the entire framework of percolation theory to flow, conduction, and other transport problems in any arbitrary network, we must be able to derive an expression for $p - p_c$ in terms of the conductances. A consistent definition of $p - p_c$ is possible using,

$$p - p_c = 1 - W(g) - [1 - W(g_c)] = W(g_c) - W(g). \tag{3.8}$$

Thus, if we substitute for the correlation length near p_c, we obtain,

$$\xi_p = \xi_0(p - p_c)^{-\nu} = \xi_0[W(g_c) - W(g)]^{-\nu}, \tag{3.9}$$

or for the cluster statistics, i.e., the number n_s of cluster of size s, where size refers to number of interconnected bonds or sites,

$$n_s = s^{-\tau}f\{s^{\sigma}[W(g_c) - W(g)]\}, \tag{3.10}$$

and for the tortuosity,

$$\Gamma_s = \Gamma_0[W(g_c) - W(g)]^{-D_{opt}}. \tag{3.11}$$

In equation (3.10), τ and σ are two universal exponents whose values depend only on the dimension of the underlying model or system, i.e., 2D or 3D, and $f(x)$ is a function whose precise nature is not known, but is well-approximated by a Gaussian distribution (Stauffer 1979). The fractal dimension D_{op} describes the tortuous length of the optimal paths (see above), and Γ_0 is a numerical constant whose value can often be determined by examining physical limits, such as when porosity is equal to 1, where the tortuosity must be 1 as well. All other results apply as well. Thus, such an identification of $p - p_c$ with the difference of the cumulative conductance distribution evaluated at g_c and at g allows one to bring all the machinery of percolation theory to bear on a problem in which the conductances are broadly distributed, rather than merely a binary distribution. Most importantly, this kind of perspective allows one to focus on the variability of the pore space, rather than on the distinction between the pore space and the solid medium. The appropriate idealization of a pore space that is fractal, with wide ranges in pore lengths, radii, and local coordination numbers, is continuum percolation, where the appropriate variable is a volume fraction and the percolation probability corresponds to a critical volume fraction. In such a case, we are not interested in the probability that a given pore has a given conductance, but, instead, are interested in the fractional volume associated with a particular conductance value. This makes the identification of the local 'conductance distribution' slightly more complex, but the rewarding results show that the extra effort is worth it.

In principle, as the percolation threshold is approached, i.e., in the context of the problems considered in this book, a porous medium moves towards drier conditions, the equation or estimate for the hydraulic conductivity obtained using the CPA should be replaced by one incorporating the universal power laws of percolation. As it turns out, however, this region of universal scaling is usually narrow, confined to moisture contents of roughly 10% or so by volume (saturations roughly 25% for a porosity of 0.4). Under certain circumstances, the regime of universal scaling disappears altogether, and the CPA formula obtained has been shown (Hunt 2004a) to agree with known non-universal exponents of percolation theory in continuum percolation (Balberg 2021). While this is a necessary result, it would be of peripheral relevance here, if the conductivity exponent were not so often used as an adjustable parameter (Mualem and Dagan 1978, Mualem 1986, Moldrup *et al* 2000a, 2000b, 2003, 2004, 2005a, 2005b). But it is important to point out that it has been shown that calculations carried out based on the CPA (Ghanbarian and Hunt 2012b) have better predictive capabilities than other theories, while justification of non-universal exponents in other contexts are artifacts of inappropriate measurement protocols (Ewing *et al* 2015).

3.9 Effective-medium approximation

When the pore-size, or pore-conductance, distribution is not of critical importance to a flow or transport property, and when the system is not close to the percolation threshold, it may be necessary to use techniques for calculating upscaled properties that are not percolative in nature. The best such alternative method is known as the effective-medium approximation, or EMA, sometimes called *poor man's percolation theory* (Sahimi 2023).

The EMA provides a simple description of flow and transport in pore and resistor networks in which a percolation disorder is relevant (Hunt and Sahimi 2017). The range of validity of EMA is normally complementary to that of percolation theory, meaning far from, rather than close to, the percolation threshold, or for small, rather than large permeability contrasts. In particular, EMA is applicable to porous media for which some of the pore throats or bonds do not conduct or allow fluid flow. In agreement with percolation theory, the prediction of the EMA for the percolation threshold of the system is nontrivial, i.e., it is neither zero nor 1, but something in between. It also predicts a power-law dependence of the permeability and conductivity of the networks on $p - p_c$ near the percolation threshold, except that the predicted exponent is incorrect. Surprisingly, the EMA can even provide an estimate of the fractal dimension of the sample-spanning percolation cluster defined earlier (Sahimi 1984). A detailed derivation of the EMA and an extensive discussion of its various aspects have been given by Sahimi (2003). Here, we simply summarize the most important results that are relevant to our discussions here.

Consider a network in which the conductance g, or the diffusivity D, of the bonds (pore throats) between the network's nodes is randomly and independently distributed according to a normalized distribution $f(D)$. The EMA replaces such a network with an effective network in which the conductance of all the bonds is D_m. Since the network is now uniform, D_m is also the overall diffusivity of the network. The lowest order EMA for the effective diffusivity D_m is obtained via physical arguments (Kirkpatrick 1971, 1973), or via lattice Green functions (Sahimi *et al* 1983b, Keffer *et al* 1996)

$$\int_0^\infty \frac{D_m - D}{D + D_m(Z/2 - 1)} f(D) dD = 0, \tag{3.12}$$

where Z is the mean coordination number or connectivity. Keffer *et al* (1996) used as a distribution of diffusivities (to ultimately estimate the diffusivity in zeolites),

$$f(D) = (1 - 0)\delta(D - D_b) + p\delta(D - D_0), \tag{3.13}$$

with D_b being a very small value, and D_0 relatively large. The solution of equation (3.12) using equation (3.13) for $f(D)$ is,

$$\frac{D_m}{D_0} = \frac{1}{2} \left[A + \left(A^2 + \frac{4k}{Z/2 - 1} \right)^{1/2} \right], \tag{3.14}$$

where $k = D_b/D_0$, and

$$A = k(1 - p) + p - \frac{k + 1 - p - k(1 - p)}{Z/2 - 1}. \tag{3.15}$$

In the limit, $k = 0$, equation (3.14) yields

$$\frac{D_m}{D_0} = \frac{p - 2/Z}{1 - 2/Z}, \tag{3.16}$$

which yields $p_c = 2/Z$ and a critical exponent, $\mu = 1$. This would be in agreement with the results of percolation theory, except that $p_c = 2/Z$ is more appropriate for 2D, rather than 3D systems, and the critical exponent is incorrect. With the identification noted, equation (3.16) becomes

$$\frac{D_m}{D_0} = \frac{p - p_c}{1 - p_c}.$$ (3.17)

Thus, according to the EMA, in any physical dimension—$d=2$ and 3—the percolation exponent μ for the power-law behavior of the effective conductivity or diffusivity is $\mu = 1$. This result is independent of $f(D)$, even if we replace it with a more general distribution. Of course, as pointed out above, near the percolation threshold, the exponent is $\mu = 2$ in three dimensions. But, otherwise, the argument of this equation is the same as that of the percolation power law, a result that allows some important simplification, and also has significant consequences.

As far back as 1971, Kirkpatrick demonstrated the need to use equation (3.17) for the conductivity of a regular network when $p \gg p_c$. In particular, the crossover value of p that he estimated was $p \approx 0.75$. For $p > 0.75$, the EMA is more accurate than the power law of percolation theory; for $p < 0.75$, the percolation power law is so. In porous media with significant disorder, the critical region in which the percolation expression $[(p - pc)/(1 - pc)]^2$ is valid, is quite wide, in accord with a choice of a crossover value of p, 0.75. But, in more ordered porous media (such as engineered media with glass beads), the crossover p is quite small. For further discussion of this point, see Ghanbarian *et al* (2015).

Figure 3.4 illustrates the predicted crossover from percolation scaling to the EMA for the case of gas diffusion at a porosity of ϵ_x. Note that we do not really expect a sharp slope break at the crossover. Such a change of slope is not seen in natural media, because porosity seldom exceeds 0.6, but the next chapter shows that it is

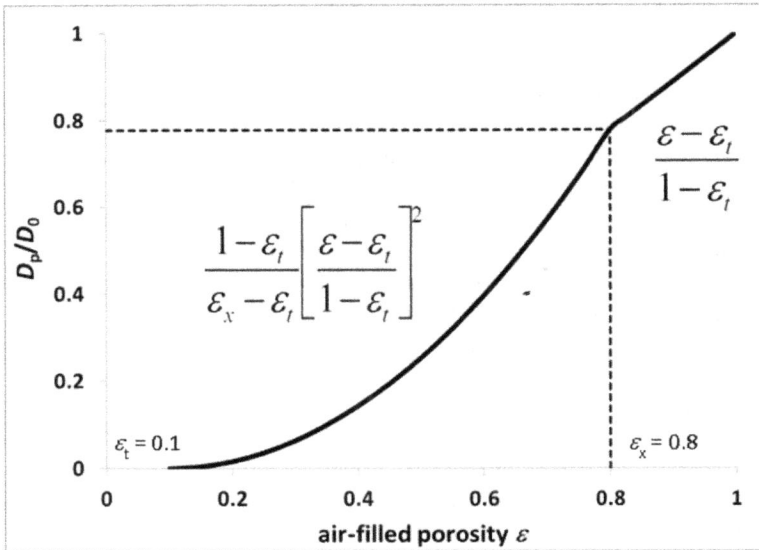

Figure 3.4. Graphical representation of the crossover from percolation power law to effective-medium approximation. This particular example is for gas diffusion, so that the role of p is played by the 'air-filled porosity' ϵ, while ϵ_t corresponds to p_c. D_p is the diffusivity of porous medium, with D_0 being its value in air.

indeed observed in some artificial media in which the crossover p is much smaller. Moreover, the slope change is observed in the thermal conductivity as a function of saturation (Hunt *et al* 2014).

References

Ambegaokar V N, Halperin B I and Langer J S 1971 Hopping conductivity in disordered systems *Phys. Rev.* B **4** 2612

Balberg I 2021 Principles of theory of continuum percolation *Complex Media and Percolation Theory* ed M Sahimi and A G Hunt (Berlin: Springer) p 89

Bernabé Y and Bruderer C 1998 Effect of the variance of pore size distribution on the transport properties of heterogeneous networks *J. Geophys. Res.* **103** 513

Blöschl G and Sivapalan M 1995 Scale issues in hydrological modelling: A review *Hydrol. Process.* **9** 251

Clement E, Baudet C, Guyon E and Hulin J P 1987 Invasion front structure in a 3D model porous medium under a hydrostatic pressure gradient *J. Phys.* D **20** 608

Clerc J P, Podolskiy V A and Sarychev A K 2000 Precise determination of the conductivity exponent of 3D percolation using exact numerical renormalization *Eur. Phys. J.* B **15** 507–16

Cushman J H, Bennethum L S and Hu B X 2002 A primer on upscaling tools for porous media *Adv. Water Resour.* **25** 1043–67

Desbarats A J 1992 Spatial averaging of hydraulic conductivity in three-dimensional heterogeneous porous media *Math. Geol.* **24** 249–67

Deutsch C 1989 Calculating effective absolute permeability in sandstone/shale sequences *SPE Form. Eval.* **4** 343–8

Ebrahimi F and Sahimi M 2002 Multiresolution wavelet coarsening and analysis of transport in heterogeneous media *Physica* A **316** 160

Ebrahimi F and Sahimi M 2004 Multiresolution wavelet scale up of unstable miscible displacements in flow through heterogeneous porous media *Transport Porous Med.* **57** 75

Ewing R P, Ghanbarian B and Hunt A G 2015 Gradients and assumptions affect interpretation of laboratory-measured gas-phase transport *Soil Sci. Soc. Am. J.* **79** 1018

Fisher M E 1971 The theory of critical point singularities, in *Critical Phenomena Proc. of 1970 Enrico Fermi Int. School on Physics* ed M S Green (Varenna, Italy: Academic) p 1

Fitzpatrick J P, Malt R B and Spaepen F 1974 Percolation theory and conductivity of random close of packed mixtures of hard spheres *Phys. Lett.* A **47** 207

Freeze R A 1975 A stochastic-conceptual analysis of one-dimensional groundwater flow in nonuniform homogeneous media *Water Resour. Res.* **11** 725

Gelhar L W 1986 Stochastic subsurface hydrology from theory to applications *Water Resour. Res.* **22** 1358

Ghanbarian B, Daigle H, Hunt A G, Ewing R P and Sahimi M 2015 Gas and solute diffusion in partially saturated porous media: Comparison with lattice-Boltzmann simulations *J. Geophys. Res. Solid Earth* **120** 182

Ghanbarian-Alavijeh B and Hunt A G 2012a Comparison of the predictions of universal scaling of the saturation dependence of the air permeability with experiment *Water Resour. Res.* **48** W08513

Ghanbarian-Alavijeh B and Hunt A G 2012b Unsaturated hydraulic conductivity in porous media: percolation theory *Geoderma* **187–88** 77

Ghanbarian-Alavijeh B, Skinner T E and Hunt A G 2012 Saturation dependence of dispersion in porous media *Phys. Rev.* E **86** 066316

Gingold D B and Lobb C J 1990 Percolative conduction in three dimensions *Phys. Rev.* B **42** 1990

Halperin B I, Feng S and Sen P N 1985 Differences between lattice and continuum percolation transport exponents *Phys. Rev. Lett.* **54** 2391

Heiba A A, Sahimi M, Scriven L E and Davis H T 1982 Percolation theory of two-phase relative permeability SPE Paper 11 015

Heiba A A, Sahimi M, Scriven L E and Davis H T 1992 Percolation theory of two-phase relative permeability *SPE Reservoir Eng.* **7** 123

Hunt A G 2004a Percolative transport and fractal porous media *Chaos Solitons Fractals* **19** 309

Hunt A G 2004b Continuum percolation theory for water retention and hydraulic conductivity of fractal soils: 2. Extension to non-equilibrium *Adv. Water Resour.* **27** 245

Hunt A G, Ewing R P and Ghanbarian B 2014 *Percolation Theory for Flow in Porous Media* 3rd edn (Berlin: Springer)

Hunt A G and Ewing R P 2015 Scaling *Handbook of Groundwater Engineering* ed J H Cushman and D Tartakovsky (CRC Press)

Hunt A G and Gee G W 2002 Application of critical path analysis to fractal porous media: comparison with examples from the Hanford Site *Adv. Water Resour.* **25** 129

Hunt A G and Sahimi M 2017 Flow, transport, and reaction in porous media: percolation scaling, critical-path analysis, and effective-medium approximation *Rev. Geophys.* **55** 993

Jerauld G R, Scriven L E and Davis H T 1984 Percolation and conduction on the 3D Voronoi and regular networks: a second case study in topological disorder *J. Phys.* C **16** 3429

Keffer D, McCormick A V and Davis H T 1996 Diffusion and percolation on zeolite sorption lattices *J. Phys. Chem.* **100** 967

Kirkpatrick S 1971 Classical transport in disordered media: scaling and effective-medium theories *Phys. Rev. Lett.* **27** 1722

Kirkpatrick S 1973 Percolation and conduction *Rev. Mod. Phys.* **45** 574

Knudby C, Carrera J, Bumgardner J D and Fogg G E 2006 Binary upscaling–the role of connectivity and a new formula *Adv. Water Resour.* **29** 590

Kogut P M and Straley J P 1979 Distribution-induced non-universality of the percolation conductivity exponents *J. Phys.* C **12** 2151

Koltermann C E and Gorelick S M 1996 Heterogeneity in sedimentary deposit: a review of structure-imitating and descriptive approaches *Water Resour. Res.* **32** 2617

Kunz H and Souillard B 1978 Essential Singularity in the Percolation Model *Phys. Rev. Lett.* **40** 133–5

Lenhard R J 1992 Measurement and modeling of 3-phase saturation pressure hysteresis *J. Contam. Hydrol.* **9** 243

Lenhard R J, Parker J C and Kaluarachchi J J 1991 Comparing simulated and experimental hysteretic 2-phase transient fluid-flow phenomena *Water Resour. Res.* **27** 2113

Lenormand R and Zarcone C 1985 Invasion percolation in an etched network: Measurement of a fractal dimension *Phys. Rev. Lett.* **54** 2226

Liu J and Regenauer-Lieb K 2011 Application of percolation theory to microtomography of structured media: percolation threshold, critical exponents, and upscaling *Phys. Rev.* E **83** 016106

Loague K and Corwin D 2006 Issues of scale *The Handbook of Groundwater Engineering* 2nd edn ed J W Delleur, F L Boca Raton and C R C Press (Boca Raton, FL: CRC Press)

Malliaris A and Turner D T 1971 Influence of particle size on the electrical resistivity of compacted mixtures of polymeric and metallic powders *J. Appl. Phys.* **42** 614

Moldrup P, Olesen T, Gamst J, Schjønning P, Yamaguchi T and Rolston D E 2000a Predicting the gas diffusion coefficient in repacked soil: water-induced linear reduction model *Soil Sci. Soc. Am. J.* **64** 1588

Moldrup P, Olesen T, Schjønning P, Yamaguchi T and Rolston D E 2000b Predicting the gas diffusion coefficient in undisturbed soil from soil water characteristics *Soil Sci. Soc. Am. J.* **64** 94

Moldrup P, Yoshikawa S, Olesen T, Komatsu T and Rolston D E 2003 Gas diffusivity in undisturbed volcanic ash soils: test of soil-water-characteristic-based prediction models *Soil Sci. Soc. Am. J.* **67** 41

Moldrup P, Olesen T, Yoshikawa S, Komatsu T and Rolston D E 2004 Three-porosity model for predicting the gas diffusion coefficient in undisturbed soil *Soil Sci. Soc. Am. J.* **68** 750

Moldrup P, Olesen T, Yoshikawa S, Komatsu T and Rolston D E 2005a Predictive-descriptive models for gas and solute diffusion coefficients in variably saturated porous media coupled to pore-size distribution. I. Gas diffusivity in repacked soil *Soil Sci.* **170** 843

Moldrup P, Olesen T, Yoshikawa S, Komatsu T and Rolston D E 2005b Predictive-descriptive models for gas and solute diffusion coefficients in variably saturated porous media coupled to pore-size distribution. II. Gas diffusivity in undisturbed soil *Soil Sci.* **170** 854

Moslehi M, de Barros F P J, Ebrahimi F and Sahimi M 2016 Upscaling of solute transport in heterogeneous porous media by wavelet transformations *Adv. Water Resourc.* **96** 180

Mualem Y 1986 Hydraulic conductivity of unsaturated soils: Prediction and formulas *Models of Soil Analysis. Part I. Physical and Mineralogical Methods* ed G S Campbell, R D Jackson, A Klute, M M Mortland and D R Nielsen (American Society of Agronomy—Soil Science Socity of America) p 799

Mualem Y and Dagan G 1978 Hydraulic conductivity of soils–unified approach to statistical models *Soil Sci. Soc. Am. J.* **42** 392

Muneepeerakul R, Bertuzzo E, Lynch H J, Fagan W F, Rinaldo A and Rodriguez-Iturbe I 2008 Neutral metacommunity models predict fish diversity patterns in Mississippi–Missouri basin *Nature* **453** 220

Neuweiler I and Vogel H-J 2007 Upscaling for unsaturated flow for non-Gaussian heterogeneous porous media *Water Resour. Res.* **43** W03443

Normand J-M and Herrmann H J 1990 Precise numerical determination of the superconducting exponent of percolation in three dimensions *Int. J. Mod. Phys.* C **1** 207

Pollak M 1972 A percolation treatment of dc hopping conduction *J. Non Cryst. Solids* **11** 1

Rasaei M R and Sahimi M 2008 Efficient simulation of water flooding in three-dimensional heterogeneous reservoirs using wavelet transformations: application to the SPE-10 model *Transport Porous Med.* **72** 311

Rasaei M R and Sahimi M 2009 Upscaling of the permeability by multiscale wavelet transformations and simulation of multiphase flows in heterogeneous porous media *Comput. Geosci.* **13** 187

Sahimi M 1984 Effective-medium approximation for density of states and the spectral dimension of percolation networks *J. Phys.* C **17** 3957

Sahimi M 1993 Flow phenomena in rocks–from continuum models to fractals, percolation, cellular automata, and simulated annealing *Rev. Mod. Phys.* **65** 1393

Sahimi M 2003 *Heterogeneous Materials* **vols I** (Berlin: Springer)

Sahimi M 2011 *Flow and Transport in Porous Media and Fractured Rock* 2nd edn (New York: Wiley-VCH)

Sahimi M 2023 *Applications of Percolation Theory* 2nd edn (Berlin: Springer)

Sahimi M, Hughes B D, Scriven L E and Davis H T 1983a Stochastic transport in disordered systems *J. Chem. Phys.* **78** 6849

Sahimi M, Hughes B D, Scriven L E and Davis H T 1983b Real-space renormalization and effective-medium approximation to the percolation conduction problem *Phys. Rev.* B **28** 307

Sahimi M and Yortsos Y C 1990 Applications of fractal geometry to porous media: a review SPE Paper 20 476

Scheibe T and Yabusaki S 1998 Scaling of flow and transport behavior in heterogeneous groundwater systems *Adv. Water Resour.* **22** 223

Seager C H and Pike G E 1974 Percolation and conductivity: a computer study II *Phys. Rev.* B **10** 1435

Sheppard A P, Knackstedt M A, Pinczewski W V and Sahimi M 1999 Invasion percolation: new algorithms and universality classes *J. Phys.* A **32** L521

Stauffer D 1979 Scaling theory of percolation clusters *Phys. Rep.* **54** 1–74

Stauffer D and Aharony A 1994 *Introduction to Percolation Theory* 2nd edn (London: Taylor and Francis)

Vereecken H, Kasteel R, Vanderborght J and Harter T H 2007 Upscaling hydraulic properties and soil water flow processes in heterogeneous souls *Vadose Zone J.* **6** 1

Wilkinson D and Willemsen J 1983 Invasion percolation: A new form of percolation theory *J. Phys.* A **16** 3365

Winterfeld P H, Scriven L E and Davis H T 1981 Percolation and conductivity of random twodimensional composites *J. Phys.* C **14** 2361

IOP Publishing

Networks on Networks (Second Edition)
Role of connectivity in physics of geobiology and geochemistry
Allen G Hunt and Muhammad Sahimi

Chapter 4

Predicting morphological, flow, and transport properties of porous media

4.1 Background

Having discussed in chapters 2 and 3 how to define and analyze pore-networks that model the morphology of a porous medium, we describe in this chapter use of such networks to compute the macroscopic properties of a pore space, which ultimately affect functioning of soil and ecosystem. Note, however, that it is typical in the community of researchers who work on porous media to distinguish between hydraulic properties, which are related to fluid flow, and conductivity—either electrical or thermal—and to regard solute transport as being in a third category, which we will describe in chapter 5. Physicists make no such fundamental distinction, and refer to all of them as *transport* properties. The steps to take in order to link properties of a pore network to ecological, hydrological, and biogeochemical functioning are as follows.

(i) Develop a pore-network model that is accurate enough to capture the essential characteristics of soil. Clearly, the model must be advanced enough to provide useful predictions for the properties of interest. In the course of this chapter, some of the criteria for evaluating the suitability of such a model will be described. Since the predictions of the models are rather far-reaching, at least for the properties and phenomena studied in this book, the proposed description of porous medium appears adequate. Whatever limitations the models may have can be more seriously tested in other contexts.

(ii) Define the physical properties of the porous medium—its porosity, statistical distributions of its pore and particle sizes, coordination or connectivity number, any possible correlations in the pore space between the pore sizs, and, for two-phase flow, wetting properties of the solid surface, and the critical volume fractions for percolation of each fluid phase.

doi:10.1088/978-0-7503-5698-5ch4

(iii) Compute the volumetric water content, the hydraulic conductivity—water-phase permeability—air permeability, solute and gas diffusion coefficients, electrical conductivity of fluid-saturated pore space, thermal conductivity, properties that are involved in coupled processes, such as heat and mass transport, phase changes, and hysteresis in wetting and drying. In addition, compute conservative, non-reacting solute transport properties by advection.

(iv) Assess the chemical properties of the medium's constituents and their correlations with its physical properties.

(v) Take into account the interaction between chemical reactions and solute transport.

(vi) Identify biological constituents of the medium, their physical properties, growth, interactions with various species, adaptability, and the changes that they introduce in the physical and chemical properties of the medium, and quantify the fluxes of heat, water, and various chemical compounds across the boundaries of the medium.

(vii) Consider the availability of water, the effects of climate and climate change, and other external factors, such as anthropogenic ones.

In this book we focus on (ii), (iii), (v), and (vi). Item (vii) is way beyond the scope of this book. Item (i), which underlies the entire discussion, is, at least in detail, also outside the present scope, but has been subject of numerous previous investigations (see Sahimi 2011, for detailed discussions) and key definitions and concepts were already provided in chapter 3. In particular, in order to explain the power-law behavior of a wide range of properties in a set of broad circumstances, fractal models of porous media have been popular over the past 30 years. While the resulting power-law distributions of the pore sizes do appear relevant to water retention models, and possibly to the hydraulic conductivity, most other properties described in (iii) exhibit universal power-law behavior near the connectivity or percolation threshold, which is not a product of any particular morphology of soil or other types of porous media. Thus, most of the power laws that we will document relate to percolation theory for fluid flow and transport, rather than the medium itself. Nevertheless, use of a concrete model is essential, and since the fractal models appear best suited to the typical water retention data, the prior work has typically utilized power-law pore-size or pore-conductance distributions, commensurate with some simpler fractal models; see, for example, models that are due to de Gennes (1985), Rieu and Sposito (1991), Tyler and Wheatcraft (1990), Perrier *et al* (1996), and Perfect (1999). Item (iv) is also somewhat peripheral to our goal in this book, as our purpose is to point out the universalities in the chemical processes due to the relevance of the underlying physics.

Since, as discussed in chapter 3, it is the pore connectivity that determines the physical properties of a pore network, the appropriate theoretical framework for understanding the dependence of the various properties of porous media on their connectivity, as well as the relevant spatial length scales, is percolation theory. When discussing fluid flow, diffusion, or hysteresis phenomenon in two-phase flow, universal power laws of percolation theory are often relevant. The application of percolation theory is, however, not simple, nor is it universal. For example, when the dominating factor is the pore-size or pore-conductance distribution rather than the connectivity of the pore space in which the fluid resides, the particular aspect of percolation theory that one needs is the critical-path analysis (CPA), described in chapter 3. Even when the pore-size or pore-conductance distribution has a minor effect on a given property, if the system is far from the percolation threshold, the effective-medium approximation (EMA) may be better suited, which is sometimes referred to as (Sahimi 2023) the *poor man's percolation*. Such complications will be considered in the following.

Five macroscopic properties of porous medium that affect the biota stand out. All are directly affected by the pore network characteristics of porous media medium and include: (a) the porosity; (b) the water content and hysteresis in the wetting and drying; (c) the water (or air) flow rate (flux); (d) solute and gas diffusion fluxes; and (e) the advective gas and solute transport fluxes. The second property is the total water content of the medium as a function of the (negative) pressure on the water, which we refer to as the tension. Tension develops due to adhesion of water molecules to the soil's internal surface, creating a negative pressure (matric potential) equal in absolute value to the energy per unit volume necessary to extract water. In chapter 3 we linked the tension to the pore radius. Soil physicists often simplify their perspective by representing the tension at a point as the height of that point above the water table. More rigorously, the tension is the energy required to remove the water from the soil, and its negative is the soil water potential. The water flow rate is defined by the permeability, or hydraulic conductivity, which is simply the water volume crossing a plane per unit area per unit time, divided by the pressure gradient. Solute and gas diffusion coefficients are measured by introducing gas or solutes at, for example, the left side of a medium in which the air and water fluxes are everywhere zero, and measuring the time-dependence of the flux out on the right boundary. Finally, the solute (or gas) transport properties of a porous medium are measured based on the time-dependence of the concentration, or the flux, of a solute at which it emerges from the right side of a medium under steady-state fluid flow, when a solute pulse is introduced on the left side of the medium.

As mentioned above, the traditional approach to describe (c) and (e) has been to treat the soil as a continuum and derive volume- or ensemble-averaged partial differential equations for water flow and transport. Such equations require flow and transport coefficients, namely, the hydraulic conductivity and the hydrodynamic dispersion coefficient, which have historically been estimated by using bundles of parallel capillary tubes (Scheidegger 1974, Mualem and Dagan 1978, Mualem 1986). Such a model is problematic because it is based on simplifying the complex soil structure to a bundle of parallel and independent tubes (Sahimi 2011, Hunt *et al* 2013). To see this, consider the following. First, such a model does not allow any transverse dispersion—dispersion in the direction perpendicular to the direction of macroscopic flow—since it has no transverse connectivity. Second, among many other difficulties, use of arithmetic averaging over tubes implies perfect connectivity and, therefore, a zero percolation threshold while, at the same time, it uses an adjustable residual water content, which is in fact a percolation threshold, hence giving rise to an internal conceptual contradiction and invalidating the entire approach. In practice, the negative exponents for the tortuosity 'corrections' to the hydraulic conductivity typically require paths that have zero length; see Kosugi (1999), Schaap and Leij (2000), Shinomaya *et al* (2001), and Borgesen *et al* (2006). In the case of solute transport, 'Fickian dispersion (a prediction of the capillary bundle model) is anomalous' (Cushman and O'Malley 2015), as also argued by Sahimi (1987), Sahimi and Imdakm (1988), and by Scher, Berkowitz and co-workers since the early 1990s (Scher *et al* 1991, Berkowitz and Scher 1995, Margolin and Berkowitz 2000, Cortis and Berkowitz 2004). Next, consider that the most commonly used phenomenologies, advanced by Moldrup and co-workers, predict power-law dependence on the pore-size distribution for diffusion coefficient that is an artifact of inconsistent measurement protocol and interpretation (Ewing *et al* 2015). The study of the water content of a porous medium is, by virtue of the principal defects in the usual simplifications, disconnected from a pore-scale theory of water (and air) transport, making it impossible to judge whether water allowed in a pore can actually reach that pore.

Thus, an understanding of every physical property of interest, including flow, advective dispersion, diffusion, conduction, and hysteresis, is essentially impossible using the capillary bundle models, but is facilitated through the percolation concepts (Sahimi 1993, 2011, 2023, Hunt *et al* 2014). In fact, percolation theory provides predictive relationships for each of such properties in terms of parameters that have physical significance, and are consistent across properties. The predictions obtained by this parsimonious approach are also as, or more accurate than, the traditional approaches. For example, the only parameter in the saturation dependence of the air permeability that is unknown, the critical air fraction for percolation of air, can be estimated through the pressure-saturation curve (Ghanbarian-Alavijeh and Hunt 2012a) and the dependence of solute arrival time distributions on saturation can be predicted using the entire pressure-saturation curve (Sahimi *et al* 1986, Ghanbarian-Alavijeh *et al* 2012). Other examples will be presented in the rest of this chapter.

The continuum models, based on the partial differential equations—the Laplace equation for water flow, and the advection-diffusion equation for solute transport—though quite similar superficially, are, in fact, quite different at their roots. For example, the Laplace equation for fluid flow in a disordered porous medium should be written as

$$\nabla \cdot (K \nabla P) = 0, \tag{4.1}$$

where K is *the local* hydraulic conductance. If we discretize this equation, we obtain a set of equations that are isomorphic to Kirchoff's law for random resistor networks. Thus, it may appear that the two formulations—continuum hydro-dynamics and pore-network models—are mathematically equivalent over the entire range of spatial scales. Solutions of the Laplace equation in strongly heterogeneous porous media are, however, typically derived using the methods of stochastic differential equations in stochastic subsurface hydrology, which is always some form of perturbation theory, which leads to closure problems on account of the fact that the real heterogeneity is not small or weak and, therefore, the perturbation-based method fails. In other words, if the effects of the heterogeneity are expressed in terms of an infinite sum of correction terms for realistic strengths of the heterogeneity, it is not possible to obtain a convergent pertubation series.

The alternative to pertubation series requires some detailed information about the pore network, as well as mathematical techniques for deriving the solution. Characterization of porous media has advanced remarkably over the past 50 years, while techniques for deriving the solutions have also advanced comparably. Neglecting for the moment the numerical approaches, the two most important analytical procedures are percolation theory and the EMA, in both of which the parameters describing the connectivity and the heterogeneity of the porous medium are explicitly incorporated. The most important aspects of the two approaches are that they reproduce, and often accurately predict, experimental data far more accurately and successfully than any other available technique, and can be utilized to analyze the macroscopic properties of real porous media. Thus, the parameters have physical meaning, and can be verified to be relevant across properties. The properties tested include solute and gas diffusion (Sahimi *et al* 1983, Ghanbarian-Alavijeh *et al* 2014a, 2014b, 2015), tortuosity (Ghanbarian *et al* 2013b), air permeability (Hunt 2005, Ghanbarian-Alavijeh and Hunt 2012a), hydraulic conductivity (Ghanbarian-Alavijeh and Hunt 2012b), and electrical conductivity (Hunt 2004b, 2004c, Ewing and Hunt 2006), particularly their saturation dependence. It is worth repeating that the same percolation framework outperforms competing phenomenologies for every property simultaneously, and uses parameters, such as a minimum moisture content for percolation, which can be determined from one property, such as the water retention curve, and applied to another property, such as the air permeability (Ghanbarian-Alavijeh and Hunt 2012a). Finally, the percolation-based theory is the first to predict successfully an entire non-Gaussian solute arrival time distribution without using adjustable parameters.

4.2 Models of porous media

Many models have been developed to provide a reasonable representation of soil or rock. An important consideration of natural porous media is that most of them have wide ranges of pore sizes, and in particular vary over at least two orders of magnitude in soils. Most models of soil are based on the assumption that pore bodies —referred to for brevity as pores—are connected with neighboring pores through pore throats (or necks). Such media models have been in existence since the 1950s (see chapter 2); were made rigorous much later (Mohanty 1981), and are what we call pore-network models. In the simplest model, one puts the pores on sites of a regular network (or lattice). In such a model, the pore throats are represented by the bonds between the sites. For example, in a square network each pore is connected by four bonds to other pores, but each bond is connected to six other bonds; see figure 3.1. Thus, the local site connectivity, described by a coordination number Z, is 4 in the square network, but 6 in the problem defined by the bonds. The larger bond connectivity provides the basis for understanding the typically lower value of bond percolation thresholds, as compared with site percolation thresholds (Stauffer and

Aharony 1994, Sahimi 2023). Since, during imbibition, pore bodies must be wetted, but during drainage water must pass through the pore throats, imbibition is a site percolation problem, while drainage is a bond percolation problem (Sahimi 1993).

Most of the water storage (and porosity) is confined to the pores, whereas most of the resistance to flow is offered by the pore throats. A considerable body of evidence exists that allows one to make the rough estimates that the typical (or at least characteristic) pore-throat diameters are about one-half that of the pore bodies (Lenhardt et al 1992). Real porous media must, of course, be modeled by more irregular networks, in which pore bodies and throats have both wide ranges of geometrical attributes, such as radius and length, and in which the coordination numbers may vary from place to place. Some of such models have been constructed using Voronoi tessellation (Jerauld et al 1984, Sahimi and Tsotsis 1997).

As pointed out by many, during wetting, the wetting fluid must fill an entire pore, but during drying the wetting fluid must be drawn out through a pore throat. The latter condition requires a greater tension (negative pressure), causing a portion of the hysteretic behavior in the pressure–volume moisture relations. The necessity to distinguish between site and bond percolation associated with wetting and drying may, in many cases, be avoided by considering continuum percolation, but different power-law exponents for advective dispersion may result, while the critical volume fractions for percolation may also be different in the two processes. In each case, however, the accessibility function of percolation theory is required to generate the actual water content from the accessible pore space (Heiba et al 1982, 1992).

In this chapter, such distinctions are not emphasized as we seek the simplest characteristics of pore-network models that allow their application to flow and transport problems of relevance to the growth and flourishing of vascular plants and soil development, i.e., over time scales that are much longer than the typical scales of soil drying and re-wetting. Over such long scales, characterizing and approximating the average behavior is sufficient. Our focus will be on continuum percolation, since such quantities as porosity and moisture content, critical to plant growth, are expressed as volume fractions. In continuum percolation, one must be able to quantify a pore-size or pore-conductance distribution and relate it to the total porosity, and one also requires the critical volume fractions of the air and water phases for percolation. At the present, given the lack of general focus in this particular direction of research, the two values are not widely known for specific soils, although some guidance is available for the proper choice, which helps us to make semi-quantitative estimates of real properties.

One way of incorporating a wide range of pore sizes is to employ a log-normal distribution. A more popular means has been to employ power-law distributions, which also arise naturally when one anaylzes such porous media as packing of particles (Halperin et al 1985). The relevance of such power-law distributions to pressure-saturation curves in porous media has certainly been demonstrated (Filgueira et al 1999, Bird et al 2000, Hunt and Gee 2002a, Ghanbarian-Alavijeh and Hunt 2012c). A convenient way of generating a power-law pore-size distribution is to employ a fractal model, popularized in the late 1980s and early 1990s (Turcotte 1986, Tyler and Wheatcraft 1990, Rieu and Sposito 1991). A frequently-used and simple model for the soil conductance distribution is the discrete random fractal model due to Rieu and Sposito (1991), henceforth called the RS model. While more advanced models (Perrier *et al* 1999) can be constructed (see, e.g., Ghanbarian *et al* 2012), the RS model suffices for the discussions in this book. In order to eliminate explicit reference to fractal structures, its discrete pore-size probability distribution is generalized to the following fractional volume distribution,

$$r^3 W(r) = \frac{(3-D)r^{2-D}}{rm^{3-D}}, \tag{4.2}$$

where r_m is the maximum pore radius, and D is the fractal dimensionality of the pores. If equation (4.2) is assumed to represent the volume fraction of the pore space with radii between r and $r + dr$, its integral between the limits r_0 and r_m, where r_0 is the minimum radius, yields the total porosity. The numerical prefactor has been chosen such that the integral of equation (4.2) between r_0 and r_m yields $1 - (r_0/r_m)^{3-D}$.

In the simplest interpretation—a porous medium with self-similar geometry—pore lengths and pore radii are proportional to each other, at least in the mean, so that the conductance of each pore is proportional to r^3, which is the Hagen–Poiseulle law for conduits of comparable length and radius. Then, the volume-normalized conductance distribution is obtained, $W(g) = g^{-\alpha}$ with $\alpha = D/3$. Note that precisely the same power-law conductance distribution arises in continuum percolation (Halperin *et al* 1985). In this form, the conductance distribution produces known prior predictions for the non-universal scaling of the hydraulic conductivity (Sahimi *et al* 1983b; Halperin *et al* 1985), but only if equation (4.2) is valid all the way to zero pore radius, which might not be a realistic assumption.

The particular pore-size distribution predicted by the RS model has been shown to provide predictions that are in good agreement with soil water retention data—pressure-saturation curves measured under conditions of drying—in a number of comparisons with experiments. Moreover, the RS model is consistent with both the wetting and drying curves in the case that the percolation-based distinction between allowability and accessibility (Heiba *et al* 1982, 1992) was used (Hunt 2004a). Since no adjustable parameters were employed, the case for use of this particular pore-size distribution is also strengthened. While this particular subject is peripheral to the focus of this book, it is certainly relevant to the needs of vegetation that must respond to different conditions, i.e., different soil water potentials, during wetting and drying, even at the same water content.

Several papers have proposed ways of determining or estimating the critical volume fractions for percolation of air and water; see chapter 3. Moldrup *et al* (2001) showed that solute diffusion coefficient tends to vanish at a moisture content related to the specific surface area of the soil. Hunt (2004a, 2004b, 2004c) found that this relationship was an indication for the tendency of water to adhere to surfaces, and the necessity to sum that contribution before counting the water belonging to the water-filled pores. An estimate of about 0.1ϕ has frequently been shown to be a good estimate of the minimal water content for percolation (Ewing and Hunt 2006,

Ghanbarian-Alavijeh *et al* 2014, 2015), where ϕ is the porosity, while Hunt (2004c) found evidence for the relevance of two terms in a critical water content, one from pore surfaces, and a second one for the pore volumes (see also Luckner *et al* 1989). It should be noted that a theoretical relationship for the critical volume, $V_c = p_c \phi$ goes back at least to Scher and Zallen (1970), while the concept of two contributions to a residual (or irreducible) water content has also been generally known in soil physics. Furthermore, the Voronoi tessellation model of Jerauld *et al* (1984) leads to the estimate $p_c = 0.1$.

4.3 Saturated hydraulic conductivity

Due to its strong dependence on pore size, shape, connectivity, and properties of surface minerals, the hydraulic conductivity is the most difficult, and at the same time most important, of the properties for modeling of various phenomena in soil, as evidenced by the characterization of the Critical Zone in terms of its permeability. The Critical Zone is the thin surface layer of Earth that extends from the top of the vegetation to the bottom of drinking water aquifers. As we explain below, the most acurate method for calculating the hydraulic conductivity of soils is the CPA, developed based on percolation theory, on account of the wide range of pore sizes occurring in soils (section 3.2). In the CPA, one uses percolation theory together with the local conductance distribution in order to determine the largest possible value g_c of the smallest conductance for percolation of the flow or transport path, i.e., one for which the associated conductance network still percolates. This particular quantity is known in soil physics as a 'bottleneck' pore. What this means in practice is that parallel disconnected paths have characteristic conductances that are so much smaller than g_c that they can be ignored, while only the smallest conductances on the connected path provide enough resistance to flow to be significant; see figure 3.3.

Indeed, as established by Bernabé and Bruderer (1998), the most accurate method for calculating the hydraulic conductivity of saturated porous media is the CPA, first proposed by Katz and Thompson (1986). Likewise, the CPA for the saturation dependence of the hydraulic conductivity was tested on a database of 107 soils and found to provide the most accurate predictions (Ghanbarian-Alavijeh and Hunt 2012a, 2012b, 2012c), as long as the value of the hydraulic conductivity does not fall by more than 5–6 orders of magnitude from its saturated value. While the latter condition is strict and a hinderance, it is likely that the difficulties of the CPA at very low saturations are due to the fact that up until the present, it has not addressed the parallel mechanism of film flow on the surface of the pores and throats. But, since this is not likely, in general, to be a mechanism important for vascular plants, it is not considered further here.

In view of the general structure of the CPA, which uses the critical percolation threshold to select the largest possible value of a limiting conductance, the two most important inputs that one needs are a conductance distribution and a threshold probability or volume. The conductance distribution depends most strongly on the pore-throat distribution, while the threshold is most succinctly expressed as a critical moisture content, or air volume fraction. Such parameters are also known in terms of their physical descriptions (Luckner *et al* 1989) as 'residual water'or 'entrapped air.'

Because Katz and Thompson (1986) applied the CPA to consolidated, rather than unconsolidated media, such as soils, they used a few concepts less familiar to soil physicists, such as mercury porosimetry curves, which are, however, analogous to pressure-saturation curves in porous media, so it is possible to draw careful analogies. We will not concentrate on this here, however, as we present only the original results in figure 4.1. The calculation of the saturated hydraulic conductivity given by Hunt and Gee (2002b) is analogous to that of Katz and Thompson (1986). The critical pore radius is determined from,

$$\int_{r_c}^{r_m} \frac{3 - D}{r_m^{3-D}} r^{2-D} dr = \theta_t, \tag{4.3}$$

where θ_t is the critical volume fraction for percolation, expressed as a residual moisture content. In equation (4.3), r_m may be replaced by A/h_A in anticipation of saturation dependence of the property. Using equation (4.3), the critical pore radius is determined, based on which the saturated hydraulic conductivity is obtained as the conductivity g_c of the bottleneck pore using the Hagen–Poiseuille law, i.e., $g_c \sim r_c^4/l$, or $g \sim r_c^3$, for pores with comparable radius r and length l.

It is important to point out that this procedure can, without adjustable parameters, make accurate predictions for over seven orders of magnitude variations in the hydraulic conductivity. Indeed, in a test using the results of numerical simulations of a known pore network, Bernabé and Bruderer (1998) confirmed the superiority of their predictions over all other formulas and approaches tested, including the Kozeny–Carman equation that actually performs the worst in highly heterogeneous media. Although the Katz–Thompson (1986) predictions were most accurate for strongly heterogeneous media, it turns out that the simplistic Kozeny–Carman model, based on a bundle of capillary tubes, is reasonably accurate in the limit of very weak heterogeneity. This is an important distinction between heterogeneous and homogeneous media, and another indication that the use of capillary bundle models is appropriate only for the latter limit. In the following, however, it will often be sufficient to use typical values of the hydraulic conductivity from observation, together with the understanding that they could have been obtained by the CPA, if sufficiently detailed information regarding the pore space was available. The most important consequence of this particular conceptual argument is that it justifies the use of percolation power laws for solute transport (Sahimi 1987, 1993) by advection through heterogeneous media.

4.4 Saturation-dependent properties

Hydraulic conductivity depends on the saturation due to the reduced pore space available for fluid flow. Chapter 3 presented the techniques used to study such a dependence, highlighting two main regimes and two families of porous media. Close to the percolation threshold, universal power laws of percolation theory are

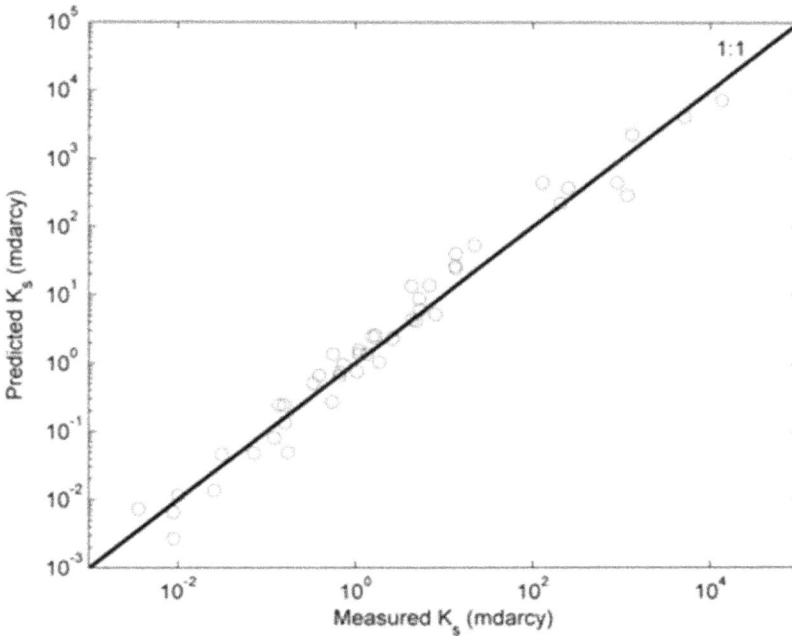

Figure 4.1. Comparison of the predictions of the Katz and Thompson for the permeability (saturated hydraulic conductivity) with experimental data over nearly seven orders of magnitude variations.

applicable, with a universal (independent of the morphology of the pore space) exponent of 2 in the conductivity–moisture relation, equation (3.2). In heterogeneous media above the threshold, the CPA provides a link between macroscopic properties and the features of the pore-size distribution, resulting in non-universal exponents that depend on the fractal dimension of the porous medium. In contrast, in homogeneous media and under the conditions close to complete saturation, the EMA is preferred to the CPA. Here, we discuss the specific cases of hydraulic and electrical conductivities, and air permeability, and gas and solute diffusivities, and how such properties are represented by universal versus medium-dependent models.

4.4.1 Hydraulic and electrical conductivities, and air permeability

It is important to point out that, at least to lowest-order approximation, the calculation of the air permeability (which is a measure of the ease of air flow) under dry conditions proceeds the same way as for saturated hydraulic conductivity. As long as the possibility that the critical air fraction for percolation can be different from the critical moisture content for percolation is ignored, the calculation would predict the same value of the air permeability under dry conditions as the permeability of water under saturated conditions. The similarity does not, however, extend to the dependence of the two properties on saturation.

Simultaneous consideration of hydraulic conductivity and air permeability illustrates their fundamental difference. Let us present general arguments regarding

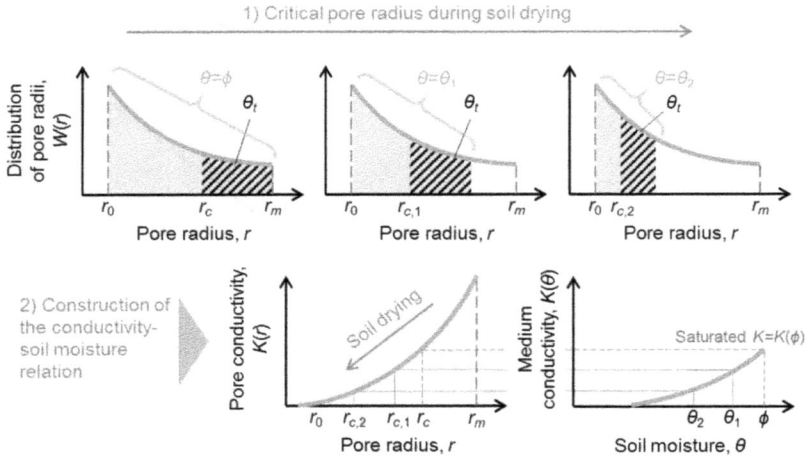

Figure 4.2. Top: schematic representation of determining the bottleneck (critical) pore size r_c under varying soil moisture (starting from saturated conditions), using the CPA. Bottom: illustration of the construction of the relation between porous medium conductivity and soil moisture.

the relevance of the pore-size distribution to their dependences on saturation, together with implications of the strong wetting condition. Consider, first, the hydraulic conductivity at full saturation, or the air permeability under perfectly dry conditions. In three-dimensional porous media, owing to the (typically) small value of the critical volume fraction for percolation, it is possible for water (air) to find an interconnected path through the medium that avoids all but the largest pores. For an arbitrary critical volume fraction, V_c, the CPA predicts the largest possible value of the smallest pore radius, r_c, on such a path. Here, V_c is expressed as a moisture content, θ_t, but for the air permeability under dry conditions, V_c would be the critical value of the air-filled porosity, ϵ_c or ϵ_t (for the threshold). The schematic of such considerations is presented in figure 4.2. The critical air fraction for percolation and the critical water content are not, in general, equal, but if, for simplicity of explanation, no distinction between the two is made, the same schematic presentation applies to both. In the case of the hydraulic conductivity, reducing the water content means that the largest pores become air-filled; see figure 4.2. This implies that the upper limit of the integral in equation (4.3) could no longer include the largest pores, as they contain no water, and it decreases from r_m to a generic radius smaller than r_m, which can be expressed as A/h. The pore space defined by $r > r_c$ and $r < A/h$ does not percolate because $h > h_A$, and the fractional volume represented by the integral with A/h as the upper limit, instead of A/h_A, is now less than θ_t. In order for the left side of equation (4.3) to remain equal to V_c, the lower limit must be reduced and r_c must diminish with decreasing saturation. Because r_c represents the bottleneck pore size that controls the conductivity, the reduction of r_c with diminishing saturation accounts for the influence of the pore-size distribution on $K(\theta)$.

In the case of the air permeability k_a, however, reducing the air content does not affect the occupation of the largest pores by air (Hunt 2005). The water—the wetting fluid—enters the smallest pores first, and the pore space defined by $r_c < r < r_m$ remains air-filled, where r_c is the critical radius. Since the air permeability at arbitrary saturation is determined by the same subset of pores that defines the air permeability for completely dry conditions, equation (4.3) for θ_t represents the critical pore radius for air flow, independent of the air content, unless the volume fraction of the pore space filled by air falls below the threshold. Thus, the CPA reveals a fundamental difference in the saturation dependences of the hydraulic conductivity K and the air permeability k_a. The bottleneck pore radius depends on the saturation for the hydraulic conductivity, but not for the air permeability. Because the pore-size distribution has no influence on the saturation dependence of k_a, the dependence of k_a on air-filled porosity should be defined by the topology of the air-filled pore space, as described by percolation theory, which expresses the universal power laws in terms of the variables appropriate to the air permeability. Figure 4.3 is a schematic representation of a water fractal water retention curve and how to relate it to the hydraulic conductivity.

By calculating the dependence of the bottleneck conductance g_c on the moisture content, given the conductance distribution $W(g)$ provided by the RS model, the unsaturated hydraulic conductivity (Hunt and Gee 2002a, 2002b) is derived and is given by

$$K(\theta) = K_S \left[\frac{1 - \phi + (\theta - \theta_t)}{1 - \theta_t} \right]^{3/(3-D)} . \tag{4.4}$$

Equation (4.4) can be modified slightly by averaging over all the resistances smaller than the threshold, which would change the power $3/(3 - D)$ to $D/(3 - D)$. The justification for the modification is that, the total resistance of the resistances that are in series is their sum. While this is not typically of great importance, it is relevant for theoretical considerations. A somewhat more complex equation was derived using a more general model (Ghanbarian-Alavijeh and Hunt 2012a). The reason for using the more general model was to estimate the model's parameters from broader sets of soil water retention curves. When its predictions were contrasted with those provided the commonly used van Genuchten model (1980) by comparison with

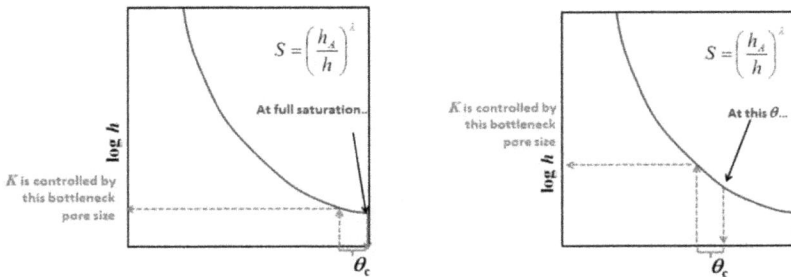

Figure 4.3. Schematic representation of the procedure for determining the bottleneck pore size for saturated (left) and unsaturated conditions (right), using the CPA.

experimental data for the hydraulic conductivity of 107 soils from the Unsaturated Soil Hydraulic Database (UNSODA) (Leij *et al* 1996), the model performed most accurately, as measured by the Akaike information criterion (AIC), which is an estimator of prediction error and, thereby, the relative quality of statistical models for a given set of data. Given a collection of models for the data, the AIC estimates the quality of each model, relative to each of the other models. Thus, AIC provides a means for model selection. One complication not discussed is the necessity to replace equation (4.4) by equation (4.2) at low enough moisture contents close to the percolation threshold. Equation (4.2) does not dependent on the properties of porous medium that equation (4.4) exhibits and is, therefore, denoted as 'universal.' For typical natural porous media, such a crossover occurs at a rather low value of the moisture content. The crossover moisture content is selected by choosing the functional form with the largest value of $dK/d\theta$, but simultaneously requiring the continuity of K and its derivative across the crossover. This procedure provides equations for both the crossover moisture content and the prefactor of the universal power-law function near the flow threshold. In contrast, in man-made media, that crossover moves to much higher values of θ, since the pore-size distribution is not so important. However, when the pore-size distribution is narrower and the system is far from the percolation threshold, the EMA will be the preferred technique.

An analogous procedure to determine the electrical conductivity as a function of saturation leads to an expression of the same form as that of equation (4.4), but with the power $1/(3 - D)$, instead of $3/(3 - D)$. Importantly, the much stronger dependence of K, when compared with the electrical conductivity, on the pore-size distribution, confines the relevance of the region of the validity of the universal power law of percolation theory to a narrow range of saturations near the threshold, a range of moisture contents investigated much less often. The weaker dependence of the electrical conductivity on pore size (and, thus, on saturation) implies, however, that the universal formulation for the electrical conductivity is appropriate over almost the entire range of accessible saturations in typical porous media. Thus, the electrical conductivity is expected to follow universal power law of percolation theory, as it, in fact, did, at least in the cases investigated. Note, however, that as Dashtian *et al* (2015) demonstrated, when there are long-range correlations in a porous medium, which usually exist in large-scale porous media, the universality of electrical conductivity with an exponent of 2 breaks down. Since solute diffusion coefficient is proportional to the electrical conductivity, it is also expected to vanish according to the same power law in the limit of the threshold moisture content for percolation; see below.

Thus, as discussed above, the air permeability as a function of saturation is expected to follow universal power law of percolation theory. The application is based on continuum percolation, so that such quantities as moisture content or air-filled porosity can be used. Although continuum percolation can also predict non-universal power laws for the permeability, it has been shown that, (a) the predictions of the non-universal power law do not apply to the air permeability, and (b) the predictions of equation (4.3), using the exponent the $D/(3 - D)$, agree with the non-universal power-law formulation in the appropriate limit. Since this limit is not, however, relevant to practical problems, at least with the particular model that is being used, it is not discussed further. The interested reader is referred to Hunt *et al* (2014).

In continuum percolation, universal scaling of the conductivity (and related properties) is expressed as a quadratic function of the difference of a volume fraction, such as the air-filled porosity, and its threshold value for percolation. In order to be most useful, the result for the air permeability is written in a power-law form that involves the value of the air permeability under dry conditions,

$$k_a(\epsilon) = k_a(\epsilon = \phi)\left(\frac{\epsilon - \epsilon_t}{\epsilon' - \epsilon_t}\right)^{\mu}. \tag{4.5}$$

In equation (4.5) ϵ' is the maximum value of the air-filled porosity, which is usually equal to the porosity. Ghanbarian-Alavijeh and Hunt (2012a) verified equation (4.5) over a wide range of experiments, as shown in figures 4.4 and 4.5. In the latter figure, the graphical representation of 16 experiments demonstrates the clear consistency

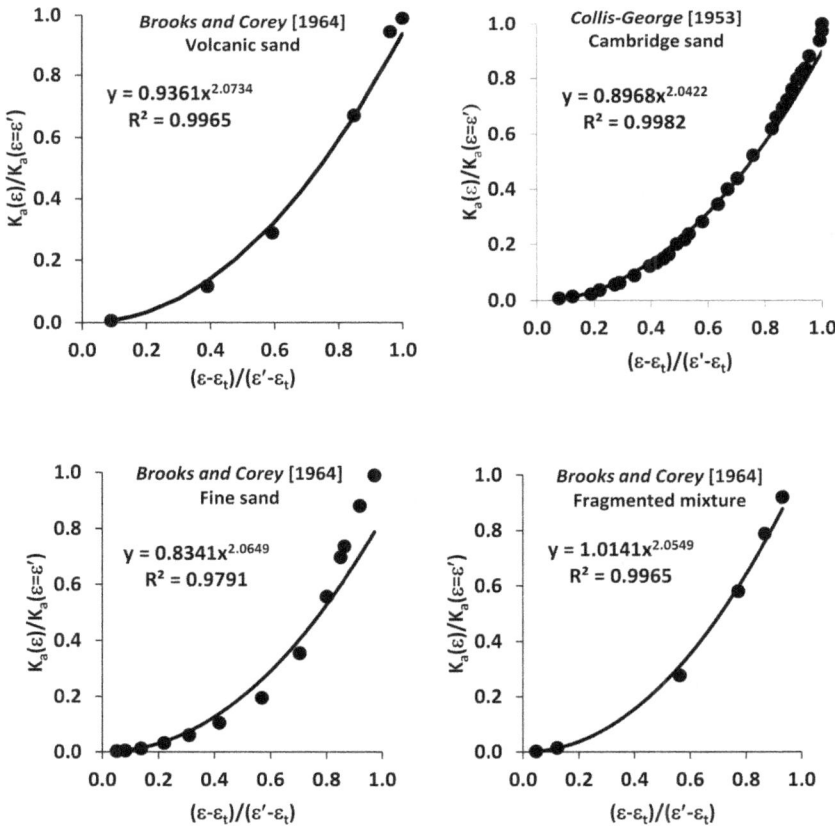

Figure 4.4. Illustration of the relation between air permeability and air-filled porosity that show clearly the relevance of the power law percolation theory with an exponent of 2.

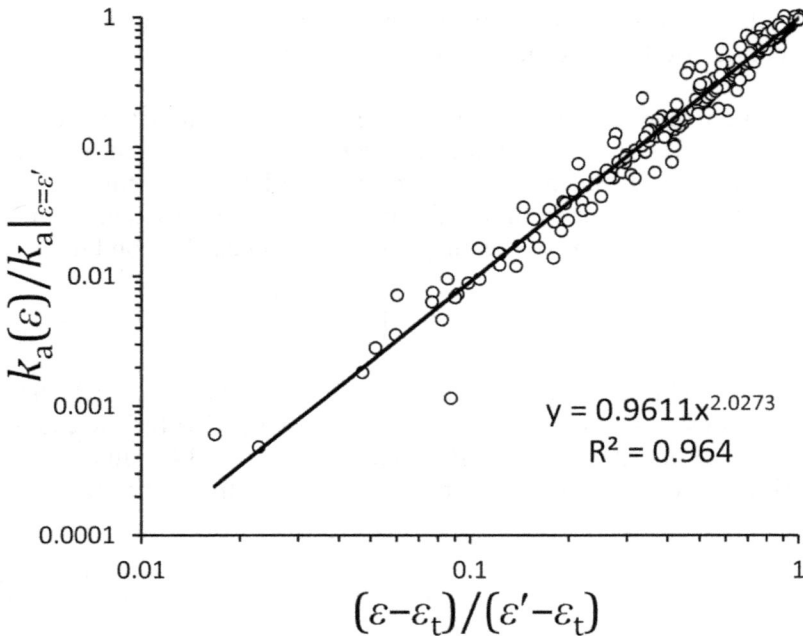

Figure 4.5. Composite representation of the accessed experimental data for the air permeability as a function of air-filled porosity.

with the predictions of percolation theory. The slope is within 1.5% of predictions, the intercept within 4%, and the scatter is small.

4.4.2 Diffusivity of gases and solutes

The same general arguments regarding the relevance of the pore-size distribution to conductivity are also applicable to gas diffusion coefficient. In other words, as the air content is reduced by adding water, the smallest pores are wetted first, so that the universal power law of percolation theory should apply to gas diffusion coefficient as a function of air-filled porosity. In a comparison of seven sets of published data reported by Per Moldrup of Aalborg University (in Denmark) and his collaborators (who advocate a non-universal power formulation that has no threshold) with the predictions of the universal power law of percolation theory, the latter was was shown to be very predictive. One small, but important, difference between gas diffusion coefficient and the air permeability is that the former is generally normalized to its value in the pure fluid, rather than to its value at full saturation (or under perfectly dry conditions). But, in any reasonable sequence of porous media media with increasing porosity that approaches 1, the heterogeneity will eventually diminish beyond a certain point, where percolation theory is no longer the most accurate approach to utilize. In such a case, an equation analogous to equation (3.17), derived by the EMA, but with a linear, rather than quadratic, dependence on the saturation difference is appropriate. Such a linear dependence prevents the diffusion coefficient from rising as much at high porosities as predicted by percolation theory. Using a crossover probability of about 0.75 suggested by Kirkpatrick (1973), well above most porosities in natural soils, Ghanbarian-Alavijeh et al (2014a, 2014b, 2014c) showed that the change to a linear dependence effectively alters the value of the numerical prefactor from 1 to 1.35, as indeed was inferred in the meta-studies of both the gas and solute diffusion coefficient. Note that the graphical representation of the gas diffusion coefficient over the entire range of saturations (thus porosities as well) should look like figure 3.3. The entire range of porosities (especially greater than 0.75, for example) is, however, essentially never observed in gas diffusion in natural porous media, though the crossover in the problem of thermal conductivity of porous media is actually observable (Hunt et al 2014), and for gas diffusion the presence of such a crossover is detected only in the normalization constant. In figures 4.6 and 4.7 we show that the percolation theory

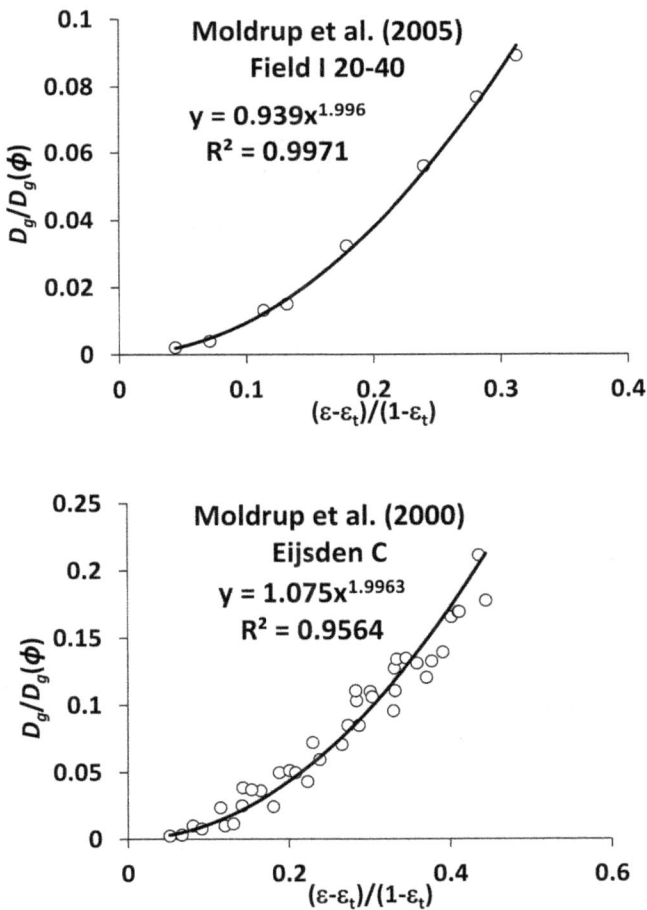

Figure 4.6. Experimental data for gas diffusion coefficient as a function of air-filled porosity for a specific soil and their comparison with the predictions of universal power law of percolation theory.

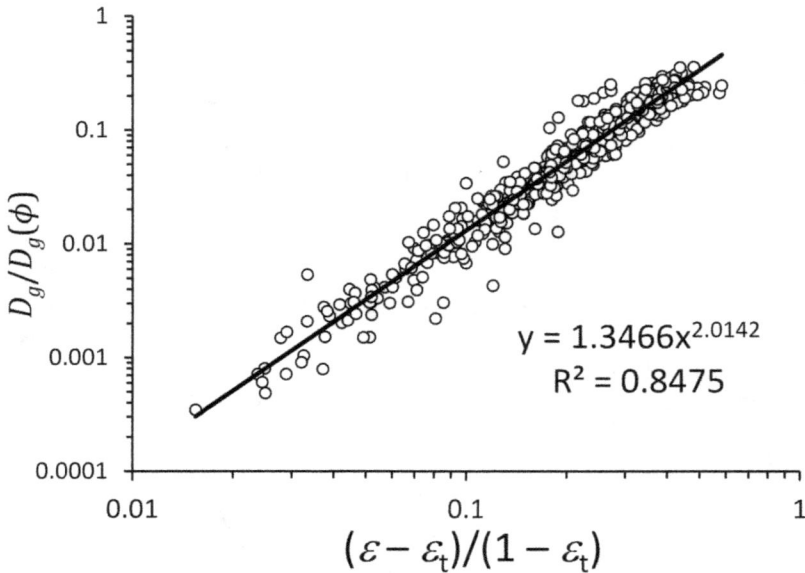

Figure 4.7. Experimental data for gas diffusion coefficient as a function of air-filled porosity, reported by the Moldrup group (Moldrup *et al* 2000a, 2000b, 2003, 2004, 2005a, 2005b, 2005c). Note that the fitted exponent differs by only 0.7% from the predicted value of 2, although the prefactor deviates by 35% from 1, which, however, is expected due to the relevance of the EMA at high porosities.

provides highly accurate predictions for the gas diffusion coefficient in individual soils, as well as across soils. Note that even though the individual soils presented did not exhibit significant evidence of a crossover to the validity of the EMA near a porosity of 1 (the numerical prefactors are mostly close to 1), the composite figure 4.8 demonstrates the relevance of the crossover (through the numerical factor of 1.35).

It should be mentioned that the non-universal power-law relationship proposed by Moldrup and co-workers (1996, 1999, 2003), and claimed by them to have been verified, has been shown (Ewing *et al* 2015) to be an artifact of a finite column height, together with the neglect of the non-zero threshold air content for the connectivity of the air phase. Interestingly, even in short columns (no more than 3.5 cm tall), where the complication from finite heights should be minimal (and it was not possible to show directly that the height of the sample generates inaccuracy), the Akaike information criterion demonstrated, nevertheless, the superiority of the percolation theory in three-fourth of the cases, even though they had previously been reported as implying the relevance of the non-universal power law of Moldrup *et al* (2003) and Masis-Meléndez (2015).

For comparison with the data for solute diffusion coefficient, an entire suite of experiments was obtained from Max Hu of Lawrence Livermore Laboratory. The same analysis for the solute diffusion coefficient leads to an exponent of 1.86, 7% less than expected, but nearly the same value of the prefactor, 1.33. It is rather straightforward, however, to simply plot all the accessible solute diffusivity data as a function of moisture content. When that is done (see below), the exponent

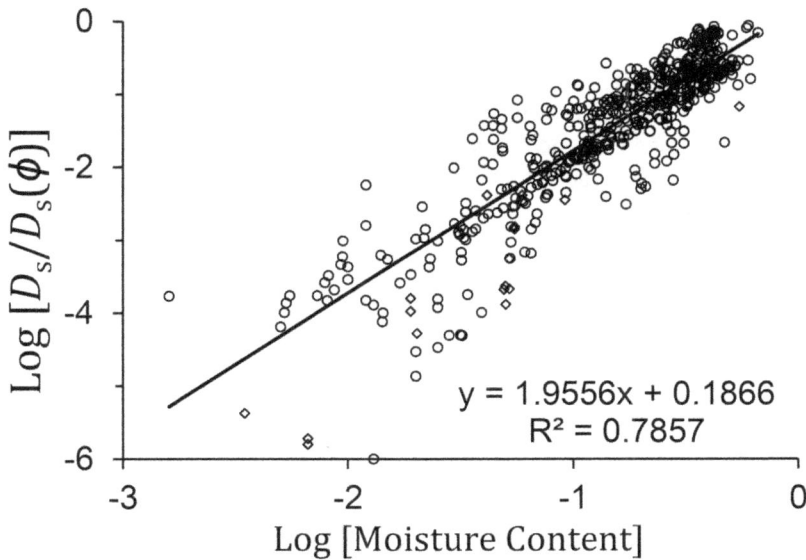

Figure 4.8. Simultaneous plot of all the data for solute diffusion coefficient as a function of soil moisture. Note that the horizontal axis is simply the volumetric moisture content, without reference to any critical value. Reprinted by permission from Springer Nature Customer Service Centre GmbH: Springer, Hunt *et al* (2014), Copyright (2014).

extracted from the data is very nearly 2, but the numerical prefactor exceeds the predictions by about 11%, being 1.53, instead of the expected 1.35. Note, however, that the prefactors are not universal.

The discrepancy between theory and experiment for the solute diffusion coefficients needs some discussion in order to clarify the effect of three complicating factors. The first is that the proportionality constant involves the volume density of conducting charges whose value, at least in an infinite system, involves the fraction of charges on the infinite cluster. In other words, in an infinte system, only the sample-spanning cluster contributes to the macroscopic diffusivity. Thus, the proper power law for the diffusivity is, $D \sim (p - p_c)^{\mu}/X_a$, where X_a is the percolation accessibility function. Since, $X_a \sim (p - p_c)^{\beta}$, one obtains, $D \sim (p - p_c)^{\mu-\beta}$, so that the actual exponent in this case $\mu - \beta = 2 - 0.41 = 1.59$, with 0.41 being the value of the exponent β. A value of 1.59 for the exponent is not, however, extracted from experimental data. The experimental value of the exponent appears to be between 1.86 and 2, which may be just close enough to 2 to be ambiguous. The pore-size distribution can also influence the value of the extracted power, either enhancing its value or reducing it. Finally, diffusion in thin water films can, at low water contents, prevent the diffusion coefficient from vanishing at the percolation threshold or critical saturation and, thus making the effective power smaller. All such complications are beyond the scope of our discussions, since they tend to be most relevant for hydraulic conductivity near zero, or in infinite systems, and at tensions so high that plants cannot extract water from the medium.

Ghanbarian *et al* (2015) considered results of lattice-Boltzmann simulations for the diffusion coefficients, both solute and gas, of more nearly homogeneous media. They demonstrated, as expected, that the region in which the quadratic power law of percolation theory is accurate is narrower in more nearly homogeneous media (similar to engineered, or artificial media, such a packing of glass beads), and the linear dependence predicted by the EMA was valid over a much wider range of air-filled porosities. Thus, in man-made media, the percolation prediction is often only visible near the percolation threshold, which might not always be seen in experiments. It is important to keep in mind, however, that the relevance of the percolation power law for the gas (or solute) diffusion coefficient would scarcely be noticed in homogeneous porous media, often investigated numerically. Figure 4.9 demonstrates the crossover explicitly in such media, but it occurs sometimes so close to the percolation threshold that one might not even guess its existence.

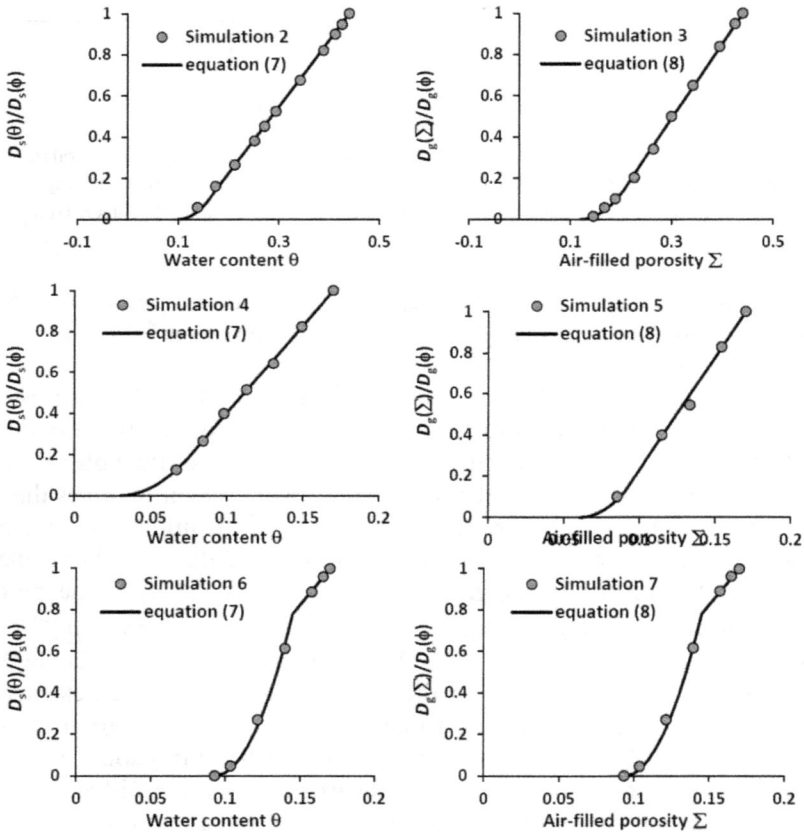

Figure 4.9. Demonstration of the importance of the linear portion of the diffusion coefficient, predicted by the EMA, to more homogeneous media used in the lattice-Boltzmann simulations of the diffusion. The quadratic part is calculated based on the power law of percolation theory. The distinction between relatively homogeneous and heterogeneous media (figures 4.7 and 4.8 for most natural porous media) is in the range of validity of the EMA, which is wide in the former, and so narrow in the latter that it is not directly seen, though its effects can nevertheless be discerned.

In three dimensions, the threshold volume fraction (water content for solute diffusion, air content for gas diffusion) is expected to be $1.5/Z$ (Vyssotsky *et al* 1961), where Z is the coordination number, or mean connectivity of the pores. The crossover to the region where the EMA is accurate (if Z is known) is expected to be $2/Z$ (Sahimi 1993). This is a relatively narrow 'critical' region. Although the two values in the above cases were used as adjustable parameters, comparison of the actual values used with the theoretical predictions yielded an approximate relation, $y = 1.03x + 0.02$ (with $R^2 = 0.94$) and $y = 1.17x - 0.02$ ($R^2 = 0.95$), an average discrepancy in the slope of 10%, and in the intercept of 0%. The reason that the critical region is wider in natural porous media is the tendency for flow (and diffusion) to be restricted to a small subnetwork of the medium, consistent with the CPA. In such a case, the typical coordination number Z of the pores on the preferential path described by percolation theory is known to be 2.7, the value required for percolation on a random network, which is much smaller than the value for Voronoi tessellation reported by Jerauld *et al* (1984), which was over 15.

4.4.3 Universality of flow and transport properties

In figure 4.10 we present the collected data for solute and gas diffusion coefficients, air permeability, and the electrical conductivity, demonstrating that each follows the

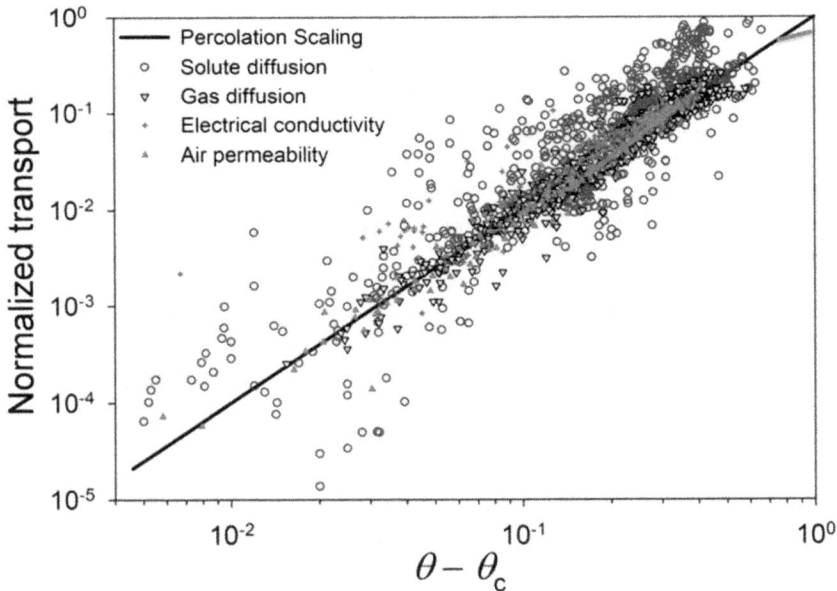

Figure 4.10. Demonstration of the near universal applicability of percolation power laws to the calculation of saturation-dependent flow and transport properties of porous media, such as air permeability, solute and gas diffusion coefficients, and electrical conductivity. Hydraulic and thermal conductivities are excluded because the former has a strong dependence on the pore-size distribution, while the latter is harder to interpret since its behavior is dominated by contact resistances between the grains, which partially obscures its correspondence to prediction.

same universal power law of percolation theory. Note, however, that the constants typically used to normalize diffusion coefficient, namely, its values in water or air, would be 1/1.35, and the appropriate universal curve would have a kink at a porosity of about 0.75, with a linear portion (the reddish brown line in the figure) that intersected the right side of the figure at $1/1.35 = 0.74$. More importantly, however, is what we learned about the physics of flow and transport in typical artificial (engineered) media, which are relatively homogeneous, applies to natural porous media in the range of the reddish brown line. It is noteworthy that none of the datasets in figure 4.10 happens to fall in that region, even though the thermal conductivity data (not shown) tend to explore the region near full connectivity. In any case, this is a second, highly significant, distinction between homogeneous (engineered) and heterogeneous (natural) porous media: the former are mostly found in the linear regime, whereas the latter are mostly in the quadratic.

References

Berkowitz B, Scher H and Silliman S E 2000 Anomalous transport in laboratory-scale, heterogeneous porous media *Water Resour. Res.* **36** 149

Berkowitz B and Scher H 1995 On characterization of anomalous dispersion in porous andfractured media *Water Resour. Res.* **31** 1461–6

Bernabé Y and Bruderer C 1998 Effect of the variance of pore size distribution on the transport properties of heterogeneous networks *J. Geophys. Res.* **103** 513

Bird N R A, Perrier E and Rieu M 2000 The water retention function for a model of soil structure with pore and solid fractal distributions *Eur. J. Soil Sci.* **51** 55

Børgesen C D, Jacobsen O H, Hansen S and Schaap M G 2006 Soil hydraulic properties near saturation, an improved conductivity model *J. Hydrol.* **324** 40

Brooks R H and Corey A T 1964 *Hydraulic Properties of Porous Media (Colorado State University Hydrology Papers vol 3)* (Fort Collins, CO: Colorado State University)

Collis-George N 1953 Relationship between air and water permeabilities in porous media *Soil Sci.* **76** 239

Cortis A and Berkowitz B 2004 Anomalous transport in 'classical' soil and sand columns *Soil Sci. Soc. Am. J.* **68** 1539

Cushman J H and O'Malley D 2015 Fickian dispersion is anomalous *J. Hydrol.* **531** 161

Dashtian H, Yang Y and Sahimi M 2015 Non-universality of the Archie exponent due to multifractality of the resistivity well logs *Geophys. Res. Lett.* **42** 10655

de Gennes P G 1985 Partial filling of a fractal structure by a wetting fluid *Physics of Disordered Materials* ed D Adler, H Fritzsche and S R Ovshinsky (New York: Plenum) p 227

Ewing R P, Ghanbarian B and Hunt A G 2015 Gradients and assumptions affect interpretation of laboratory-measured gas-phase transport *Soil Sci. Soc. Am. J.* **79** 1018

Ewing R P and Hunt A G 2006 Dependence of the electrical conductivity on saturation in real porous media *Vadose Zone J.* **5** 731

Filgueira R R, Ya A P, Fournier L L, Sarli G O and Aragon A 1999 Comparison of fractal dimensions estimated from aggregate mass-size distribution and water retention scaling *Soil Sci.* **164** 217

Ghanbarian B, Daigle H, Hunt A G, Ewing R P and Sahimi M 2015 Gas and solute diffusion in partially saturated porous media: comparison with lattice-Boltzmann simulations *J. Geophys. Res. Solid Earth* **120** 182

Ghanbarian B, Hunt A G, Ewing R P and Sahimi M 2013a Tortuosity in porous media: a critical review *Soil Sci. Soc. Am. J.* **77** 1461

Ghanbarian B, Hunt A G, Sahimi M, Ewing R P and Skinner T E 2013b Percolation theory generates a physically based description of tortuosity in saturated and unsaturated porous media *Soil Sci. Soc. Am. J.* **77** 1920

Ghanbarian-Alavijeh B and Hunt A G 2012a Comparison of the predictions of universal scaling of the saturation dependence of the air permeability with experiment *Water Resour. Res.* **48** W08513

Ghanbarian-Alavijeh B and Hunt A G 2012b Unsaturated hydraulic conductivity in porous media: percolation theory *Geoderma* **187-88** 77

Ghanbarian-Alavijeh B and Hunt A G 2012c Estimation of soil-water retention from particle-size distribution: fractal approaches *Soil Sci.* **177** 321

Ghanbarian-Alavijeh B and Hunt A G 2014a Saturation dependence of solute diffusion in porous media: universal scaling compared with experiments *Vadose Zone J.* **13** 1920

Ghanbarian-Alavijeh B and Hunt A G 2014b Universal scaling of gas diffusion in porous media *Water Resour. Res.* **50** 2242

Ghanbarian-Alavijeh B, Hunt A G, Skinner T E and Ewing R P 2014 Universal scaling of the formation factor in porous media derived by combining percolation and effective-medium theories *Geophys. Res. Lett.* **41** 3884

Ghanbarian-Alavijeh B, Skinner T E and Hunt A G 2012 Saturation dependence of dispersion in porous media *Phys. Rev. E* **86** 066316

Halperin B I, Feng S and Sen P N 1985 Differences between lattice and continuum percolation transport exponents *Phys. Rev. Lett.* **54** 2391

Heiba A A, Sahimi M, Scriven L E and Davis H T 1982 Percolation theory of two-phase relative permeability *SPE Paper 11 015*

Heiba A A, Sahimi M, Scriven L E and Davis H T 1992 Percolation theory of two-phase relative permeability *SPE Reservoir Eng.* **7** 123

Hunt A G 2004a Comparing van Genuchten and percolation theoretical formulations of the hydraulic properties of unsaturated media *Vadose Zone J.* **3** 1483

Hunt A G 2004b Continuum percolation theory and Archie's law *Geophys. Res. Lett.* **31** L19503

Hunt A G 2004c Continuum percolation theory for water retention and hydraulic conductivity of fractal soils: 1. Estimation of the critical volume fraction for percolation *Adv. Water Resour.* **27** 175

Hunt A G 2005 Continuum percolation theory for saturation dependence of air permeability *Vadose Zone J.* **4** 134

Hunt A G, Ewing R P and Ghanbarian B 2014 *Percolation Theory for Flow in Porous Media* 3rd edn (Berlin: Springer)

Hunt A G, Ewing R P and Horton R 2013 What's wrong with soil physics? *Soil Sci. Soc. Am. J.* **77** 1877

Hunt A G and Gee G W 2002a Water retention of fractal soil models using continuum percolation theory: tests of Hanford Site soils *Vadose Zone J.* **1** 252

Hunt A G and Gee G W 2002b Application of critical path analysis to fractal porous media: comparison with examples from the Hanford site *Adv. Water Resour.* **25** 129

Jerauld G R, Hatfield J C, Scriven L E and Davis H T 1984 Percolation and conduction on Voronoi and triangular networks: a case study in topological disorder *J. Phys.* C **17** 1519

Katz A J and Thompson A H 1986 Quantitative prediction of permeability in porous rock *Phys. Rev.* B **34** 8179

Kirkpatrick S 1973 Percolation and conduction *Rev. Mod. Phys.* **45** 574–88

Kosugi K 1999 General model for unsaturated hydraulic conductivity for soils with lognormal pore-size distribution *Soil Sci. Soc. Am. J.* **63** 270

Leij F J, Alves W J, van Genuchten M T and Williams J R 1996 Unsaturated soil hydraulic database, UNSODA 1.0 User's Manual, Rep. EPA/600/R-96/095 (Ada, OK: U.S. Environmental Protection Agency)

Lenhard R J 1992 Measurement and modeling of 3-phase saturation pressure hysteresis *J. Contam. Hydrol.* **9** 243

Luckner L, van Genuchten M T and Nielsen D R 1989 A consistent set of parametric models for the two-phase flow of immiscible fluids in the subsurface *Water Resour. Res.* **25** 2187

Margolin G and Berkowitz B 2000 Application of continuous-time random walks to transport in porous media *J. Phys. Chem.* B **104** 3942

Masís-Meléndez F, de Jonge L W, Chamindu Deepagoda T K K, Tuller M and Moldrup P 2015 Effects of soil bulk density on gas transport parameters and pore-network properties across a sandy field site *Vadose Zone J.* **14** 1

Mohanty K K 1981 Fluids in porous media: two-phase distribution and flow *PhD Thesis* University of Minnesota

Moldrup P, Kruse C W, Rolston D E and Yamaguchi T 1996 Modeling diffusion and reaction in soils: III. Predicting gas diffusivity from the Campbell soil-water retention model *Soil Sci.* **161** 366

Moldrup P, Olesen T, Yamauchi T, Schjønning P and Rolston D E 1999 Modeling diffusion and reaction in soils. IX. The Buckingham-Burdine-Campbell equation for gas diffusivity in undisturbed soil *Soil Sci.* **164** 542

Moldrup P, Olesen T, Gamst J, Schjønning P, Yamaguchi T and Rolston D E 2000a Predicting the gas diffusion coefficient in repacked soil: water-induced linear reduction model *Soil Sci. Soc. Am. J.* **64** 1588

Moldrup P, Olesen T, Schjønning P, Yamaguchi T and Rolston D E 2000b Predicting the gas diffusion coefficient in undisturbed soil from soil water characteristics *Soil Sci. Soc. Am. J.* **64** 94

Moldrup P, Oleson T, Komatsu T, Schjøning P and Rolston D E 2001 Tortuosity, diffusivity, and permeability in the soil liquid and gaseous phases *Soil Sci. Soc. Am. J.* **65** 613

Moldrup P, Yoshikawa S, Olesen T, Komatsu T and Rolston D E 2003 Gas diffusivity in undisturbed volcanic ash soils: test of soil-water-characteristic-based prediction models *Soil Sci. Soc. Am. J.* **67** 41

Moldrup P, Olesen T, Yoshikawa S, Komatsu T and Rolston D E 2004 Three-porosity model for predicting the gas diffusion coefficient in undisturbed soil *Soil Sci. Soc. Am. J.* **68** 750

Moldrup P, Olesen T, Yoshikawa S, Komatsu T and Rolston D E 2005a Predictive-descriptive models for gas and solute diffusion coefficients in variably saturated porous media coupled to pore-size distribution. I. Gas diffusivity in repacked soil *Soil Sci.* **170** 843

Moldrup P, Olesen T, Yoshikawa S, Komatsu T and Rolston D E 2005b Predictive-descriptive models for gas and solute diffusion coefficients in variably saturated porous media coupled to pore-size distribution. II. Gas diffusivity in undisturbed soil *Soil Sci.* **170** 854

Moldrup P, Olesen T, Yoshikawa S, Komatsu T, McDonald A M and Rolston D E 2005c Predictive-descriptive models for gas and solute diffusion coefficients in variably saturated porous media coupled to pore-size distribution. III. Inactive pore space interpretations of gas diffusivity *Soil Sci.* **170** 867

Mualem Y 1986 Hydraulic conductivity of unsaturated soils: Prediction and formulas *Methods of Soil Analysis. Part I. Physical and Mineralogical Methods* ed G S Campbell, R D Jackson, A Klute, M M Mortland and D R Nielsen (American Society of Agronomy—Soil Science Socity of America) p 700

Mualem Y and Dagan G 1978 Hydraulic conductivity of soils–unified approach to statistical models *Soil Sci. Soc. Am. J.* **42** 392

Perfect E 1999 Estimating mass fractal dimensions from water retention curves *Geoderma* **88** 221

Perrier E, Bird N and Rieu M 1999 Generalizing the fractal model of soil structure: the pore-solid fractal approach *Geoderma* **88** 137

Perrier E, Rieu M, Sposito G and de Marsily G 1996 Models of water retention curve for soils with a fractal pore size distribution *Water Res. Res.* **32** 3025

Rieu M and Sposito G 1991 Fractal fragmentation, soil porosity, and soil water properties. I. Theory *Soil Sci. Soc. Am. J.* **55** 1231

Sahimi M 1987 Hydrodynamic dispersion near the percolation threshold: scaling and probability densities *J. Phys.* A **20** L1293

Sahimi M 1993 Flow phenomena in rocks–from continuum models to fractals, percolation, cellular automata, and simulated annealing *Rev. Mod. Phys.* **65** 1393

Sahimi M 2011 Flow and Transport in Porous Media and Fractured Rock 2nd edn (New York: Wiley-VCH)

Sahimi M 2023 *Applications of Percolation Theory* 2nd edn (Berlin: Springer)

Sahimi M, Heiba A A, Davis H T and Scriven L E 1986 Dispersion in flow through porous media: II. Two-phase flow *Chem. Eng. Sci.* **41** 2123

Sahimi M, Hughes B D, Scriven L E and Davis H T 1983 Stochastic transport in disordered systems *J. Chem. Phys.* **78** 6849

Sahimi M and Imdakm A O 1988 The effect of morphological disorder on hydrodynamic dispersion in flow through porous media *J. Phys.* A **21** 3833

Sahimi M and Tsotsis T T 1997 Transient diffusion and conduction in heterogeneous media: beyond the classical effective-medium approximation *Ind. Eng. Chem. Res.* **36** 3043

Schaap M and Leij F 2000 Improved prediction of unsaturated hydraulic conductivity with the Mualem-van Genuchten model *Soil Sci. Soc. Am. J.* **64** 843

Scheidegger A E 1974 *The Physics of Flow Through Porous Media* 3rd edn (Toronto: University of Toronto Press)

Scher H, Shlesinger M F and Bendler J T 1991 Time-scale invariance in transport and relaxation *Phys. Today* Jan., 26–34

Scher H and Zallen R 1970 Critical density in percolation processes *J. Chem. Phys.* **53** 3759

Shinomiya Y, Takahashi K, Kobiyama M and Kubota J 2001 Evaluation of the tortuosity parameter for forest soils to predict the unsaturated hydraulic conductivity *J. For. Res.* **6** 221

Stauffer D and Aharony A 1994 *Introduction to Percolation Theory* 2nd edn (London: Taylor and Francis)

Tuli A and Hopmans J W 2004 Effect of degree of saturation on transport coefficients in disturbed soils *Eur. J. Soil Sci.* **55** 147

Turcotte D L 1986 Fractals and fragmentation *J. Geophys. Res.* **91** 1921

Tyler S W and Wheatcraft S W 1989 Application of fractal mathematics to soil water retention estimation *Soil Sci. Soc. Am. J.* **53** 987

Tyler S W and Wheatcraft S W 1990 Fractal processes in soil water retention *Water Resour. Res.* **26** 1047

van Genuchten M T 1980 A closed-form equation for predicting the hydraulic conductivity of unsaturated soils *Soil Sci. Soc. Am. J.* **44** 892

Vyssotsky V A, Gordon S B, Frisch H L and Hammersley J M 1961 Critical percolation probabilities (bond problem) *Phys. Rev.* **123** 1566

IOP Publishing

Networks on Networks (Second Edition)
Role of connectivity in physics of geobiology and geochemistry
Allen G Hunt and Muhammad Sahimi

Chapter 5

Solute transport and reaction rate in heterogeneous porous media

5.1 Background

The relevance and significance of percolation theory, and in particular the backbone of percolation clusters (the current-carrying part of the cluster; see chapter 3), to solute transport in heterogeneous porous media was first pointed out in a series of papers by Sahimi and co-workers (Sahimi *et al* 1982, 1983, 1986a, 1986b), and confirmed later by the experiments of Charlaix *et al* (1987, 1988), Hulin *et al* (1988), and Gist *et al* (1990). Pore-network simulations of Sahimi *et al* (1982, 1983, 1986a) had indicated that, as the connectivity or percolation threshold of a porous medium is approached, the dispersivity—the ratio of solute dispersion coefficient and the mean fluid velocity—should increase without bounds, and the aforementioned experimental studies provided strong experimental evidence for the prediction. The power-law behavior of the dispersion coefficient near the connectivity threshold was first studied by Sahimi (1987) and Sahimi and Imdakm (1988), and extended later by Sahimi (2012).

The statististic of percolation clusters were then used to derive a probabilistic theory of solute transport in heterogeneous porous media. The fundamental means to derive the equation (Hunt 1988, Hunt and Skinner 2008) was based on the critical-path analysis (CPA) (Ambegoakar *et al* 1971, Pollak 1972, Friedman and Seaton 1998, Hunt 2001), described and used in chapters 3 and 4. Typical solute transport experiments of relevance to test the theory include the introduction of solute to water flowing through a column under steady-state conditions. At time $t = 0$ at position $x = 0$, the solute is introduced at the upstream, which is discontinued after some time. The duration of the injection may vary, but the probabilistic theory was developed for the case in which an instantaneous pulse of solute is injected into the system. What is reported by the experimentalist, and what is to be predicted, is the breakthrough curve (BTC), i.e., the difference between the introduced solute concentration and the background level. In what follows we present the derivation of the equation, which is then used to derive the distribution of arrival times of the solute at the boundary of the system, and its related statistics as a function of space and the properties of a porous medium. In order to apply the results to the more typical inlet step function concentration used in experiments, either a Green's function technique must be utilized, or the predictions must be compared with the negative of the time derivative of the experimental solute BTC.

5.2 Percolation theory for solute transport in heterogeneous porous media

The basic approach uses cluster statistics of percolation theory (Stauffer and Aharony 1994), melded with the CPA, to derive a relative probability $W(g, x)$ that a continuously interconnected path through conductances, not less than an arbitrary given value g, will connect opposite faces of a three-dimensional (3D) system with side lengths equal to the Euclidean dimensions x of the system. For the local conductance distribution, the Rieu and Sposito (RS) (1991) model is used, though Ghanbarian-Alavijeh *et al* (2012) extended the model so as to allow extraction of parameters describing the pore space by a wider range of soil water retention curves (SWRC). The local conductance distribution is (Hunt and Skinner 2008):

$$W(g, x) \propto \beta^{-1} Ei\left[\alpha\left(\frac{x}{L}\right)^{\beta} \right], \tag{5.1}$$

where Ei is the exponential integral, $Ei(z) = \int_z^\infty \exp(-y)y^{-1}dy$, and the parameters α and β are defined by, $\alpha = |1 - (g/g_c)^{(3-D)/3}|^2$ and $\beta = 2/\nu$, with ν being the critical exponent of percolation correlation length described in chapter 3. In this expression, L is a representative elementary length, i.e., the same as the correlation length ξ_p of percolation theory introduced in chapter 3. The distance or length x represents, for example, the length of a laboratory sample of a porous medium, or the distance from an entry surface, such as the top of a soil column. While a solute traveling through such a cluster of pores does not arrive all at once at the exit, the variability in the system's arrival times due to the variability across a particular cluster is not evaluated. Instead, only the most likely pore cluster crossing time is evaluated. Consequently, the distribution first-passage times for crossing the system arise from the variations in the controlling conductances. The most likely time $t(g, x)$ for solute transport across such a pore cluster is then determined, which involves two separate contributions. One is the contribution of the various pore sizes, or pore conductances, along the backbone of the percolation cluster of pores, while the second has to do with the topology of the backbone. The first contribution is determined by summing the travel times along a 1D path with conserved flux through conductances whose values are between the limiting conductance for that particular cluster, and the largest conductance in the system. This is analogous to the calculation of a total resistance of such a 1D path. The contribution due to the topology of the backbone expresses the relevance of the fractal dimension of the backbone to the pore cluster's arrival time. Thus, in this way, the contributions of the pore connectivity (topology) and pore sizes (geometry) are decoupled.

$t(g, x)$ contains a power-law divergence at $g = g_c$, where g_c is the critical conductance determined by the CPA (see chapter 4), together with a factor that describes the fractal slowing with distance (Hunt and Skinner 2008):

$$t = \left(\frac{x}{L}\right)^{D_{bb}} \left(\frac{t_0}{3-D}\right)\left(\frac{g_c}{g}\right)\left[\frac{1}{1-\theta_t}\left(\frac{g_c}{g}\right)^{(1-D)/3} - 1\right]\left\{(1-\theta_t)\left[\left(\frac{g}{g_c}\right)^{(1-D)/3} - 1\right]\right\}^{\nu(D_{bb}-1)}$$

$$\equiv t_g\left(\frac{x}{L}\right)^{D_{bb}}. \tag{5.2}$$

Here, D_{bb} is the fractal dimensionality of the backbone; t_0 is a typical pore-crossing time (which is inversely proportional to a macroscopic flow velocity); D is the fractal dimensionality in the RS model (kept at 2.95 for the comparison with the experimental data for heterogeneous porous media, for which its value was not known from the WRC), and θ_t is the critical volume fraction for percolation, expressed as a moisture fraction, which is unvaried throughout the comparisons with the data. Moreover, the parameters on the right side of equation (5.2) have been grouped in the characteristic time t_g, which implies that the role of both the porous medium conductivity and pore-crossing time are accounted for.

The explicit proportionality of $t(g, x)$ to $x^{D_{bb}}$ for heterogeneous media makes solute velocities, to a good approximation, a diminishing power of the time (Sahimi 1987, 2012). Therefore, solute transport becomes progressively slower than water flow as time increases and the space grows. Moreover, as the conductivity g approaches the critical value g_c, t_g diverges due to the compound effect of increasing $[(g/g_c)^{(1-D)/3} - 1]^{-\nu(D_{bb}-1)}$ and t_0. As a consequence, drier soils exhibit longer and longer travel times at a given spatial scale, an effect first pointed out by Sahimi *et al* (1982). The same effect occurs in fine-textured soils when compared to coarse-textured ones, at given t, x, and moisture.

Only four values of D_{bb} are relevant to flow and transport in most porous media and under the most general conditions, as long as certain long-range correlations can be neglected, or are not relevant (Sahimi and Mukhopadhyay 1996). The four values were listed in table 3.2, and correspond to 2D or 3D flow, and saturated or unsaturated conditions that represent, respectively, random and invasion percolation. In the applications to solute transport, it is assumed that drying conditions are relevant for unsaturated conditions most of the time, so that one can focus on trapping bond invasion percolation (see chapter 3). The assumption is supported by the observation that, during rainfall, dry-down durations are typically much longer than rewetting events. Therefore, the relevant values of D_{bb} are, respectively, 1.643 and 1.217 for saturated and unsaturated conditions in 2D, and 1.870 and 1.458 in 3D; see table 3.2.

The exponent D_{bb} in equation (5.2) could be interpreted as controlling a temporal tortuosity on the backbone, i.e., how long the transport time is delayed by the topology of the flow paths. It does not describe, however, the tortuous distance of the solute travel, which is given by a different exponent, the optimal paths exponent, D_{op}, introduced in chapter 3, such that plumes travelling only on the optimal paths will be associated with the exponent D_{op}, rather than with D_{bb}. D_{op} takes on the same value for each percolation problem in either 2D or 3D, though the 2D value, 1.21, is different from its 3D value, 1.43, listed in table 3.2. This concept and the values of D_{op} will turn out to be relevant in what we will describe later in this book.

The explicit fractal scaling will be dropped for nearly homogeneous systems; thus, $t = t_g(x/L)$, but the divergence at the critical conductance $g = g_c$ still slows to some extent the overall solute transport. Note, however, that without the explicit fractal scaling, the velocity of non-interacting solutes, at least up to fairly long time intervals, does not deviate greatly from that of the fluid. These are known based on the studies of Canadian Forces Base Borden aquifer, in Ontario, Canada, and Cape Cod aquifer in Massachusetts (Roberts *et al* 1986), but equation (5.2) without the factor $(x/L)^{D_{bb}}$ reproduces very well the data from the two studies, hence highlighting the major difference between strongly and weakly heterogeneous porous media.

The expression for the cluster crossing time t contains contributions from both the local conductance distribution, and the topology of large clusters near the percolation threshold. Its derivation, as a sum of times crossing individual pores that sample the distribution of resistances less than the critical resistance $1/g_c$ makes it analogous to a total path resistance. The distribution of arrival times is determined within such a framework (Hunt and Skinner 2008) using, $gW(g, x)/[dt(g, x)/dg] = W(t, x)$. Use of an analogous relationship allows determination of the spatial distribution of the solute at a given time, from which the moments of the distribution are determined (Hunt and Skinner 2010). The moments are important for a variety of purposes, as explained in the following.

5.3 Dispersivity

The first moment of the distribution is the mean position of the solute 'plume,' while the difference between the second moment and the square of the first is the variance. Although such distributions do not always have a finite mean arrival time, it is still possible to determine a mean solute velocity (or a flux) from the first spatial moment. In figure 5.1 an example of the total solute displacement as a function of time for a strongly heterogeneous medium, as described by equations (5.1) and (5.2) is shown. It should be kept in mind that the solute flux that we wish to determine below is simply the time derivative of this function.

The dependence of the dynamic evolution of the mean travel distance $\langle x(t) \rangle$ on the spatial heterogeneity, as represented by the fractal dimension D of the RS model, are relatively weak, with the main factor influencing the result being the fractal dimension D_{bb} of the backbone, whose importance arises from the power-law prefactor of equation (5.2). The results shown in figure 5.1 predict fluxes that agree with chemical reaction rates over up to ten orders of magnitude of time scale

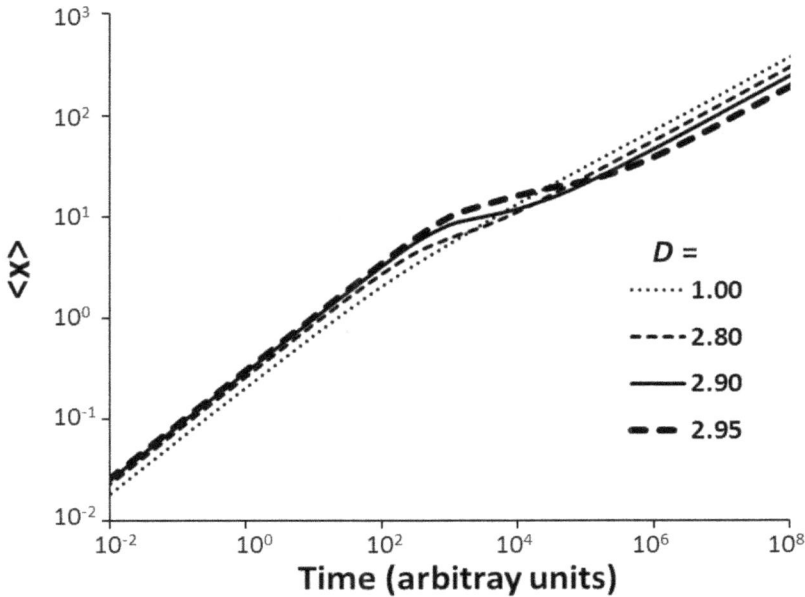

Figure 5.1. Dynamic evolution of the total mean transport distance $\langle x \rangle$ as predicted by equations (5.1) and (5.2) and its dependence on the fractal dimension D. Note that approximately two orders of magnitude of increase in time leads to approximately one order of magnitude of increase in $\langle x \rangle$, as is appropriate for the 3D random percolation (saturated conditions) value of $D_{bb} \approx 1.87$, or almost 2.

(the maximum range that has been calculated). In general, the initial slope of the logarithmic plot is very nearly $1/D_{bb}$.

It has been shown that, at larger distances, the predicted ratio of the variance and the mean distance travelled by the solute, the aforementioned dispersivity α, scales linearly with the transport distance, and that at smaller distances the equation predicts 2–3 orders of magnitude variations in the dispersivity, both of which are borne out in a comparison with the experimental data from nearly 2200 experiments; see figure 5.2. Not only does the envelope of experimental data follow the prediction, but most of the individual experiments also do so as well (Hunt *et al* 2011). One notable exception is the data reported by Sternberg *et al* (1996), where the experiment concerned a series configuration of media whose flow characteristics alternated between fast and slow. Interestingly, the kind of reaction-rate scaling that we predict does not hold in such systems either (Salehikhoo *et al* 2013).

5.4 Distribution of solute arrival times

The predictions for the solute arrival-time distribution have also been verified by comparison with several sets of data without using any adjustable parameters (Ghanbarian-Alavijeh *et al* 2012). The data include the saturation-dependent data of Cherrey *et al* (2003), for which the medium-specific parameters were extracted from the pressure-saturation curve, with the remaining parameters specified by

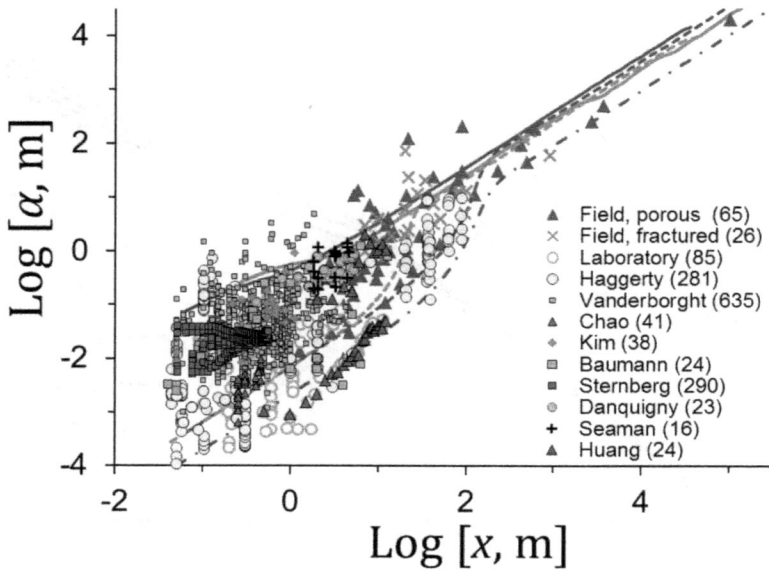

Figure 5.2. Comparison of experimental dispersivitiess, as a function of length, with the predictions using equations (5.1) and (5.2) Experimental data are from from Seaman *et al* (2007), Huang *et al* (2006), Vanderborght and Vereecken (2007), Haggerty (2002), Haggerty *et al* (2004), Neuman and di Federico (2003), Danquigny *et al* (2004), Sternberg *et al* (1996), Kim *et al* (2002), Chao *et al* (2000), and Bauman *et al* (2002). Values reported by Neuman and di Federico (2003) had been obtained by them from other sources, and are labelled 'field, porous' and 'field, fractured.' The numbers in front of the authors refer to the number of experimental points. The various theoretical results are for two values of the fractal dimension D, 1.5 and 2.95, in the RS model, which represent relatively homogeneous and strongly heterogeneous porous formations, respectively, Other distinctions are the values of the backbone fractal dimension D_{bb}, corresponding to different spatial dimensions, namely two for fractures, and three for porous media, as well as the relevance of invasion percolation (e.g., drainage curves) or random percolation (full saturation).

percolation theory. This is shown in figure 5.3. In particular, the 3D value of the exponent ν of percolation correlation length, $\nu \simeq 0.89$, as well as the 3D value of the appropriate backbone exponent (see table 3.2) were used. The particular value that was used was, $D_{bb} \approx 1.46$, for desaturating conditions modeled by bond invasion percolation with trapping. While the shape of the distribution was not adjusted, the peak time of the distribution was, which is inversely proportional to the hydraulic conductivity. All but one of the experimental peak times were, however, within 20% of the calculated values, implying that the calculation of the hydraulic conductivity, while reasonably accurate, was probably less accurate, and more sensitive to details than that of the arrival time distribution. Similar trends were produced when compared with the data reported by Jardine *et al* (1993), except that, in this case, there were no data for the pore-size distribution of the porous medium. Given the higher sensitivity of the hydraulic conductivity to the pore-size distribution, it is noteworthy that the system parameters that produced the closest agreement with the scaling of the peak time with saturation also produced excellent agreement with the arrival time distribution.

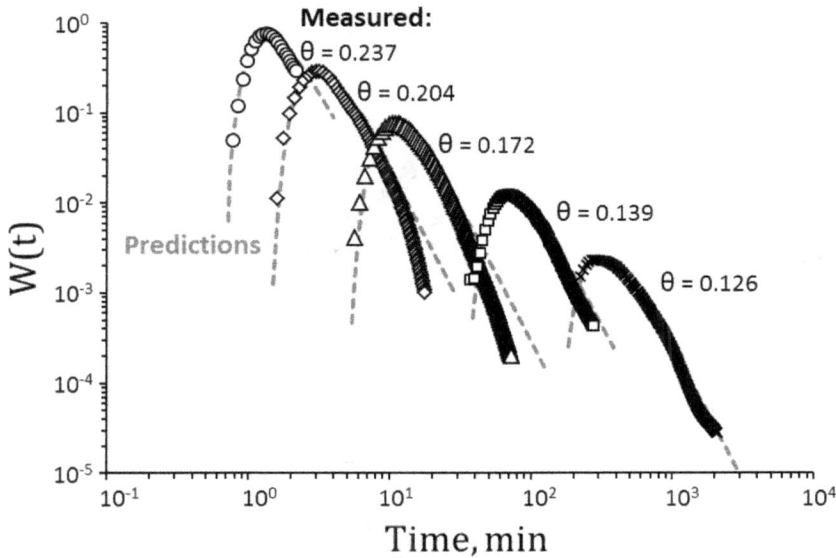

Figure 5.3. Comparison of the predicted arrival-time distributions, as a function of saturation, with experimental data. Reprinted with permission from Ghanbarian *et al* (2012), Copyright (2012) by the American Phsycial Society.

5.5 Scaling of the reaction rate

To predict the scaling of reaction rates in porous media, the solute flux is used as a proxy. In other words, it is assumed that, $v_s = d\langle x \rangle / dt$ is proportional to the reaction rate, R, such that R/R_0 is proportional to v_s/v_0, where v_0 is the fluid velocity, and R_0 is the initial reaction rate. The results could then be expressed as, $R = R_0 f(t/t_0)$, where $f(t/t_0)$ is obtained as the temporal derivative of the function shown in figure 5.1. The assumption requires that the reaction be transport-limited, which is often true and is believed to be the case for the range of flow velocities (Molins *et al* 2012, Maher 2010) that we consider here, and which is typical of the subsurface (Blöschl and Sivapalan 1995). Comparisons have been made with many different experiments, as well as with field studies, of which three comparisons are presented here.

The first comparison is with the experiment of Du *et al* (2012) in which elution of uranium from a (stirred) slurry was measured for various particle-size distributions. The data are shown in figure 5.4. Du *et al* (2012) could have scaled all the six experimental curves onto a single curve by using different proportionalities of pore volume to time. The original interpretation was that the changes in slope were due to a crossover from intra-particle diffusion to inter-particle transport (convection) dynamics. Here, however, we note that the experimental data, if not the procedure or apparatus, have more in common with the crossover in the dispersivity data at the scale of 100 m. Although in this case, t_0 had to be adjusted to produce the observed fit, all the R_0 values were identical, and it should be kept in mind that the two

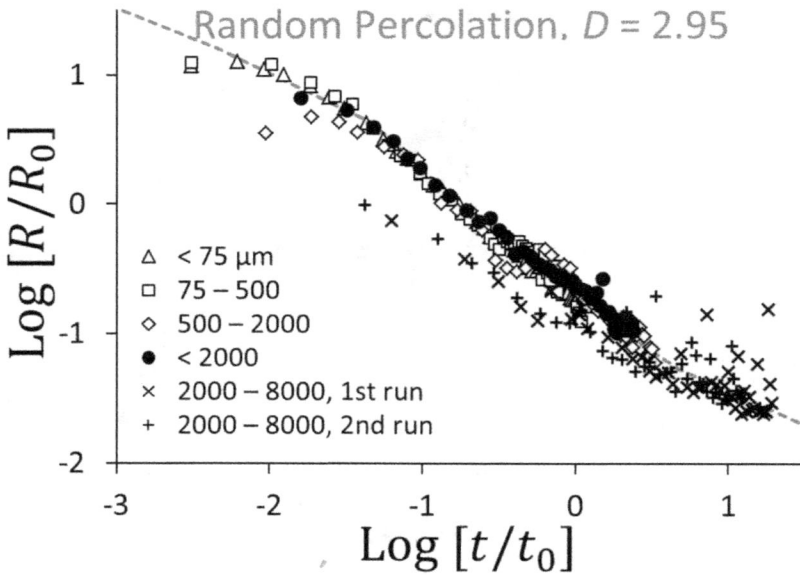

Figure 5.4. Comparison of dynamic evolution of reaction rates in six elution experiments, reported by Du *et al* (2012). The fundamental relationship between pore volumes and time scale was set by a flow velocity, but the velocity was not precisely constant between experiments, as indicated by comparing the two different runs for the same particle-size distribution, i.e, the same porous medium. The relationship, $t_0 = x_0/v_0$, was used together with the best estimates of inter-particle distances x_0, in order to determine the best estimate for the fluid and solute 'pore-crossing' times, which was then scaled up using the experimental column length x to determine the time of the experiment through, $t = t_0(x/x_0)^{D_{bb}}$, with $D_{bb} \approx 1.87$ (table 3.1). These values were always within a factor of approximately 3 of the actual experimental times. Then, the time scales was adjusted by up to a factor 3, while the reaction rate was not rescaled, implying that the curves were translated horizontally, in order to place all the six experiments on the same curve. Two adjustable parameters, t_0 and R_0, were used to fit all the six experimental curves.

parameters were able to yield estimates of the values of four different slopes in four distinct time decades, as well as the three positions for each crossover, in each of the six experiments. Furthermore, the various values of t_0 were checked to show that they scale approximately as predicted (Hunt *et al* 2015) using $t_0 = x_0/v_0$, where v_0 is the known pore-scale fluid velocity, x_0 is the inter-particle separation, and $t = t_0(x/x_0)^{D_{bb}}$ for the various particle-size distributions.

Comparisons were also made with silicate weathering data (Hunt *et al* 2015). Consider field data for weathering rates reported by Maher (2010) and White and Brantley (2003); see figure 5.5. Since weathering rates were shown to be proportional to the fluid flow rates greater than 0.5 μm s^{-1} (Maher 2010) and because Blöschl and Sivapalan (1995) assumed that such flow rates in the subsurface can vary over roughly an order of magnitude in either direction from 1 μm s^{-1}, a corresponding two orders of magnitude uncertainty in weathering rates is explicitly shown in the figure. R_0 was specified for Maher's data, but in the case of the data reported by White and Brantley, the largest values measured were utilized. Maher suggested that the time-dependence set on at approximately 10^{-2} years, which was used for t_0. Note

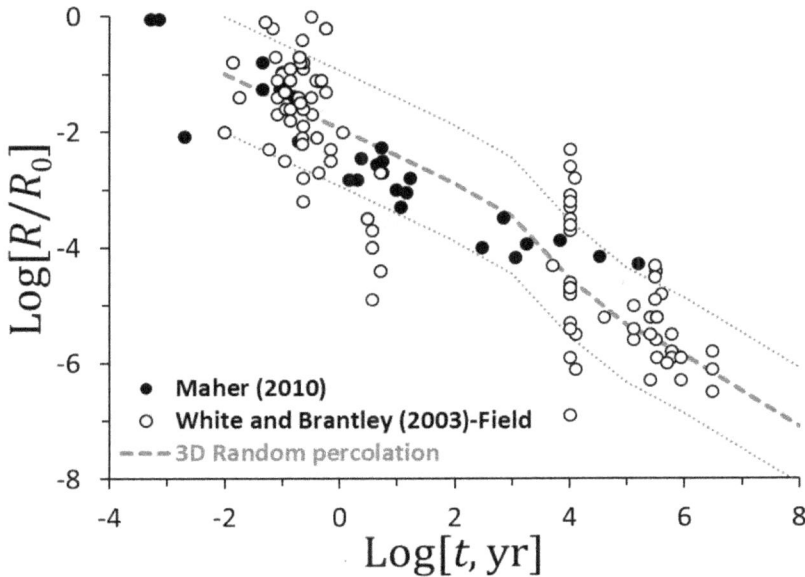

Figure 5.5. Comparison of experimental field data for the dynamic evolution of reaction rate, reported by White and Brantley (2003) and Maher (2010), for the weathering of silicate minerals, with the predicted solute fluxes using equations (5.1) and (5.2)

that the largest value of R in the White and Brantley data set were measured at times close to 10^{-2} years. For the sake of interpretation of the variability in the data, consider the fact that White and Brantley stated that when variations in temperature were accounted for, approximately two orders of magnitude of variability remained, implying that the data that do not fit within the curves, were probably influenced strongly by temperature.

Use of a given t_0, which may well be greater than a pore-crossing time, is in accord with Maher's (2010) assertion that, at shorter time intervals, the reaction rates were not transport-limited. Such analysis is typically performed using the Damköhler number, which is the ratio of advection and reaction time scales (Salehikhoo *et al* 2013). We do not use this rigorously here, although in the intervening time period, Yu and Hunt (2017) developed a precise treatment of incorporating the Damköhler number in the analysis of solute transport and reaction.

Next, consider the experiments of Zhong *et al* (2005), Liu *et al* (2008), and White and Brantley (2003), shown in figure 5.6. In the case of Zhong *et al*, neither t_0 nor R_0 could be estimated and, thus, the predictions were simply scaled to their initial values, which explains why they appear at the beginning of the curve in figure 5.6. In the case of the data reported by White and Brantley, it was possible to use the experimental conditions to estimate t_0, but not R_0 and, therefore, R_0 was determined using the scaling prediction for v_s. As for the data of Liu *et al* (2008), both t_0 and R_0 were known. These analyses were also extracted from Hunt *et al* (2015). The discussion of silicate weathering rates will be revisited later in this book when the formation of soil is considered.

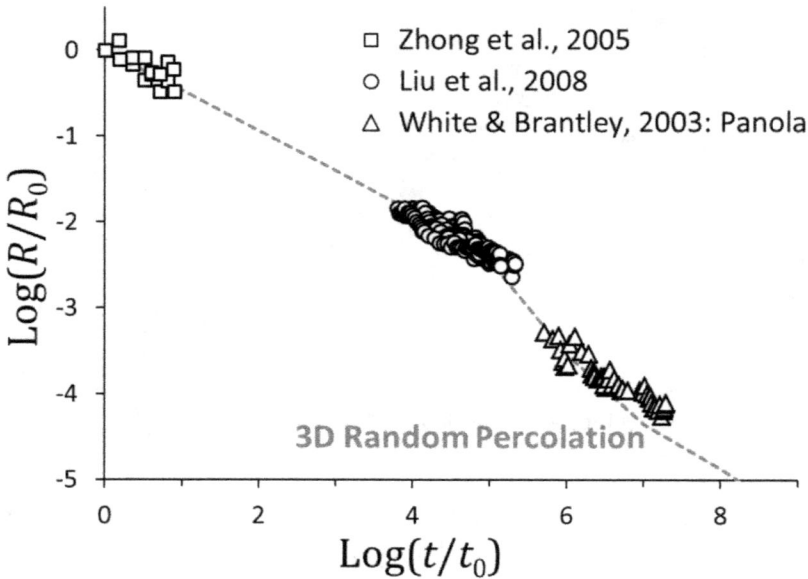

Figure 5.6. Experimental data for scaling of the reaction rates as a function of time, reported by Zhong *et al* (2005), Liu *et al* (2008), and White and Brantley (2003). The latter experiment was reported in the same source as their summary of the field data, and the scaling relationship, equation (5.2), was tested under the conditions of their Panola plagioclase experiment, which yielded a range of times, from 0.19–2.1 years. Their experiment was run for a duration of 0.21–6.3 years.

Summarizing this chapter, it is useful to highlight some of the distinctions between flow, diffusion, and solute transport in weakly and strongly and heterogeneous porous media, as well as to indicate where the percolation variables are relevant. The distinctions are important, as some techniques for growing vascular plants, such as hydroponics, or certain uniform media, may make the models for weakly disordered porous media more nearly appropriate. It is possible to use the capillary bundle models for weakly heterogeneous porous media and derive reasonably accurate expressions for most quantities of interest, except for the transverse dispersion of solutes. In particular, the hydraulic conductivity may be calculated based on the Kozeny–Carman relationship, the solute and gas diffusion coefficients are linear in the relevant phase concentration over most of the range of accessible saturations—a result similar to simple linear mixing models—and for advective flow the solute and flow velocities are equal, while the dispersivity does not increase with transport distance in the medium.

In strongly heterogeneous porous media, on the other hand, pore-network models are preferred over the capillary bundle models. The hydraulic conductivity is calculated by the CPA of Katz and Thompson (1986), or a related method (Banavar and Johnson 1987, Le Doussal 1989, Hunt and Gee 2002). Research in this field is ongoing and, therefore, a more precise analysis that incorporates the effects of the surface roughness on the local conductances is already available (Madadi *et al* 2003, Madadi and Sahimi 2003, Ghanbarian *et al* 2016). Near the connectivity threshold, gas and solute diffusion coefficients are quadratic functions of the air-filled porosity and water content, respectively, and contain an explicit threshold below which the diffusion coefficient vanishes. Finally, solutes carried along passively by flowing fluids have, as a rule, velocities less than that of the average flow velocity, while the dispersivity increases, on the average, proportionate to the transport distance. The most serious issue to address is in the lack of clear guidance to choose what level of heterogeneity is required to consider a porous medium truly heterogeneous. We borrow from the physics of supercooled liquids (Hunt 1992), for which several order of magnitude spread in the local conductance distribution was required. Such a value, as suggested by Seager and Pike (1974), was roughly in accord with the data that indicated the deviation from the proportionality of viscosity and resistivity in viscous liquids (Hunt 1992).

As discussed in chapter 4, it is worth recalling that the nature of effective-medium approximation (EMA) is conducive to an understanding of the flow and transport properties of a porous medium that is weakly heterogeneous. Consider that the theoretical basis in the EMA is in writing down a single equation for the pressure or potential difference across an arbitrary bond—pore throat—in the medium, which effectively allows the bond to interact with the remaining medium in a way that samples the entire medium variability at that particular bond, while forcing the bond to behave, on average, similar to the effective medium itself surrounding it (see Sahimi 2003, for a comprehensive discussion). The assumption transforms spatial heterogeneity to an analogue of an ensemble-averaged heterogeneity that, if ergodicity holds, is equivalent to temporal heterogeneity. But a temporal heterogeneity is quite the opposite of quenched heterogeneity whose effect is addressed by percolation theory. Such a temporal heterogeneity is more akin to the diffusion problem, a stochastic process that is irreproducible temporally, but spatially homogeneous. The original percolation problem, however, is reproducible temporally, but spatially heterogeneous. Thus, it makes sense that such a transformation as is associated with the use of the EMA would work best if the heterogeneity is not too strong. Bernabé and Bruderer (1998) concluded that, for strong heterogeneity, the entire pressure field is so strongly controlled by a few, large pressure drops, that stochastic theories cannot capture it. Such few but large pressure drops occur at the critical conductances on the critical paths, which are the subject of critical-path analysis developed based on percolation theory.

References

Ambegaokar V N, Halperin B I and Langer J S 1971 Hopping conductivity in disordered systems *Phys. Rev.* B **4** 2612

Banavar J R and Johnson D L 1987 Characteristic pore sizes and transport in porous media *Phys. Rev.* B **35** 7283

Baumann T, Müller S and Niessner R 2002 Migration of dissolved heavy metal compounds and PCP in the presence of colloids through a heterogeneous calcareous gravel and a homogeneous quartz sand–pilot scale experiments *Water Res.* **36** 1213

Bernabé Y and Bruderer C 1998 Effect of the variance of pore size distribution on the transportproperties of heterogeneous networks *J. Geophys. Res.* **103** 513–25

Blöschl G and Sivapalan M 1995 Scale issues in hydrological modelling: a review *Hydrol. Process.* **9** 251

Chao H, Rajaram H and Illangasekare T 2000 Intermediate-scale experiments and numerical simulations of transport under radial flow in a two-dimensional heterogeneous porous medium *Water Resour. Res.* **36** 2869

Charlaix E, Hulin J-P, Leroy C and Zarcone C 1988 Experimental study of tracer dispersion in flow through two-dimensional networks of etched capillaries *J. Phys.* D **21** 1727

Charlaix E, Hulin J-P and Plona T J 1987 Experimental study of tracer dispersion in sintered glass porous materials of variable compaction *Phys. Fluids* **30** 1690

Cherrey K D, Flury M and Harsh J B 2003 Nitrate and colloid transport through coarse Hanford sediments under steady state, variably saturated flow *Water Resour. Res.* **39** 1165

Danquigny C, Ackerer P and Carlier J P 2004 Laboratory tracer tests on three-dimensional reconstructed heterogeneous porous media *J. Hydrol.* **294** 196

Du J, Bao J, Hu Q and Ewing R P 2012 Uranium release from different size fractions of sediments in Hanford 300 area, Washington, USA *J. Env. Rad.* **107** 92

Friedman S P and Seaton N A 1998 Critical path analysis of the relationship between permeability and electrical conductivity of three-dimensional pore networks *Water Resour. Res.* **34** 1703

Ghanbarian B, Hunt A G and Daigle H 2016 Fluid flow in porous media with rough pore-solid interface *Water Resour. Res.* **52** 2045

Ghanbarian-Alavijeh B, Skinner T E and Hunt A G 2012 Saturation dependence of dispersion in porous media *Phys. Rev.* E **86** 066316

Gist G A, Thompson A H, Katz A J and Higgins R L 1990 Hydrodynamic dispersion and pore geometry in consolidated rock *Phys. Fluids* A **2** 1533

Haggerty R 2001 Matrix diffusion: heavy-tailed residence time distributions and their influence on radionuclide retention *Radionuclide Retention in Geologic Media, Workshop Proc. (Oskarshamn, Sweden)* (Radioactive Waste Management GEOTRAP Project (Paris: Organisation for Economic Cooperation and Development, 2002)) p 81

Haggerty R, Harvey C F, Freiherr von Schwerin C and Meigs L C 2004 What controls the apparent time-scale of solute mass transfer in aquifers and soils? A comparison of experimental results *Water Resour. Res.* **40** W01510

Huang G, Huang Q and Zhang H 2006 Evidence of one-dimensional scale-dependent fractional advection-dispersion *J. Contam. Hydrol.* **85** 53

Hulin J-P, Charlaix E, Plona T J, Oger L and Guyon E 1988 Tracer dispersion in sintered glass beads with a bidisperse size distribution *AIChE J.* **34** 610

Hunt A G 1992 Dielectric and mechanical relaxation; transition from effective medium to percolation theories *Solid State Commun.* **84** 701

Hunt A G 1998 Upscaling in subsurface transport using cluster statistics of percolation *Transp. Porous Media* **30** 177

Hunt A G 2001 Applications of percolation theory to porous media with distributed local conductances *Adv. Water Resour.* **24** 279

Hunt A G and Gee G W 2002 Application of critical path analysis to fractal porous media: comparison with examples from the Hanford Site *Adv. Water Resour.* **25** 129

Hunt A G, Ghanbarian B, Skinner T E and Ewing R P 2015 Scaling of geochemical reaction rates via advective solute transport *Chaos* **25** 075403

Hunt A G and Skinner T E 2008 Longitudinal dispersion of solutes in porous media solely by advection *Phil. Mag.* **88** 2921

Hunt A G and Skinner T E 2010 Predicting dispersion in porous media *Complexity* **16** 43

Hunt A G, Skinner T E, Ewing R P and Ghanbarian-Alavijeh B 2011 Dispersion of solutes in porous media *Eur. Phys. J.* B **80** 411

Jardine P M, Jacobs G K and Wilson G V 1993 Unsaturated transport processes in undisturbed heterogeneous porous media: 1. Inorganic contaminants *Soil Sci. Soc. Am. J.* **57** 945

Katz A J and Thompson A H 1986 Quantitative prediction of permeability in porous rock *Phys. Rev.* B **34** 8179

Kim D-J, Kim J-S, Yun S-T and Lee S-H 2002 Determination of longitudinal dispersivity in an unconfined sandy aquifer *Hydrol. Process.* **16** 1955

Le Doussal P 1989 Permeability versus conductivity for porous media with wide distribution of pore sizes *Phys. Rev.* B **39** 4816

Liu C, Zachara J M, Qafoku N P and Wang Z 2008 Scale-dependent desorption of uranium from contaminated subsurface sediments *Water Resour. Res.* **44** W08413

Madadi M and Sahimi M 2003 Lattice Boltzmann simulation of fluid flow in fracture networks with rough, self-affine surfaces *Phys. Rev.* E **67** 026309

Madadi M, Van Siclen C D and Sahimi M 2003 Fluid flow and conduction in fractures with rough, self-affine surfaces: a comparative study *J. Geophys. Res.* **108** 2396

Maher K 2010 The dependence of chemical weathering rates on fluid residence time *Earth Plan. Sci. Lett.* **294** 101

Molins S, Trebotich D, Steefel C I and Shen C 2012 An investigation of the effect of pore scale flow on average geochemical reaction rates using direct numerical simulation *Water Resour. Res.* **48** W03527

Neuman S P and di Federico V 2003 Multifaceted nature of hydrogeologic scaling and its interpretation *Rev. Geophys.* **41** 1014

Pollak M 1972 A percolation treatment of dc hopping conduction *J. Non Cryst. Solids* **11** 1

Rieu M and Sposito G 1991 Fractal Fragmentation, Soil Porosity, and Soil Water Properties: I. Theory *Soil Sci. Soc. Am. J.* **55** 1231–8

Roberts P V, Goltz M N and Mackay D M 1986 A natural gradient experiment on solute transport in a sand aquifer. 3. Retardation estimates and mass balances for organic solutes *Water Resour. Res.* **22** 2047

Sahimi M 1987 Hydrodynamic dispersion near the percolation threshold: scaling and probability densities *J. Phys.* A **20** L1293

Sahimi M 2003 *Heterogeneous Materials* I (Berlin: Springer)

Sahimi M 2012 Dispersion in porous media, continuous-time random walks, and percolation *Phys. Rev.* E **85** 016316

Sahimi M 2023 *Applications of Percolation Theory* 2nd edn (Berlin: Springer)

Sahimi M, Davis H T and Scriven L E 1983 Dispersion in disordered porous media *Chem. Eng. Commun.* **23** 329

Sahimi M, Heiba A A, Davis H T and Scriven L E 1986b Dispersion in flow through porous media: II. Two-phase flow *Chem. Eng. Sci.* **41** 2123

Sahimi M, Heiba A A, Hughes B D, Scriven L E and Davis H T 1982 Dispersion in flow through porous media SPE Paper 10 969

Sahimi M, Hughes B D, Scriven L E and Davis H T 1986a Dispersion in flow through porous media: I. One-phase flow *Chem. Eng. Sci.* **41** 2103

Sahimi M and Imdakm A O 1988 The effect of morphological disorder on hydrodynamic dispersion in flow through porous media *J. Phys.* A **21** 3833

Sahimi M and Mukhopadhyay S 1996 Scaling properties of a percolation model with long-range correlations *Phys. Rev.* E **54** 3870

Salehikhoo F, Li L and Brantley S L 2013 Magnesite dissolution rates at different spatial scales: the role of mineral spatial distribution and flow velocity *Geochim. Cosmochim. Acta* **108** 91

Scher H, Shlesinger M F and Bendler J 1991 Time-scale invariance in transport and relaxation *Phys. Today* **44** 26

Seager C H and Pike G E 1974 Percolation and conductivity: A computer study II *Phys. Rev.* B **10** 1435

Seaman J C, Bertsch P M, Wilson M, Singer J, Majs F and Aburime S A 2007 Tracer migration in a radially divergent flow field: longitudinal dispersivity and anionic tracer retardation *Vadose Zone J.* **6** 373

Sheppard A P, Knackstedt M A, Pinczewski W V and Sahimi M 1999 Invasion percolation: new algorithms and universality classes *J. Phys.* A **32** L521

Stauffer D and Aharony A 1994 *Introduction to Percolation Theory* 2nd edn (London: Taylor and Francis)

Sternberg S P K, Cushman J H and Greenkorn R A 1996 Laboratory observation of nonlocal dispersion *Transp. Porous Media* **23** 135

Vanderborght J and Vereecken H 2007 Review of dispersivities for transport modeling in soils *Vadose Zone J.* **6** 29

White A F and Brantley S L 2003 The effect of time on the weathering rates of silicate minerals: why do weathering rates differ in the lab and in the field? *Chem. Geol.* **202** 479

Yu F and Hunt A G 2017 Damköhler number input to transport-limited chemical weathering and soil production calculations *ACS Earth Space Chem.* **1** 30

Zhong L, Liu C, Zachara J M, Kennedy D W, Scezcody J E and Wood B 2005 Oxidative remobilization of biogenic uranium (IV) precipitates: effects of iron (II) and pH *J. Environ. Qual.* **34** 1763

IOP Publishing

Networks on Networks (Second Edition)
Role of connectivity in physics of geobiology and geochemistry
Allen G Hunt and Muhammad Sahimi

Chapter 6

Water transport and storage

6.1 Background

Water transport and storage in the unsaturated zone are of fundamental importance to plant growth and ecosystem function. The unsaturated zone is of particular importance for biota because for saturations S, over a relatively wide range from near 10% to about 90% (Hunt and Gee 2003, Hunt 2004), both air and water phases are typically continuously interconnected, allowing both fluids to flow, including their associated nutrients or dissolved constituents. Typically, both phases need to be present in such concentrations that flow rates allow optimal growth of plants and their associated communities, i.e., bacteria and fungi.

Water storage is a property of the unsaturated zone. It represents the amount of water as a function of the porosity, i.e., the saturation, rather than storage within fully saturated aquifers. With certain limits, stored water is accessed by plants for use in transpiration. The limits arise due to pore sizes and the phase continuity of air and water. Plants cannot draw water out of pores smaller than about 0.3 μm, while pores larger than about 30 μm tend to drain too rapidly under the gravitational force to store significant water (Watt *et al* 2006). If either the water or air phase connectivity is lost, as percolation theory has taught us, it becomes almost impossible to remove significant amounts of that phase from the medium. Water is also transported into and out of the unsaturated zone of the soil. The vast majority of water transported into the vadose zone is from above closer to the surface, though in desert climates, deeper water can be accessed by plants and drawn up into the vadose zone. On the other hand, water is transported out of the vadose zone towards the surface through evapotranspiration, as well as towards the water table under the influence of gravity. The implications for plants of the mobility of these fluid phases as a function of saturation is considerable. When the water content is too low, for example, plants can no longer draw water from the soil and reach the permanent wilting point.

By water transport in this chapter we mean not simply the movement of water under steady-state conditions, but such movement that actually changes its amount in various regions of the soil. A classical equation, known as the Richards equation (Richards 1951), is usually used to describe the changes in water content of the porous medium, and is a nonlinear diffusion equation due to the nonlinear relationship between pressure head (suction pressure) and water content. In pore networks, the Richards equation is replaced by a process described by invasion percolation algorithm, already described in chapters 3 and 4. For the case of wetting from one side (as soil during a rainstorm), a site-percolation problem, the two models, namely, site and bond invasion percolation, provide nearly the same predictions, though the work by Sharma *et al* (1980), designed to test the classic Philip (1957, 1967, 1969) model and analysis, suggested (Hunt *et al* 2017) superior agreement with the percolation model of infiltration (Wilkinson and Willemsen 1983, Sahimi 1993, 2011, Hunt *et al* 2017). For drying (Shokri *et al* 2024), which is a bond percolation problem, differences are more significant.

6.2 Water transport: Richards equation, Philip infiltration, and invasion percolation

Consider a small volume of soil, dV. The rate at which water is transported out of the volume is given by the divergence of the moisture flux $\mathbf{J}_\theta$, leading to,

$$\boldsymbol{\nabla} \cdot \mathbf{J}_\theta = -\frac{\partial \theta}{\partial t}. \tag{6.1}$$

Next, we invoke Darcy's law for slow flow of a fluid through a porous medium to describe unsaturated flow,

$$\mathbf{J}_\theta = -K\boldsymbol{\nabla}P, \tag{6.2}$$

with K being the hydraulic conductivity, and P the pressure. As discussed in the previous chapters, the hydraulic conductivity represents the ease with which a given pressure gradient can move water. In contrast to saturated media for which, by definition, the moisture content does not change, equation (6.1) is written with a right side that allows such changes explicitly. Equation (6.2), analogous to Ohm's law of electromagnetism, is a linear equation, since the water flux may depend also on a higher power of the gradient of P. If we substitute equation (6.2) into equation (6.1), we obtain

$$\boldsymbol{\nabla} \cdot \mathbf{J}_\theta + \frac{\partial \theta}{\partial t} = -\boldsymbol{\nabla} \cdot (K\boldsymbol{\nabla}P) + \frac{\partial \theta}{\partial t} = -\boldsymbol{\nabla} \cdot \left(K\frac{\partial P}{\partial \theta}\boldsymbol{\nabla}\theta \right) + \frac{\partial \theta}{\partial t}. \tag{6.3}$$

Equation (6.3) is rewritten as

$$\frac{\partial \theta}{\partial t} = \boldsymbol{\nabla} \cdot (d_h \boldsymbol{\nabla}\theta), \tag{6.4}$$

where, $d_h = K(\theta)(dP/d\theta)$. Equation (6.4) is in the form of a diffusion equation, with d_h representing a diffusion coefficient, which is a function of the moisture content θ, due to its dependence on the pressure-saturation curve, $dP/d\theta$, and the hydraulic conductivity $K(\theta)$. The dependence of d_h on θ implies that equation (6.4) is, in general, a complex equation that involves non-integer powers of a nonlinear function of moisture content. Thus, more often than not, its solution typically requires numerical simulation. Some analytical solutions are, however, known to represent reasonably well the various physical systems under certain conditions. One, which is of particular interest here, is due to Philip (1957, 1969) for wetting of soils, given by

$$J_\theta = A + Bt^{-1/2}, \tag{6.5}$$

which was intended for the moisture entering a porous medium from the surface, when water is introduced to an initially dry surface at a constant rate, A. The time integral of the moisture current yields the total moisture transported into the medium,

$$\Delta\theta \equiv I = At + St^{1/2}. \tag{6.6}$$

Here, A represents the ratio of the saturated hydraulic conductivity and the porosity, while S is proportional to $A^{1/2}$. Using the power law provided by invasion percolation, described in chapters 3 and 4 for wetting conditions, yields,

$$I = \left(\frac{z_0}{t_0}\right)t + z_0\left(\frac{t}{t_0}\right)^{1/D_{bb}}, \tag{6.7}$$

where the exponent $D_{bb} \simeq 1.86$ is the fractal dimension of the backbone in three-dimensional (3D) site invasion percolation with trapping, since wetting is inherently a site-percolation problem. In equation (6.7), both the coefficients and the power require some discussion, particularly as the distinction between equations (6.7) and (6.8) can be considered from either perspective.

The non-trivial time-dependent term of equation (6.7) is nearly the same as the equation for solute transport in chapter 5. However, for solute transport, only the unsteady term was given and a slightly different exponent (fractal dimension), $D_{bb} \simeq 1.87$, was used. The absence of the steady-state term in the equation for solute transport in chapter 5 was consistent with the assumption of small (point) solute sources in a steady flow, rather than a continuous moisture front, as here. The use of $D_{bb} \simeq 1.87$, appropriate for random percolation, was consistent with assuming that weathering is most pronounced under saturated conditions. In fact, most chemical weathering with transport of products towards the water table likely takes place under either wetting or steady-state conditions. But the corresponding distinction in the backbone fractal dimensions (1.86 versus 1.87), which leads to scaling exponents of 0.538 and 0.532, was not considered significant enough in chapter 5, so that as long as weathering was occurring with primarily downward water flow, whether saturated or wetting, the same exponent, 0.53, was applicable. Experiments on wetting at field scale are, however, able to detect the difference between exponents of 0.54 and 0.5, with the latter proposed by the Philip equations.

In addition, z_0 in equation (6.7) is a particle size, and t_0 is the ratio of a particle size and the flow rate that, in the case of a flooded field, is the saturated hydraulic conductivity, K_S, divided by the porosity ϕ. Thus, z_0/t_0 is the same as A in equation (6.6). The evaluation of the second coefficient requires further knowledge of K_S, which is ordinarily considered proportional to be the square of a characteristic (or critical) particle size, though the proportionality constant is not generally agreed on. Thus, $(z_0/t_0)^{0.54} = z_0/[z_0/(K_S\phi)]^{0.54} = z_0(z_0)^{0.54} = (K_S^{1.54})^{1/2}$, which turns out to be proportional to $K^{0.77}$ in the simplest approximation, implies that the constants S and A, extracted from experiments if they are actually in accord with equation (6.7), will be related by $S \approx A^{0.77}$.

It is instructive to discuss how two such similar-looking equations can be distinguished. Sharma *et al* (1980) addressed this issue by using a transformation to scaled variables,

$$\beta = I\left(\frac{A}{S^2}\right), \quad \tau = t\left(\frac{A}{S}\right)^2. \tag{6.8}$$

In the absence of systematic deviations from the Philip's scaling relationship (or significant experimental error), all experimental values for $\beta(\tau)$ will fall on a universal curve, $\beta = \tau + \tau^{1/2}$. Systematic deviations indicate a discrepancy between theory and experiment. Comparisons may, of course, be distorted by inaccurate determination of A and S, for the estimation of which many methods exist. One can, for example, divide the data by either t or $t^{1/2}$, in order to convert one term to a constant. Then, asymptotic methods can be utilized to determine A at large times in the former case, or S at small times in the latter. If, however, equation (6.7) is the appropriate description of the infiltration process, then use of the transformations given by equation (6.8) will lead to, $\beta = \tau + t_0^{0.04}\tau^{0.54}$, which introduces scatter into the predicted relationship, as reflected by the variability of the parameter t_0, which was not given in the original reference. Its variability can, however, be estimated from the largest and smallest values of the data in their given range, and used to guide a range of predictions that equation (6.7) would provide.

A scaled plot of $\beta(\tau)$ (Hunt *et al* 2017), which used digitized data of Sharma *et al* (1980), is given in figure 6.1. A clear spread in the data is revealed, which is not in accord with the prediction of equation (6.4). Note that the blue curve, which corresponds to the smallest value of t_0, is also nearly the the prediction of equation (6.6). Similarly, one can investigate the relationship between what is called the sorptivity S and the parameter A of the Philip model. According to Sharma *et al* (1980), such a relationship should be, $S \propto A^{1/2}$. The data are shown in figure 6.2. The best fit, however, is given by a power of 0.727, rather than 1/2, significantly closer to the percolation prediction of 0.77.

Notably, data from over 5000 field infiltration experiments, compiled by Dr Mehdi Rahmati (of the University of Maragheh, Iran, and a Senior Researcher at Forschungszentrum Jülich, Germany) and co-workers (personal communication; see also Rahmati *et al* (2018)), indicate a spike of exponents very close to 0.5, but many more near 0.54, and a sizable component with values larger than 0.54. This is not incompatible with the present interpretation, as it is not asserted that all porous media are sufficiently disordered for the percolation

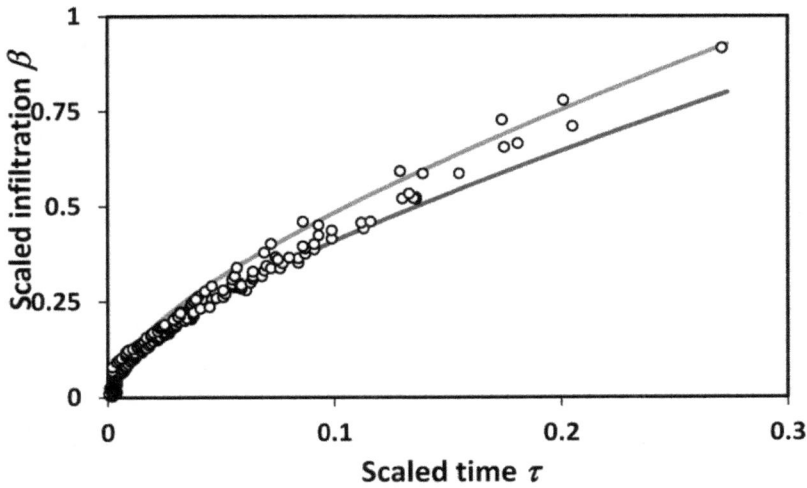

Figure 6.1. Scaled infiltration parameter β and its dependence on the scaled time τ. Reprinted from Hunt *et al* (2017).

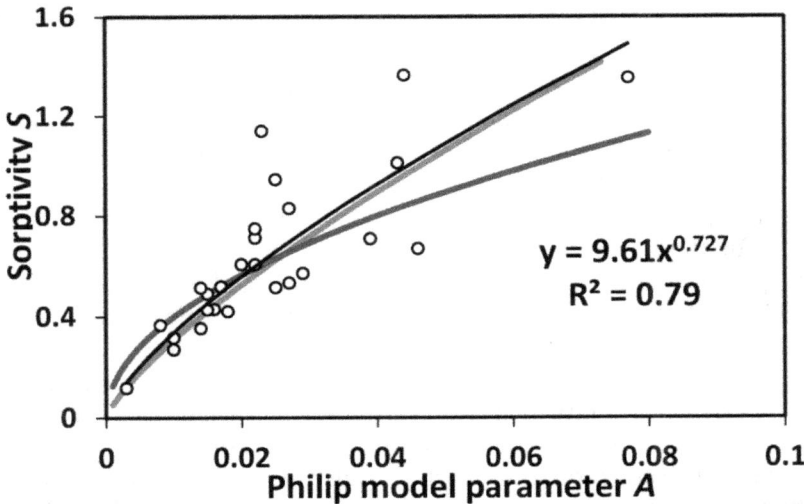

Figure 6.2. The relationship between sorptivity and the parameter A of Philip's model. Reprinted from Hunt *et al* (2017).

concepts to predominate. If a porous medium is sufficiently homogeneous, it would certainly be plausible that a description that does not specifically consider disorder would be sufficient. Furthermore, particularly at shorter time (and, thus, length) scales, the specifics of the disorder in the pore space will become more important. As noted by Hunt (2001), transport coefficient exhibit a greater spread at small spatial scales, and the distribution is skewed to lower values, whereas the mean value is

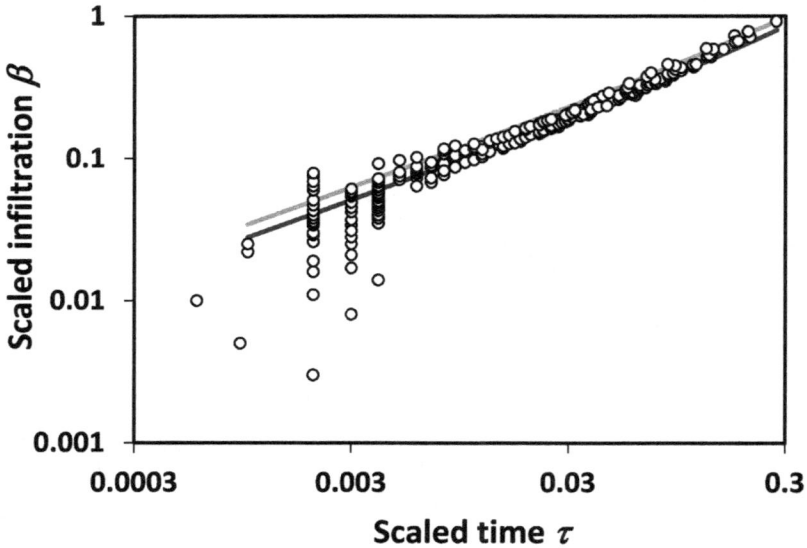

Figure 6.3. Plot of the data reported by Sharma *et al* (1980), indicating expotentially wide distribution. Reprinted from Hunt *et al* (2017).

enhanced by a distribution that is exponentially wide. The data of Sharma *et al* (1980) demonstrate this well, when plotted logarithmically; see figure 6.3.

Our general conclusions are compatible with other reports in the literature. For example, Ferer *et al* (2004) reported both simulations and experimental data support a crossover from capillary fingering–invasion percolation with trapping with $D_{bb} \simeq 1.86$ for the wetting fluid, and $D_{bb} \simeq 1.43$ for drying—at low velocities to viscous fingering—described by a model, diffusion-limited aggregation (Witten and Sander 1981, Patterson 1984) with $D_f \simeq 1.71$ at high velocities—and referenced by Fernandez *et al* (1991) for the proposed behavior. However, the conclusions of other researchers on experiments in disordered porous media differ somewhat. For example, Toussaint *et al* (2005) state, 'We study viscous fingering during drainage experiments in linear Hele-Shaw cells filled with a random porous medium. The central zone of the cell is found to be statistically more occupied than the average, and to have a lateral width of 40% of the system width, irrespectively of the capillary number Ca. A crossover length, $w_f \propto Ca^{-1}$, separates the lower scales where the invader's fractal dimension, 1.83, is identical to capillary fingering, and the larger scales where the fractal dimension is found to be 1.53. The lateral width and the large-scale dimension are lower than that of diffusion-limited aggregation, but can be explained in terms of the dielectric breakdown model (see Sahimi 2003, for the description of the model). Indeed, we show that when averaging over the quenched disorder in capillary thresholds, an effective law, $v \propto (\nabla P)^2$ relates the average interface growth rate and the local pressure gradient.' It may be interesting to note that 1.53 is closer to 1.46 than to 1.71, as well as (possibly) the dielectric breakdown model that generates a rather wide range of fractal dimensions (Wiesmann and

Pietronero 1986, Satpathy 1986, Tsonis 1991). Drainage experiments, however, represent bond invasion percolation with trapping, rather than site invasion percolation with trapping for imbibition. While the latter has a fractal dimension of 1.82, the former has, $D_{bb} \simeq 1.46$.

6.3 Water storage and its implications for plants

Water storage in the unsaturated zone is critical for plants. When water becomes too scarce, plant uptake of water ceases and they reach the permanent wilting point. When thermal gradients can be ignored, the water potential is higher at higher elevations in the soil, while the water content is lower. This water is found more exclusively in the smaller pores of the medium with increasing water potential (and height) and, thus, decreasing water content. However, since water is so strongly held in the smallest pores (with high water potential values, expressed as a height h above the water table), even though it may be there in significant amounts, plants cannot access it. We wish to be able to address a few of the physical properties of soils that are relevant to the function of plant root systems and plant water uptake.

The model proposed by Rieu and Sposito (1991), already described and utilized in chapters 4 and 5, is, in essence, a fractal pore-size distribution, though its origin and purpose were conceived slightly differently. Given, (a) the strong wetting condition that a given radius, say $r \geqslant A/h$, which is a function of the water (matric) potential into pores with $r > \langle r \rangle$, in which only air is allowed, and pores with $r < \langle r \rangle$ in which only water is allowed, and (b) that water (and air) has connected pathways of water-allowed pores, as well as sufficient time to flow out of (or into) such pores, respectively, it is possible to write down an equation for the volumetric water content, θ—the volume fraction of the pore space occupied by water—in terms of the porosity ϕ:

$$\phi = \theta - \left[1 - \left(\frac{h_A}{h} \right)^{3-D} \right], \tag{6.9}$$

where D is the fractal dimension of the pore space, h_A is the air-entry pressure (discussed previously), which is the value of h when air is allowed into the largest pores by $r = A/h$. Frequently, $A = 0.149$ is used, which would be appropriate for the height (in centimeters) that water would rise in a cylindrical tube of radius r (in centimeters). It is commonly assumed that a value of $h = 15\,000$ cm corresponds to the permanent wilting point of plants, which corresponds to a pore radius of 0.1 μm. In agreement with this, Watt *et al* (2006) stated that the range of pore sizes that contain water and can be extracted by plants is from about 0.2 μm to about 30 μm. In an ideal porous medium in which pore sizes are between the two limits, and with a typical porosity of 0.4, the application of the Rieu and Sposito model leads to a fractal dimension of 2.89. Applying the critical-path analysis (see chapters 3 and 4) to such a porous medium leads to a critical pore size under perfectly saturated conditions of 20 μm, but under the condition of 90% saturation, which is the highest common soil saturation, the critical pore size would be only 13 μm. For such a pore radius, an equation for predicting the hydraulic conductivity of a weakly heterogeneous porous medium, such as the Kozeny–Carman correlation, leads to a hydraulic conductivity of about $1~\mu\text{m s}^{-1} = 10^{-4}~\text{cm s}^{-1}$. The analogous calculation for a pore of 1/10 of the radius will, however, lead to a hydraulic conductivity 100 times smaller, $10^{-6}~\text{cm s}^{-1}$. At values just 10 times smaller, flow rates of $10^{-7}~\text{cm s}^{-1}$ correspond to 100 μm/day or smaller, and the experiments begin to fall out of equilibrium, simply because it takes weeks for measurable amounts of water to drain.

We will return to these in subsequent chapters.

References

Ferer M, Ji C, Bromhal G S, Cook J, Ahmadi G and Smith D H 2004 Crossover from capillary fingering to viscous fingering for immiscible unstable flow: experiment and modeling *Phys. Rev.* E **70** 016303

Fernandez J F, Rangel R and Rivero J 1991 Crossover length from invasion percolation to diffusion-limited aggregation in porous media *Phys. Rev. Lett.* **67** 2958

Hunt A G 2001 Applications of percolation theory to porous media with distributed local pore conductances *Adv. Water Resour.* **24** 279

Hunt A G 2004 Continuum percolation theory for water retention and hydraulic conductivity of fractal soils: Estimation of the critical volume fraction for percolation *Adv. Water Resour.* **27** 175–83

Hunt A G and Gee G W 2003 Wet-end deviations from scaling of the water retention characteristics of fractal porous media *Vadose Zone J.* **2** 759

Hunt A G, B Ghanbarian B and Holtzmann R 2017 A percolation-based approach to scaling of infiltration and evapotranspiration *Water* **9** 104

Paterson L 1984 Diffusion-limited aggregation and two-fluid displacement in porous media *Phys. Rev. Lett.* **52** 1621

Philip J R 1957 The theory of infiltration: 4. Sorptivity and algebraic infiltration equations *Soil Sci.* **84** 257

Philip J R 1969 Moisture equilibrium in the vertical in swelling soils. I. Basic theory *Soil Res.* **7** 99

Philip J R 1967 Sorption and infiltration in heterogeneous media *Aust. J. Soil Res.* **5** 1

Rahmati M *et al* 2018 Development and analysis of the Soil Water infiltration Global database *Earth Syst. Sci. Data (ESSD)* **10** 1237–63

Richards L A 1951 Capillary conduction of liquids through porous mediums *Physics* **1** 318

Rieu M and Sposito G 1991 Fractal fragmentation, soil porosity, and soil water properties. I. Theory *Soil Sci. Soc. Am. J.* **55** 1231

Sahimi M 1993 Flow phenomena in rocks: from continuum models to fractals, percolation, cellular automata and simulated annealing *Rev. Mod. Phys.* **65** 1395

Sahimi M 2003 *Heterogeneous Materials II* (Berlin: Springer)

Sahimi M 2011 *Flow and Transport in Porous Media and Fractured Rock* 2nd edn (New York: Wiley-VCH)

Satpathy S 1986 Dielectric breakdown in three dimensions: results of numerical simulation *Phys. Rev.* B **33** 5093

Serra T and Casamitjana X 1998 Structure of the aggregates during the process of aggregation and breakup under a shear flow *J. Colloid Interface Sci.* **206** 505

Sharma M L, Gander G A and Hunt C G 1980 Spatial variability of infiltration in a watershed *J. Hydrol.* **45** 101

Shokri N, Hassani A and Sahimi M 2024 Soil salinization, from pore to global scale: mechanisms, modeling and outlook *Rev. Geophys.* **62** e2023RG000804

Toussaint R, Løvoll G, Méheust Y, Maløy K J and Schmittbuhl J 2005 Influence of pore-scale disorder on viscous fingering during drainage *Europhys. Lett.* **71** 583

Tsonis A 1991 A fractal study of dielectric breakdown in the atmosphere *Non-linear Variability in Geophysics* ed D Schertzer and S Lovejoy (Berlin: Springer) p 167

Watt M, Silk W K and Passioura J B 2006 Rates of root and organism growth, soil conditions and temporal and spatial development of the rhizosphere *Ann. Bot.* **97** 839

Wiesmann H J and Pietronero L 1986 Properties of Laplacian fractals for dielectric breakdown in 2 and 3 dimensions *Fractals in Physics* ed L Pietronero and E Tosatti (Amsterdam: Elsevier) p 151

Wilkinson D and Willemsen J 1983 Invasion percolation: a new form of percolation theory *J. Phys.* A **16** 3365

Witten T A and Sander L M 1981 Diffusion-limited aggregation, a kinetic critical phenomenon *Phys. Rev. Lett.* **47** 1400

IOP Publishing

Networks on Networks (Second Edition)
Role of connectivity in physics of geobiology and geochemistry
Allen G Hunt and Muhammad Sahimi

Chapter 7

Water transport in plants

7.1 Background

This chapter is dedicated to the description of water transport in plant tissues. Differences and similarities between plant conductive tissue (xylem) and soils are described, and detailed models of flow through xylem conduits are used to explain inherent trade-offs in plant water use. In fact, xylem tissues cannot transport water efficiently under dry conditions—either soil drought or high evaporative fluxes— thereby constraining plant formation and function.

Terrestrial plants require evaporation of water from the stomatal pores to acquire CO_2 to fuel photosynthesis and growth. Globally, about 70% of water evaporated from land is transpired by plants (Dingman 1994) and, therefore, passes through the plant xylem tissues connecting the roots to leaves and stomata. Water transport is driven by declining water potential levels between the soil and the leaves, based on the cohesion–tension theory (Dixon and Joly 1895, van den Honert 1948, Tyree and Ewers 1991, Tyree 1997, Cruiziat et al 2002). The cohesion–tension theory, sometimes referred to as the C–T theory, postulates that the water ascent in trees is exclusively due to the transpirational pull from continuous water columns in the xylem conduit, running from the roots to the leafs. The pull creates tension gradients large enough that overcome the gravitational force and frictional resistances. The gradient is maintained by evaporation from the stomata, and the steepness of the gradient is controlled by the properties of the xylem and stomatal regulation of transpiration. Here, we focus on transport of liquid water. Previous work described stomatal controls on transpiration and the interactions between liquid- and vapor- phase transport properties (Jarvis and McNaughton 1986, Buckley et al 2003, Mencuccini 2003, Manzoni et al 2013b).

Xylem tissues are specialized to transport water efficiently by offering the least possible resistance to flow and allowing leaf water potential to remain as close to the soil water potential as possible. The goal is met by various designs as the earlier vascular plants evolved towards extant gymnosperms and angiosperms, including

7-1

unicellular structures (tracheids) and multicellular ones (vessels), connected through porous pits of varying complexity (Sperry 2003); see figure 7.1. As a result of such different designs, xylem conduits vary in size, shape, and function across plant types, shown in figure 7.2. Their diameters are on the order of 10^{-6} to 10^{-4} m, whereas their length is much larger, ranging between 10^{-3} to 100 m. Therefore, compared to soil pores, xylem conduits are more elongated and their diameter varies less. As a result, different from soils, their geometry can be approximated by a bundle of parallel capillaries.

Efficient transport of water is not, however, always possible. As the pressure gradients steepen, when the evaporative demand increases or soil moisture declines, liquid water in the xylem reaches increasingly more negative water potentials. Eventually, this meta-stable state breaks down and vapour or air bubbles form in the conduits (Tyree 1997, Cruiziat et al 2002). The presence of bubbles inhibits flow of liquid water, hence reducing the hydraulic conductivity. Should refilling not occur,

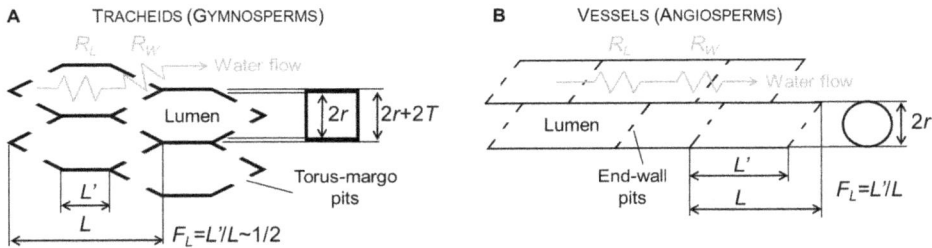

Figure 7.1. Schematic representation of xylem conduits in (A) gymnosperms, and (B) angiosperms, and definition of symbols [following the notation of Wheeler et al (2005)].

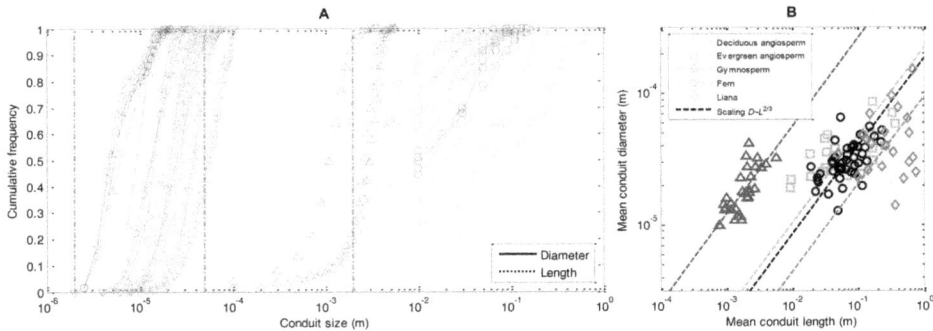

Figure 7.2. Illustration of typical xylem conduit sizes across plant functional types. Left panel: cumulative distributions of vessel diameters (solid lines) and lengths (dotted lines), compared to soil particle sizes, as indicated by vertical dot-dashed lines that separate clay from silt and sand (from left to right). The data are from Sperry and Sullivan (1992); Sperry et al (1994); Lipp and Nilsen (1997); Herman et al (1998); Kocacinar and Sage (2004); Cai and Tyree (2010); Mayr et al (2010); Peguero-Pina et al (2011); Pittermann et al (2011), and Torres-Ruiz et al (2013). Right panel: correlation between mean conduit length and diameter and predicted scaling laws [equation (7.5)]. Data for various organs from the same species (roots and stems) are pooled (Sperry et al 2006, Jacobsen et al 2012).

for example, at night or after a rainfall event, lowered conductivity may further steepen the pressure gradients, leading to further bubble formation. Eventually, hydraulic failure, or runaway cavitation, ensues (Tyree and Sperry 1988, Manzoni *et al* 2014a). The curve describing the decline in conductivity as a function of water potential is commonly referred to as cavitation curve. In analogy with soils, tissue-scale hydraulic conductivity and cavitation curves are emerging macroscopic results of micro-scale features of the xylem conduits. As shown below, trade-offs exist that required xylem conduits to evolve toward an optimal design, capable of balancing transport efficiency through relatively large pores and safety from cavitation through narrow connections between them.

Despite the obvious size, as indicated in figure 7.2, and architectural differences— directed versus random networks—xylem tissues present some analogies to soils. As such, they are modelled as porous media with a set of macroscopic properties. Flow rates are computed using the equivalent of Darcy's law for xylem tissues (Tyree and Sperry 1988, Bohrer *et al* 2005, Manzoni *et al* 2013b), where the driving force is imposed by declining water potential, and the resistance to transport is caused by frictional forces along the conduits. How plant-scale transport is linked to micro-scale properties is explored in detail next.

7.2 Pore scale

In this section, the properties of xylem pores are discussed and their theoretical descriptions are presented based on previous work (Wheeler *et al* 2005, Pittermann *et al* 2006, Sperry *et al* 2006). Because of the clear geometrical and functional separation between the main conductive conduit and the connecting pits, as shown in figure 7.1, the overall xylem resistivity to water transport per unit lumen cross sectional area (R) is expressed as the sum of the lumen R_L and end-wall (pit) contributions, R_W,

$$R = R_L + R_W. \tag{7.1}$$

Lumen resistivity follows the Hagen–Poiseuille law for laminar flow, while the end-wall resistivity is expressed as the resistance r_W distributed over the average lumen length, i.e., the distance L' between subsequent end walls. In turn, r_W is written as the ratio of the total pit resistance r_P and the pit area, A_P, yielding

$$R_L = \frac{8\eta}{\pi r^4}, \quad R_W = \frac{2r_P}{A_P L'}, \tag{7.2}$$

where r is the radius of the conduit lumen, and the factor 2 accounts for the fact that end walls are in contact with two conduit units. Equation (7.2) holds for both tracheid and vessel structures; the two xylem types differ in how A_P and L' are linked to the diameter and length of the conduits; see figure 7.1.

Vessels have circular cross-sections, so that the pit area is given by, $A_P = 2\pi r L F_P$, where F_P is the pit fraction on the vessel surface. As a result, the total resistivity is obtained,

$$R = \frac{8\eta}{\pi r^4} + \frac{r_P}{\pi r L^2 F_L F_P}. \tag{7.3}$$

In contrast to angiosperm vessels, gymnosperm tracheids have square cross-sections, so that $A_P = 8rLF_P$, and $F_P \approx 1/2$, yielding,

$$R = \frac{8\eta}{\pi r^4} + \frac{r_P}{2rL^2 F_P}. \tag{7.4}$$

As conduit size varies, the resistances in lumens and pits in equations (7.3) and (7.4) need to change in a coordinated manner so as to maintain R to a minimum, in order to avoid creating bottlenecks. Therefore, in both vessels and tracheids, conduit radius should scale with conduit length as

$$r \sim L^{2/3}. \tag{7.5}$$

Scaling relation (7.5) is consistent with observations across species within a plant functional type, but not across functional types due to their different geometries and scaling coefficients, as indicated by dashed lines in figure 7.2(B). The observations agree particularly well with equation (7.5) in gymnosperms and self-supporting angiosperms, whereas in lianas and vines the relation tends to break down. It should also be pointed out that the description holds for a representative, average conduit, but a range of conduit sizes typically characterizes the xylem; see figure 7.1.

As r and L and their distribution in a given stem change among species and plant functional types, the likelihood that cavitation occurs in the conduits also varies. It had originally been hypothesized that air seeding develops from pit membranes and, therefore, starts at less negative pressures when the membrane pores are larger. Contrary to expectations, however, vulnerability to cavitation was not related to pit porosity, but rather to the pit area per vessel. This observation suggests that cavitation is related to the presence of rare, large pores (Wheeler *et al* 2005). In turn, the size of the largest pore is expected to increase with the total pit area, A_P, which is proportional to vessel size. While this hypothesis was later revisited (Christman *et al* 2009, Lens *et al* 2011), evidence points to conduits with high hydraulic efficiency, requiring large r, being more prone to cavitation and, therefore, being less 'hydraulically safe,' as safety requires low values of r (Wheeler *et al* 2005, Hacke *et al* 2006). This trade-off emerges from the physics of water transport and phase transition at the pore scale. As discussed in the following sections, the trade-off propagates not only to large scales in the xylem, but also to a suite of other plant functional traits.

7.3 Tissue scale

Integrating the hydraulic resistance per unit lumen area R over the entire range of conduit diameters yields the tissue-level resistivity. Most commonly, its inverse, the hydraulic conductivity K_X is reported (with units of flux divided by pressure gradient). As indicated by equation (7.2), hydraulic conductivity scales as the forth power of conduit lumen radius, so that the mean sapwood conductivity under saturated conditions, i.e., without cavitation, is estimated as

$$K_{X,\text{sat}} \sim \int_0^\infty r^4 p_r(r) dr, \tag{7.6}$$

where p_r is the probability density of conduits with radius r, and the proportionality accounts for any parameter that is independent of r, including sapwood area. Note that we assumed that the end-wall conductivity scales as the lumen conductivity, as discussed above, and that its effects can be incorporated in the proportionality constants between $K_{X,\text{sat}}$ and the integral on the right side of equation (7.6).

The saturated hydraulic conductivity is reduced when cavitation ensues. To account for such a reduction and derive an equation for the xylem vulnerability curve, the probability that a conduit is impaired must be considered (Cai and Tyree 2010). The probability of hydraulic failure of a conduit with radius r at a water potential ψ_P is denoted by $p_{\psi_P}(\psi_P, r)$. For each conduit radius r, a vulnerability curve is constructed as the non-exceedance probability of hydraulic failure at water potential Ψ_P,

$$F_{\psi_P}(\psi_P, r) = \int_0^{|\psi_P|} p_{\psi_P}(\psi'_P, r) d\psi'_P. \tag{7.7}$$

In this framework, however, linkages between hydraulic failure and the properties of the xylem are not explicit yet. The 'rare pit hypothesis' has been proposed to fill this gap, suggesting that hydraulic failure is more likely in conduits with large surface area and more pits. According to this hypothesis, the mean water potential at cavitation becomes more negative with decreasing r. The hypothesis links the mean of p_{Ψ_P} to r, thereby introducing an explicit dependence of F_{Ψ_P} upon conduit size. The functional dependence between the statistics of p_{Ψ_P} and conduit radius is not known in general, but some empirical correlations have been suggested. For example, Cai and Tyree (2010) reported that the absolute value of the median water potential at cavitation scaled as $r^{-0.31}$ in the 5–25 μm range in conduit radius. Weighing each diameter class in equation (7.6) by the probability of hydraulic failure, equation (7.7), and normalizing the conductivity by its maximum value, a vulnerability curve for the entire sapwood is obtained:

$$\frac{K_X(\psi_P)}{K_{X,\text{sat}}} = \frac{\int_0^\infty r^4 p_r(r) F_{\psi_P}(\psi_P, r) dr}{\int_0^\infty r^4 p_r(r) dr}. \tag{7.8}$$

Due to the nonlinearity of equation (7.8), a fully analytical solution has not been derived so far. However, numerical solutions based on discrete diameter classes (Cai and Tyree 2010) and detailed simulations of the xylem network (Loepfe *et al* 2007) yielded 'emerging' vulnerability curves that capture the main features of the observed ones. In the absence of theoretical cavitation curves, most plant-scale hydraulic models use empirical formulations (Tyree and Sperry 1988, Bohrer *et al* 2005). Among the numerous expressions that have been proposed, the inverse polynomial form has the advantage of simplicity and analytical tractability; see section 7.4; see also figure 7.3. As an example,

$$\frac{K_X(\psi_P)}{K_{X,\text{sat}}} = \left[1 + \left(\frac{\psi_P}{\psi_{50}} \right)^{\alpha} \right]^{-1}, \tag{7.9}$$

where ψ_{50} is the water potential at 50% loss of conductivity, and α is a shape exponent. Typical values of ψ_{50} vary between -0.5 to -15 MPa, depending on plant functional type, species, climatic conditions, and position along the root-to-leaf xylem system. Despite significant overlap, gymnosperms tend to have more negative ψ_{50} values than angiosperms, whereas species from more mesic climates tend to have less negative values of ψ_{50} (Hacke *et al* 2001, Maherali *et al* 2004, Choat *et al* 2012, Manzoni *et al* 2013b). Moreover, ψ_{50} consistently decreases, i.e., becomes more negative, from the roots to the terminal twigs in a given tree, consistent with tapering conduits, and more negative water potential values (Johnson *et al* 2012). The shape exponent α varies comparatively less, with typical values between 2 and 8, and

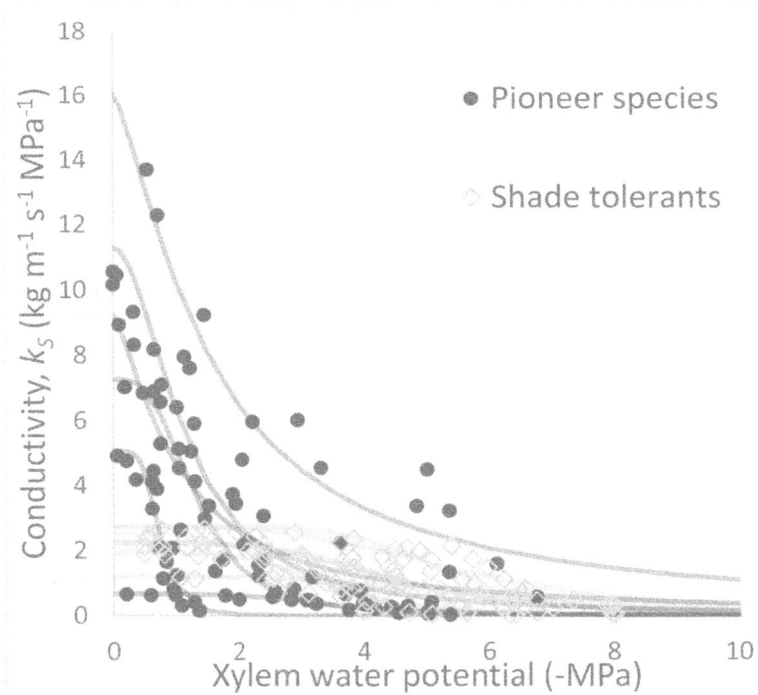

Figure 7.3. An example of vulnerability curves for pioneer and shade tolerant (late successional) species from the same ecosystem, showing an inter-specific trade-off between hydraulic efficiency (high xylem conductivity; blue data points and curves) and safety against cavitation (sensitivity to changes in xylem water potential; orange data points and curves) (Markesteijn *et al* 2011). Curves represents fitting of equation (7.9) to species-specific data.

higher values in gymnosperms than in angiosperms, and in more mesic climates, i.e., steeper decline in K_X at $\psi_P = \psi_{50}$ (Manzoni *et al* 2013b).

7.4 Ecological implications of the safety-efficiency trade-off

Because larger conduits are more prone to cavitation, a tissue-level trade-off between safety and efficiency emerges. The trade-off manifests itself in the inverse correlation between $K_{X,\text{sat}}$ and $|\psi_{50}|$ that is often observed across organs within individuals (Meinzer *et al* 2010), across species within communities (Hacke *et al* 2006, Lens *et al* 2011, Markesteijn *et al* 2011), as shown in figure 7.3, along climatic gradient (Maherali *et al* 2004, Manzoni *et al* 2013b), and across plant functional types, and in particular between conifers and angiosperms (Maherali *et al* 2004, Maherali *et al* 2006, Sperry *et al* 2006, Meinzer *et al* 2009). Nevertheless, no trade-off emerged in other studies, indicating that it is not universal and may be masked by high trait variability and co-variation of hydraulic traits and allocation patterns (Maseda and Fernandez 2006). From an ecological perspective, the safety-efficiency trade-off delineates two end-member water use strategies. On the one hand, species with highly conductive xylem reduce the water potential gradient between the soil and the leaves, but are prone to cavitation, as soon as the xylem water potential starts declining. Deciduous angiosperms growing in moist environments typically exhibit this behaviour. At the other end of the spectrum, gymnosperms and evergreen angiosperms from dry environments that tend to have low transport efficiency, causing steep water potential gradients, but also exhibit high resistance to cavitation. Despite such differences in trait values, similar rates of transpiration can be achieved by the end-members for a given plant size, as shown in figure 7.5(A), because in the efficient species the driving force for transpiration is small, whereas in less efficient species the driving force is large, compensating for the low conductance.

Notably, the continuum between hydraulically safe and efficient species is mirrored by a similar continuum of stomatal regulation strategies. Species that are most hydraulically efficient, i.e., those with high K_X, tend to have proportionally high maximum stomatal conductance, while the opposite is true for less efficient species with conservative water use (Mencuccini 2003, Manzoni *et al* 2013b). Moreover, species with vulnerable xylem, i.e., less negative ψ_{50}, tend to close their stomata earlier during a dry period, exhibiting an isohydric strategy. In contrast, less vulnerable species maintain gas exchange under drought at the expense of lowered leaf water potential, thus exhibiting an anisohydric strategy (Cruiziat *et al* 2002, Skelton *et al* 2015). Most conifers exhibit high resistance to cavitation and tight stomatal control, constraining the minimum leaf water potentials to less negative values than ψ_{50}, which represents wide safety margins, whereas most (but not all) angiosperms maintain transpiration in dry conditions. Therefore, the minimum leaf water potential can reach ψ_{50}, implying a large degree of conductivity loss (narrow safety margins) (Choat *et al* 2012). Maintaining coordinated stomatal closure and loss of conductivity allows plants to maximize their long-term transpiration rate, which is associated with carbon uptake and fitness (Manzoni *et al* 2013b), while also avoiding catastrophic hydraulic failure (Manzoni *et al* 2013a).

7.5 Plant scale

The hydraulic conductivity K_X measured in laboratory conditions on wood samples of known size is normalized by sapwood area A_S or leaf area A_L, and reported as the area-specific conductivities, $K_S = K_X/A_S$ and $K_L = K_X/A_L$, respectively, (with typical units of kg m^{-1} s^{-1} MPa^{-1}). In turn, these conductivities can be scaled up to the entire plant level, employing the total stem sapwood area and total plant leaf area (Whitehead *et al* 1984, Mencuccini 2003). This is shown in figure 7.4. The procedure neglects trends in hydraulic traits from roots to terminal twigs, but it provides a simple upscaling approach. The scaled-up conductivities, combined with representative cavitation curves, are then be employed to estimate transpiration rates using Darcy's law at the plant scale (Bohrer *et al* 2005, Novick *et al* 2009, Manzoni *et al* 2013b),

$$E = K_S(\psi_P)\left(\frac{A_S}{h}\right)(\psi_R - \psi_L - \rho g h), \qquad (7.10)$$

where the canopy height h approximates the length of the hydraulic path, ψ_R, ψ_P, and ψ_L are, respectively, the root, shoot xylem, and leaf water potentials, and the last term in brackets accounts for gravitational effects. The role of xylem water potential

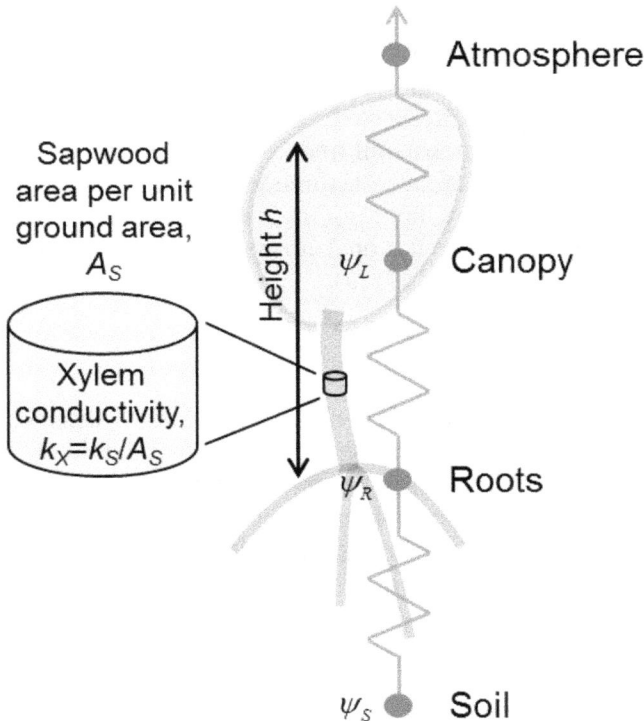

Figure 7.4. Schematic of the lumped plant hydraulic model used in section 7.3.

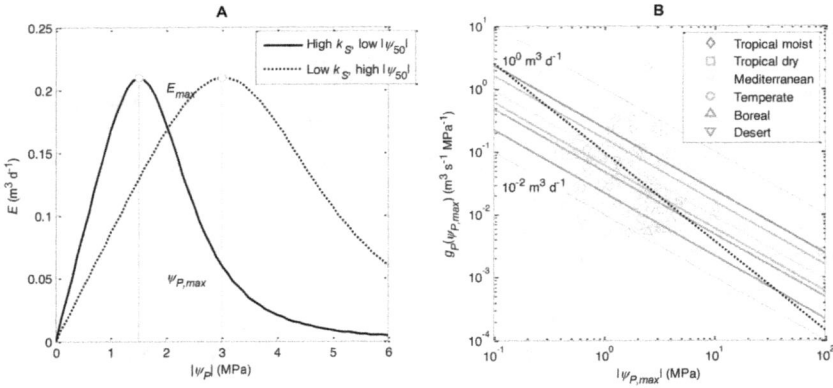

Figure 7.5. (A) Relation between plant-scale transpiration rate and canopy water potential for species with contrasting hydraulic traits, leading to the same maximum transpiration rate, E_{max}. (B) Hydraulic conductance $g_P(\psi_{P,max})$, when transpiration rate is maximum, scaling inversely with the driving force $-\psi_{P,max}$, resulting in relatively constrained maximum transpiration rates for a given plant size, despite large variation of plant hydraulic traits. Symbols refer to woody species from various ecosystems; sapwood area-specific conductivities have been converted to whole-plant conductances assuming $A_S = 10^{-2}$ m^2 and $h = 20$ m. The dotted black line is a reduced major axis regression of all data points, $g_P(\psi_{P,max}) \sim |\psi_{P,max}|^{-1.41}$; the solid colour lines represent the median E_{max} for each ecosystem type (E_{max} increases in the order, boreal < desert < temperate < Mediterranean < tropical dry < tropical moist); the dashed lines indicate two rates of transpiration on a ground area basis for reference. Modified after Manzoni *et al* (2013a), John Wiley & Sons, copyright 2013 The Authors, copyright 2013 New Phytologist Trust.

in the conductance K_S is emphasized in equation (7.10) to remind us of the cavitation effects.

Under well-watered conditions, soil and root water potential levels are nearly zero, so that $\psi_S \approx \psi_R \approx 0$. Moreover, assuming that leaf water potential is close to the xylem water potential that limits g_P, i.e., assuming $\psi_L \approx \psi_P$—a conservative assumption—and neglecting gravitational effects, equation (7.10) is reformulated as (Meyra *et al* 2011, Manzoni *et al* 2013a),

$$E \approx K_S(\psi_P)|\psi_P|\left(\frac{A_S}{h}\right), \tag{7.11}$$

where $|\psi_P|$ represents the driving force for water flow, and K_S depends on the canopy water potential according to a vulnerability curve, such as that in equation (7.9). The combined effects of a monotonically decreasing vulnerability curve and a linearly increasing driving force determine a maximum transpiration rate at an intermediate water potential; see figure 7.5(A). Using equations (7.9) and (7.11), the maximum rate is determined to be,

$$E_{max} = g_P(\psi_{P,max})|\psi_{P,max}| = K_{S,sat}\left(\frac{A_S}{h\alpha}\right)|\psi_{50}|(\alpha - 1)^{1-1/\alpha} = g_{P,sat}|\psi_{50}|$$
$$\alpha^{-1}(\alpha - 1)^{1-1/\alpha}, \tag{7.12}$$

where, in the last equality, all morphological features have been included in the plant-scale hydraulic conductance $g_{P,\text{sat}} = K_{X,\text{sat}}/h = K_{S,\text{sat}}A_S/h$.

The presence of a maximum was also emphasized earlier, but using a numerical approach (Sperry *et al* 1998). The derivation and the result, equation (7.12), analytically demonstrate that, to sustain a given maximum transpiration rate, the saturated hydraulic conductivity and the water potential at 50% loss of conductivity must be inversely correlated. Based on this result, it can be hypothesized that the hydraulic safety-efficiency trade-off at the plant scale emerges as a way of guaranteeing high rates of transpiration during the warmer and most physiologically active time of the day (Manzoni *et al* 2013a). For a given plant size, the estimated $E_{\max}$ tends to increase with temperature (and, thus, vapour pressure deficit), from boreal to temperate and tropical ecosystems; see the solid colour line in figure 7.5(B). This trend is consistent with the idea that plants strive to meet the atmospheric demand for evaporation by evolving xylem tissues with larger $E_{\max}$ in the tropics than in colder ecosystems. A notable exception is represented by deserts, where $E_{\max}$ is comparable to temperate moist ecosystems, while higher values could be expected to meet a high evaporation demand. It is possible that perennial woody desert species have not evolved to intensely exploit water during the short periods, when it is available, but rather to use water conservatively to guarantee survival to prolonged dry periods.

It is important to point out that, we assumed the same plant size for all species in figure 7.5—clearly an unreasonable assumption—since both tree and shrub species have been considered. Therefore, the analysis needs to be interpreted as an 'all else being equal' comparison across traits, rather than actual plants exhibiting a range of sizes. When changes in A_S and h are accounted for in datasets where the parameters are available, the $E_{\max}$ predicted by equation (7.12) closely matched observed peak transpiration rates in well-watered conditions (Manzoni *et al* 2013a).

The description of water transport presented in this chapter focused on physiological function, but neglected plant form. The next chapter links the microscopic features of xylem conduits to the macroscopic morphological features of vascular plants, and to plant- and ecosystem-scale metabolism.

References

Bohrer G, Mourad H, Laursen T A, Drewry D, Avissar R, Poggi D, Oren R and Katul G G 2005 Finite element tree crown hydrodynamics model (FETCH) using porous media flow within branching elements: a new representation of tree hydrodynamics *Water Resour. Res.* **41** W11404

Buckley T N, Mott K A and Farquhar G D 2003 A hydromechanical and biochemical model of stomatal conductance *Plant Cell Environ.* **26** 1767

Cai J and Tyree M T 2010 The impact of vessel size on vulnerability curves: data and models for within-species variability in saplings of aspen *Populus tremuloides Michx Plant Cell Environ.* **33** 1059

Choat B *et al* 2012 Global convergence in the vulnerability of forests to drought *Nature* **491** 752

Christman M A, Sperry J S and Adler F R 2009 Testing the 'rare pit' hypothesis for xylem cavitation resistance in three species of Acer *New Phytol.* **182** 664

Cruiziat P, Cochard H and Ameglio T 2002 Hydraulic architecture of trees: main concepts and results *Ann. For. Sci.* **59** 723

Dingman S L 1994 *Physical Hydrology* (Englewood Cliffs, NJ: Prentice-Hall)

Dixon H H and Joly J 1895 On the ascent of sap *Phil. Trans. R. Soc.* B **186** 563

Hacke U G, Sperry J S, Pockman W T, Davis S D and McCulloch K A 2001 Trends in wood density and structure are linked to prevention of xylem implosion by negative pressure *Oecologia* **126** 457

Hacke U G, Sperry J S, Wheeler J K and Castro L 2006 Scaling of angiosperm xylem structure with safety and efficiency *Tree Physiol.* **26** 689

Herman M, Dutilleul P and Avella-Shaw T 1998 Intra-ring and inter-ring variations of tracheid length in fast-grown versus slow-grown Norway spruces (*Picea abies*) *Iawa J.* **19** 3

van den Honert T H 1948 Water transport in plants as a catenary process *Discuss. Faraday Soc.* **3** 146

Jacobsen A L, Pratt R B, Tobin M F, Hacke U G and Ewers F W 2012 A global analysis of xylem vessel length in woody plants *Am. J. Bot.* **99** 1583

Jarvis P G and McNaughton K G 1986 Stomatal control of transpiration–scaling up from leaf to region *Adv. Ecol. Res.* **15** 1

Johnson D M, McCulloch K A, Woodruff D R and Meinzer F C 2012 Hydraulic safety margins and embolism reversal in stems and leaves: Why are conifers and angiosperms so different? *Plant Sci.* **195** 48

Kocacinar F and Sage R F 2004 Photosynthetic pathway alters hydraulic structure and function in woody plants *Oecologia* **139** 214

Lens F, Sperry J S, Christman M A, Choat B, Rabaey D and Jansen S 2011 Testing hypotheses that link wood anatomy to cavitation resistance and hydraulic conductivity in the genus Acer *New Phytol.* **190** 709

Lipp C C and Nilsen E T 1997 The impact of subcanopy light environment on the hydraulic vulnerability of Rhododendron maximum to freeze-thaw cycles and drought *Plant Cell Environ* **20** 1264

Loepfe L, Martinez-Vilalta J, Pinol J and Mencuccini M 2007 The relevance of xylem network structure for plant hydraulic efficiency and safety *J. Theor. Biol.* **247** 788

Maherali H, Moura C F, Caldeira M C, Willson C J and Jackson R B 2006 Functional coordination between leaf gas exchange and vulnerability to xylem cavitation in temperate forest trees *Plant Cell Environ* **29** 571

Maherali H, Pockman W T and Jackson R B 2004 Adaptive variation in the vulnerability of woody plants to xylem cavitation *Ecology* **85** 2184

Manzoni S, Katul G and Porporato A 2014a A dynamical-system perspective on plant hydraulic failure *Water Resour. Res.* **50** 5170

Manzoni S, Vico G, Katul G, Palmroth S, Jackson R B and Porporato A 2013a Hydraulic limits on maximum plant transpiration and the origin of the safety-efficiency tradeoff *New Phytol.* **198** 169

Manzoni S, Vico G, Katul G and Porporato A 2013b Biological constraints on water transport in the soil-plant-atmosphere system *Adv. Water Resour.* **51** 292

Manzoni S, Vico G, Katul G, Palmroth S and Porporato A 2014b Optimal plant water-use strategies under stochastic rainfall *Water Resour. Res.* **50** 5379

Markesteijn L, Poorter L, Paz H, Sack L and Bongers F 2011 Ecological differentiation in xylem cavitation resistance is associated with stem and leaf structural traits *Plant Cell Environ* **34** 137

Maseda P H and Fernandez R J 2006 Stay wet or else: three ways in which plants can adjust hydraulically to their environment *J. Exp. Bot.* **57** 3963

Mayr S, Beikircher B, Obkircher M A and Schmid P 2010 Hydraulic plasticity and limitations of alpine Rhododendron species *Oecologia* **164** 321

Meinzer F C, Johnson D M, Lachenbruch B, McCulloh K A and Woodruff D R 2009 Xylem hydraulic safety margins in woody plants: coordination of stomatal control of xylem tension with hydraulic capacitance *Funct. Ecol.* **23** 922

Meinzer F C, McCulloh K A, Lachenbruch B, Woodruff D R and Johnson D M 2010 The blind men and the elephant: the impact of context and scale in evaluating conflicts between plant hydraulic safety and efficiency *Oecologia* **164** 287

Mencuccini M 2003 The ecological significance of long-distance water transport: short-term regulation, long-term acclimation and the hydraulic costs of stature across plant life forms *Plant Cell Environ.* **26** 163

Meyra A G, Zarragoicoechea G J and Kuz V A 2011 A similarity law in botanic. The case of hydraulic conductivity of trees *Eur. Phys. J.* D **62** 19

Novick K, Oren R, Stoy P, Juang J Y, Siqueira M and Katul G 2009 The relationship between reference canopy conductance and simplified hydraulic architecture *Adv. Water Resour.* **32** 809

Peguero-Pina J J, Sancho-Knapik D, Cochard H, Barredo G, Villarroya D and Gil-Pelegrin E 2011 Hydraulic traits are associated with the distribution range of two closely related Mediterranean firs, *Abies alba Mill.* and *Abies pinsapo Boiss Tree Physiol.* **31** 1067

Pittermann J, Limm E, Rico C and Christman M A 2011 Structure-function constraints of tracheid-based xylem: a comparison of conifers and ferns *New Phytol.* **192** 449

Pittermann J, Sperry J S, Hacke U G, Wheeler J K and Sikkema E H 2006 Inter-tracheid pitting and the hydraulic efficiency of conifer wood: the role of tracheid allometry and cavitation protection *Am. J. Bot.* **93** 1265

Skelton R P, West A G and Dawson T E 2015 Predicting plant vulnerability to drought in biodiverse regions using functional traits *Proc. Natl Acad. Sci. USA* **112** 5744

Sperry J S 2003 Evolution of water transport and xylem structure *Int. J. Plant Sci.* **164** S115

Sperry J S, Adler F R, Campbell G S and Comstock J P 1998 Limitation of plant water use by rhizosphere and xylem conductance: results from a model *Plant Cell Environ.* **21** 347

Sperry J S, Hacke U G and Pittermann J P 2006 Size and function in conifer tracheids and angiosperm vessels *Am. J. Bot.* **93** 1490

Sperry J S, Nichols K L, Sullivan J E M and Eastlack S E 1994 Xylem embolism in ring-porous, diffuse-porous, and coniferous trees of northern Utah and interior Alaska *Ecology* **75** 1736

Sperry J S and Sullivan J E M 1992 Xylem embolism in response to freeze-thaw cycles and water-stress in ring-porous, diffuse-porous, and conifer species *Plant Physiol.* **100** 605

Torres-Ruiz J M, Diaz-Espejo A, Morales-Sillero A, Martin-Palomo M J, Mayr S, Beikircher B and Fernandez J E 2013 Shoot hydraulic characteristics, plant water status and stomatal response in olive trees under different soil water conditions *Plant Soil* **373** 77

Tyree M T 1997 The Cohesion-Tension theory of sap ascent: current controversies *J. Exp. Bot.* **48** 1753

Tyree M T and Ewers F W 1991 The hydraulic architecture of trees and other woody-plants *New Phytol.* **119** 345

Tyree M T and Sperry J S 1988 Do woody-plants operate near the point of catastrophic xylem dysfunction caused by dynamic water-stress–answers from a model *Plant Physiol.* **88** 574

Wheeler J K, Sperry J S, Hacke U G and Hoang N 2005 Inter-vessel pitting and cavitation in woody Rosaceae and other vesselled plants: a basis for a safety versus efficiency trade-off in xylem transport *Plant Cell Environ.* **28** 800

Whitehead D, Edwards W R N and Jarvis P G 1984 Conducting sapwood area, foliage area, and permeability in mature trees of Picea sitchensis and Pinus contorta *Can. J. For. Res.-Rev. Can. Rec. For.* **14** 940

IOP Publishing

Networks on Networks (Second Edition)
Role of connectivity in physics of geobiology and geochemistry
Allen G Hunt and Muhammad Sahimi

Chapter 8

Allometric scaling and metabolism

8.1 Background

While the lumped-parameter approximation employed in chapter 7 is useful for interpreting coarse-scale data, and is sufficiently simple to allow for developing analytically tractable transport models, including cavitation, it clearly fails to describe plant morphology and its relation to function. Exploring branch furcation and vertical variations in conduit size places the conduit network within the context of the branch transport network, hence linking micro- to macro-scale in a more realistic and mechanistic framework. This is the focus of the present chapter.

Universal allometric scaling laws emerge from self-similar network architectures evolving towards the highest water transport efficiency or, in other words, the minimum energy dissipation. Some well-known examples are river networks (Rodriguez-Iturbe and Rinaldo 1997) and vascular systems (West *et al* 1997). In the case of vascular plants, an efficient transport network can deliver more water to the leaves, allowing larger CO_2 uptake through the stomata and, thus, higher rates of photosynthesis and ultimately plant fitness (Bejan *et al* 2008, Manzoni *et al* 2014). From an evolutionary standpoint, higher fitness confers a competitive advantage and is, therefore, selected for. Therefore, it becomes clear that the water flow rate, i.e., the transpiration rate E, due to its tight coupling to plant metabolism, is likely to be maximized, and the network system delivering water is likely to evolve towards efficient, self-similar forms. As discussed earlier in this book, however, hydraulic efficiency always comes at a cost, such as increased risk of damage when water is scarce. Based on such contrasting requirements for efficiency and safety, a suite of ecological strategies based on coordinated hydraulic and biochemical traits have evolved, so that fitness is maximized subject to a set of constraints accounting for water limitations (Cowan 1986, Manzoni *et al* 2013b, 2014). As shown in this chapter, optimization criteria also impose constraints on plant form, leading to general allometric relations linking the size of plant organs and their function in terms of water transport.

doi:10.1088/978-0-7503-5698-5ch8

Based on these premises, we will, in the following, use E as a proxy for metabolic rate. To derive the relations between water flow rate and plant architecture and size, we first describe the general results obtained by Banavar *et al* (1999) for a generic directed network that is also applicable to vascular plants, but lacks physiological detail. Next, we consider the more detailed model by West *et al* (1999), which considers a fractal network with conduits approximated as tapering pipes, linking the base of the network to the terminal units. Finally, the more realistic approach by Savage *et al* (2010) is described in which conduit furcation is accounted for. Hence, this chapter follows a hierarchy of models of increasing physiological and mechanistic detail, and investigates their implications at plant-to-ecosystem scales. Moreover, the results derived here (in particular, the scaling relations of root extent with the tree height, and height with age) will be useful to interpret the data and results in chapter 9 and to a lesser extent chapter 11.

8.2 A general model for scaling of metabolic rates

A general derivation of the scaling relation between metabolic rate, i.e., the transpiration E, and organism size starts from the heuristic consideration that the fluid volume should scale as the number of links in the network, i.e., ℓ^d in a space-filling network in D-dimensional space, where ℓ is a linear scale, as shown in figure 8.1, times the mean distance between the source and the terminal units (Banavar *et al* 1999). The latter may vary, depending on the geometry of the network, ranging from ℓ when all links are directed outward from the source, as in figure 8.1, to ℓ^D in a space-filling D-dimensional object, such as a spiral. Therefore, the fluid volume may follow the two limiting scaling laws,

$$V_f \sim \begin{cases} \ell^{D+1} \rightarrow \ell \sim V_f^{1/(D+1)} \text{ for directed networks in a } D\text{-dimensional space,} \\ \ell^{2D} \rightarrow \ell \sim V_f^{1/2D}, \qquad \text{for 1D objects in a } D\text{-dimensional space.} \end{cases} \tag{8.1}$$

The fluid volume V_f is assumed to scale isometrically with the mass of the organism, $V_f \sim M$, and the metabolic rate scales as the number of terminal units that use the supplied resources, ℓ^D. It thus follows from equation (8.1) that,

$$E \sim \begin{cases} M^{D/(D+1)} \text{ for directed networks in a } D\text{-dimensional space,} \\ M^{1/2} \qquad \text{for one-dimensional networks} \end{cases} \tag{8.2}$$

For three-dimensional directed networks, equation (8.2) yields the well-known 3/4-power scaling relation between metabolic rate and body mass.

8.3 Plant allometry emerging from fractal branching networks

Let us now consider a more specific network that captures the main features of plant architecture, as shown in figure 8.1. Branches are assumed to have length l_k and radius $r_{s,k}$ at branching level k, with k varying from 0, representing the main stem, to N, the terminal twigs or petioles. Note that the radii of structural components are denoted by the subscript s to differentiate them from conduit radii, denoted by r_k.

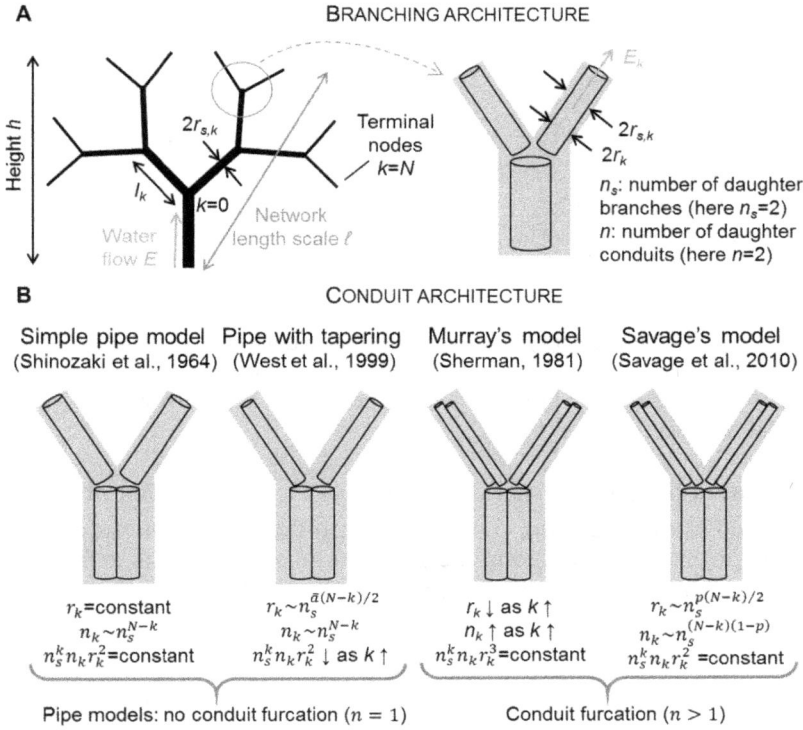

A   BRANCHING ARCHITECTURE

B   CONDUIT ARCHITECTURE

Simple pipe model   Pipe with tapering   Murray's model   Savage's model
(Shinozaki et al., 1964)   (West et al., 1999)   (Sherman, 1981)   (Savage et al., 2010)

r_k=constant

$n_k \sim n_s^{N-k}$

$n_s^k n_k r_k^2$=constant

$r_k \sim n_s^{\bar{a}(N-k)/2}$

$n_k \sim n_s^{N-k}$

$n_s^k n_k r_k^2 \downarrow$ as $k \uparrow$

$r_k \downarrow$ as $k \uparrow$

$n_k \uparrow$ as $k \uparrow$

$n_s^k n_k r_k^3$=constant

$r_k \sim n_s^{p(N-k)/2}$

$n_k \sim n_s^{(N-k)(1-p)}$

$n_s^k n_k r_k^2$ =constant

Pipe models: no conduit furcation ($n = 1$)   Conduit furcation ($n > 1$)

Figure 8.1. Schematic representation of a plant branching network, explaining the notations used in the text. (A) Branching architecture in a fractal tree model, and (B) conduit geometry according to various theories (Shinozaki *et al* 1964, Sherman 1981, West *et al* 1999, Savage *et al* 2010).

At each furcation, n_s daughter branches emanate from a parent branch, where subscript *s* again refers to a structural component; see figure 8.1.(A). Each branch contains n_k conduits with radius r_k, which decreases with *k* to describe conduit tapering. The conduits span the entire network, from the main stem to the petioles, which constitutes the pipe model, but in this simple model do not furcate, i.e., $n = 1$. Moreover, the total number of conduits is invariant with order *k*, and is equal to the number of conduits in a terminal unit, n_N. The characteristics of the terminal units are denoted by subscript *N*. The following order-invariant ratios are defined (West *et al* 1999):

$$\frac{n_{k+1}}{n_k} = n_s^{-1}$$

$$\frac{\ell_{k+1}}{\ell_k} = \gamma$$

$$\frac{r_{s,k+1}}{r_{s,k}} = \beta \frac{r_{k+1}}{r_k} = \bar{\beta}$$

(8.3)

Trends in these quantities as the branching order increases are shown in figure 8.2. Let us then consider, as a reference, the scale-invariant terminal units with subscript

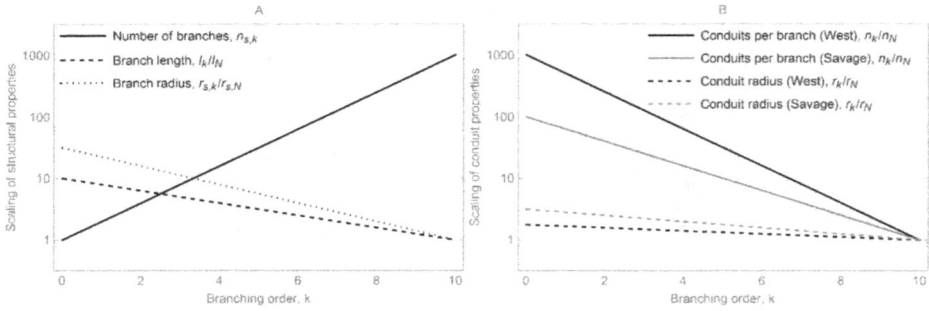

Figure 8.2. Trends in key structural (A) and conduit (B) sizes and numbers as a function of the branching order k. The value of each property (except for branch number) is normalized with respect to the value at the terminal units. In (B), the predictions by the allometric models of West *et al* (1999) and Savage *et al* (2010) are depicted in black and grey, respectively.

N, assume a volume-filling network, i.e., $\gamma n_s^{-1/3}$, and define tapering ratio, $\bar{\beta} \approx n_s^{-\bar{a}/2}$, and branching ratio, $\beta \approx n_s^{-a/2}$. With these assumptions and definition, it is possible to express the geometric features at any order as a function of the features at order N and the furcation number for the structural elements, n_s:

$$n_k = n_N n_s^{N-k}$$
$$l_k = l_N n_s^{(N-k)/3}$$
$$r_{s,k} = r_{s,N} n_s^{(N-k)a/2} \qquad (8.4)$$
$$r_k = r_N n_s^{(N-k)\bar{a}/2}$$

Moreover, the sapwood area of a branch scales as the product of conduit number and its cross-section, so that using the above relations, we obtain

$$A_{S,k} \sim n_s^{(N-k)(1+\bar{a})}. \qquad (8.5)$$

Two more constraints are imposed in order to estimate the exponents a and $\bar{a}$ (West *et al* 1999). First, to guarantee tree structural stability, the branching of the structural elements is area-preserving, i.e., the branch cross-sectional areas are conserved, so that $\beta = n_s^{(-1/2)}$ and, thus, $a = 1$. Second, the resistance to water flow throughout the network is minimized, while also minimizing tapering—this condition leads to an estimate of the tapering exponent $\bar{a}$. It can be shown that the total resistance along a flow path from $k=0$ to N is, for large N, given by,

$$Z = \sum_{k=0}^{N} Z_k = Z_N \frac{1 - \left[\left(\frac{l_t}{l_N}\right)(n_s^{1/3} - 1)\right]^{1-6\bar{a}}}{1 - n_s^{1/3 - 2\bar{a}}} \approx \frac{Z_N}{1 - n_s^{1/3 - 2\bar{a}}}, \quad \bar{a} > 1/6, \qquad (8.6)$$

where the resistance of the kth segment is computed using the Hagen–Poiseuille law for laminar flow in pipes, and l_t is the total path length, $l_t = \sum_{k=0}^{M} N l_k \approx l_0/(1 - n_s^{-1/3})$. For $\bar{a} > 1/6$, Z becomes independent of the path length—an important property that allows tall trees to transport water efficiently from the roots to the leaves, despite their long hydraulic pathways. Large values of $\bar{a}$ imply, however, rapidly increasing conduit size in the lower branches and the main stem, for a given r_N. In turn, wide conduits at the tree base may impair the plant structural stability by occupying a large fraction of the stem cross-sectional area, and are more prone to cavitation. Therefore, to minimize the total resistance while limiting tapering, $\bar{a}$ should tend towards a value of 1/6. This threshold is used in the following.

Starting from equation (8.4), the total volume of the woody elements is estimated as the sum of the volumes of all segments (again assuming large N):

$$V = \sum_{k=0}^{N} n_s^k n^k V_k = V_N \frac{n_s^{N(1/3+a)}}{1 - n_s^{2/3-a}}.$$ (8.7)

Because wood density can be assumed to be independent of plant size, the plant mass M is proportional to its volume, $M \propto V$. As a consequence, using equation (8.7), the number of terminal units or leaves is linked to plant mass by,

$$n_s^N \sim M^{3/(1+3a)} \sim M^{3/4}.$$ (8.8)

Combining equations (8.4), (8.5), and (8.8), the scaling relations between plant biomass and plant height $h \sim l_t$, the main stem length l_0, the main stem radius $r_{s,0}$, conduit radius at the plant base r_0, and the main stem sapwood area $A_{s,0}$ are given by

$$\begin{aligned} h &\sim M^{1/(3+a)} \sim M^{1/4}, \\ l_0 &\sim M^{1/(3+a)} \sim M^{1/4}, \\ r_{s,0} &\sim M^{3a/2(3+a)} \sim M^{3/8}, \\ r_0 &\sim M^{3\bar{a}/[2(3+a)]} \sim M^{1/16}, \\ A_{s,0} &\sim M^{3(1+\bar{a})/(3+a)} \sim M^{7/8}, \end{aligned}$$ (8.9)

where, on the far right side, the scaling laws correspond to $a = 1$ and $\bar{a} = 1/6$. The next step to derive an allometric equation linking plant mass and metabolic rate is to express the water flow rate E through the entire network as a function of the flow rate in a terminal unit, E_N, which is assumed to be scale-invariant,

$$E = n_s^N E_N.$$ (8.10)

It follows from equation (8.8) that metabolism scales as the 3/4 power of biomass [as in equation (8.2)],

$$E \sim M^{3/(1+3a)} \sim M^{3/4}.$$ (8.11)

The assumption of invariant E_N is justified by the argument that it is proportional to the number of conduits in a terminal unit (which is invariant) times the flow rate in a single tube (also invariant) (West et al 1999). Other scaling relations are easily derived by combining equations (8.9) and (8.11), and are reported in table 8.1. It should be noted that the above derivation has some internal inconsistencies

Table 8.1. Scaling exponents in allometric relations $y(x)$ linking parameters y in each row to parameters x in each column, obtained by combining equations (8.9) and (8.11) and assuming $a = 1$ and $\bar{a} = 1/6$ (after West *et al* (1999)).

$y \downarrow x \rightarrow$	E	M	h, l_0	r_0	$r_{s,0}$	$A_{S,0}$
E	1	3/4	3	12	2	6/7
M	4/3	1	4	16	8/3	8/7
h, l_0	1/3	1/4	1	4	2/3	2/7
r_0	1/12	1/16	1/4	1	1/6	1/14
$r_{s,0}$	1/2	3/8	3/2	6	1	3/7
$A_{S,0}$	7/6	7/8	7/2	14	7/3	1

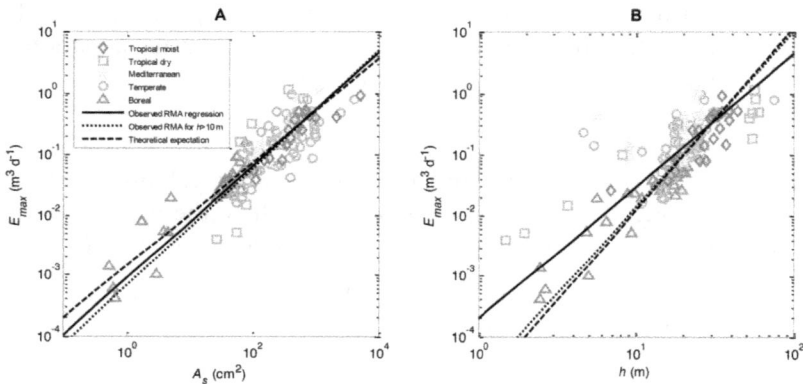

Figure 8.3. Scaling relations between peak transpiration rate, $E_{\max}$ and (A) sapwood area A_s, and (B) tree height h. Dashed lines are the expected scaling relations (table 8.2); solid lines are observed reduced major axis (RMA) regressions; the dotted lines represent the reduced major axis regressions for tall trees only. Data points from trees growing in a range of ecosystems are due to Schiller and Cohen (1995); Granier *et al* (2000); Pataki and Oren (2003); Chang *et al* (2006); Tang *et al* (2006); Manzoni *et al* (2013a), and references therein. Regression statistics are reported in table 8.4.

(Kozlowski and Konarzewski 2004), and some of the predicted exponents have been shown to differ from the experimental data. For example, the expected exponent of the height-diameter relation is 2/3 (see table 8.1). However, both lower (Muller-Landau *et al* 2006, Feldpausch *et al* 2011) and higher exponents have been reported (Givnish *et al* 2014). Despite the shortcoming, the above theory remains a generally accepted approximation, supported by numerous pieces of evidence.

The observed maximum transpiration rates are compared to A_s and h in figure 8.3. Instead of daily transpiration, the peak rate is used here, because it is less affected by stem capacitance and might better represent metabolic potential when transpiration depends more strongly on water transport, rather than storage. The observed scaling between transpiration and sapwood area for all data points, as shown in figure 8.3 (A), and between transpiration and height in tall trees, as presented in figure 8.3(B), are reasonably well captured by the equations in table 8.1. Including small trees decreases significantly the slope of the E versus h relation. The decrease is caused jointly by some relatively short trees from tropical dry forests, and three individuals with unusually high transpiration rates for their height. The latter three were all grown in weighing lysimeters (Edwards 1986), probably with more space available than in natural conditions, leading to a low, spreading canopy, supported by large sapwood area. Moreover, the scaling exponents in table 8.1 are only approximated for short trees, since equation (8.7) holds for mature trees with large N, so that deviations from the theoretical scaling are to be expected (West *et al* 1999).

Figure 8.4 illustrates the scaling of daily transpiration rate and various measures of tree size, i.e., $r_{s,0}$, A_s, and M, from the datasets of Meinzer *et al* (2005) and Enquist *et al* (1998). Angiosperm species closely follow the theoretical predictions

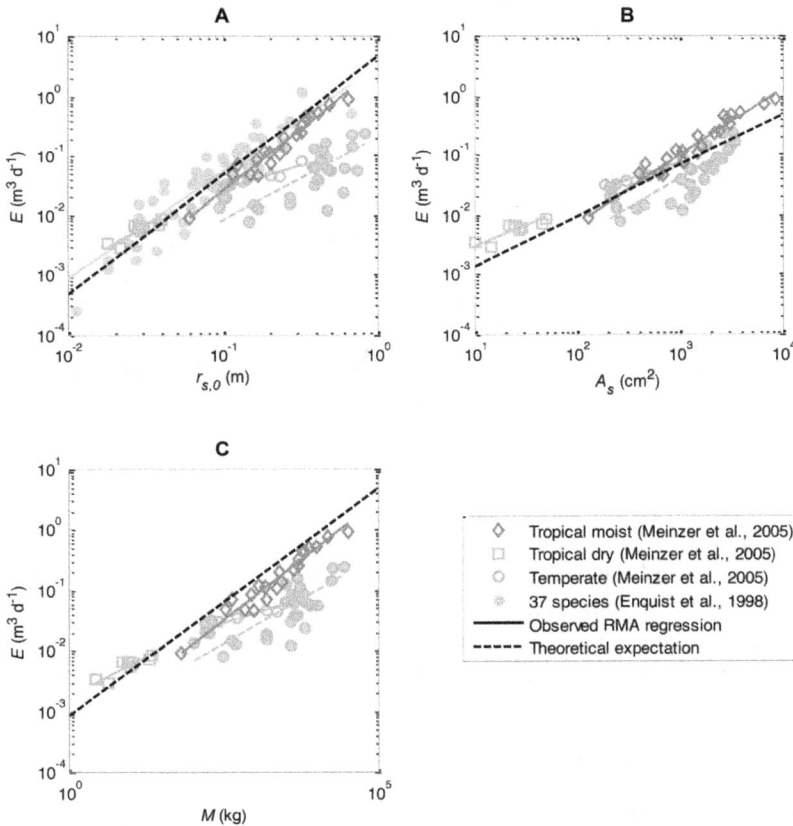

Figure 8.4. Scaling relations between daily transpiration rate E and (A) stem radius $r_{s,0}$, (B) sapwood area A_s, and (C) tree mass M. Open and solid coloured symbols refer to angiosperms and gymnosperms, respectively [the dataset was provided by Frederick C Meinzer (reported by Meinzer *et al* 2005)]. In panel (A), grey symbols refer to the dataset by Enquist *et al* (1998). The black dashed lines are the expected scaling relations (table 8.2), plotted for reference (not fitted). Solid and dashed colour lines are observed reduced major axis regressions for angiosperms and gymnosperms, respectively, in each ecosystem. Regression statistics are reported in table 8.3.

(black dashed line), whereas conifers exhibit shallower slopes. Previous work also indicated that conifers are expected to have an exponent of 0.45, rather than 0.75 (McCulloh *et al* 2010). Perhaps the elongated morphology of several conifer species explains their lower scaling exponent, compared to the less regular canopies of angiosperms

By combining equations (8.6), (8.9) and (8.11), a set of interrelated scaling relations linking plant biomass, size, and hydraulic properties (conductivity K_X and conductance g_P) are derived (West *et al* 1999). The hydraulic conductance of a single conduit spanning the entire plant, i.e., $1/Z$ from equation (8.6), is expected to be independent of size. Since in any individual conduit water flows at a size-invariant rate E_N, it follows that the water potential difference between the base of the stem and the leaves should also be conserved. The total plant conductance, g_P, is computed as the sum of the conductances in parallel of $n_N \times n_s^N$ conduits. Because n_N and Z are size-invariant, as indicated by equation (8.6), and $n_s^N \sim M^{3/4}$, equation (8.8), it follows that,

$$g_P = \frac{n_N n_s^N}{Z} = \frac{E}{\Delta \psi_t} \sim M^{3/4}. \tag{8.12}$$

In other words, the model predicts that the water potential difference driving water flow is independent of the plant size, consistent with the need to maintain a stable leaf water potential ψ_P during ontogeny that is suitable for leaf functioning. Despite the stability of ψ_P, as plants grow taller and larger, the total transpiration rate increases proportionally to the number of conduits, n_s^N, which in turn scales as the 3/4 power of plant mass.

At the level of a single conduit in a branch of order k, the hydraulic resistance is obtained using the Hagen–Poiseuille law and equation (8.4),

$$Z_k = \frac{E_N}{\Delta \psi_k} \sim \frac{l_k}{r_k^4} \sim M^0 \rightarrow \Delta \psi_k \sim M^0, \tag{8.13}$$

indicating that the water potential drop in a single branch is also scale-invariant. As a consequence, employing the definition of hydraulic conductivity of a single conduit, l_k/Z_k, the scaling relation between branch length and radius, $l_k \sim r_{s,k}^{2/3a}$ [see equation (8.4)], and accounting for the number of conduits in a branch of order k, $n_k \sim r_{s,k}^{2/a}$ [see equation (8.4)], the entire branch conductivity is determined to be

$$K_{X,k} = n_k \frac{l_k}{Z_k} \sim r_{s,k}^{8/(3a)} \sim r_{s,k}^{8/3}. \tag{8.14}$$

As a last corollary, the relation between sapwood area $A_{s,k}$ and the total cross-sectional area $A_{t,k}$ is determined based on equations (8.4) and (8.5):

$$A_{s,k} \sim A_{t,k}^{(1+\bar{a})/a} \sim A_{t,k}^{7/6}. \tag{8.15}$$

The relations between hydraulic properties and plant size have been tested, yielding mixed results. Mencuccini (2002) considered three spatial scales. First, he examined tissue-level (segment) conductivity-diameter relations, showing that they are consistent with theory, both within and across species. Second, at the branch and tree scale, conductance-diameter relations were shallower—with lower scaling exponent—than expected, suggesting that intra-specific ontogenetic changes are not well captured by allometric theory. Third, inter-specific patterns of tree conductance versus mass and diameter were not statistically distinguishable from theoretical predictions, but mainly due to the large uncertainty. Meinzer *et al* (2005) found lower than expected exponents for the sapwood area versus cross-sectional area relation, but ordinary least square fitting was used, which tends to underestimate the scaling exponents.

In light of equation (8.12), it is useful to comment on the relations between the scaling theory presented here and the equations discussed earlier. Allometric theory provides a static description of water transport and its relation to the architecture of trees. In contrast, equation (8.11) represents a dynamic description of water transport rate as it emerges from the interaction of highly variable driving force and hydraulic conductance. Upon averaging over long periods, the two approaches are reconciled. Equation (8.11) can be expressed in terms of the product of whole-plant conductance and water potential difference between the soil and the leaves. The former scales as $M^{3/4}$, whereas the latter scales as M^0, as indicated by equation (8.13), so that the 3/4 scaling exponent of equation (8.11) is recovered.

8.4 Conduit furcation

As figure 8.1 indicates, pore sizes exhibit significant variability, which can in part be explained in the context of whole-plant water transport by tapering and furcation. The fractal network outlined above assumes that conduits taper, but do not furcate. That is, each conduit links the plant base to the petioles following a rigid pipe analogy. The assumption leads to a decrease in conduit density as k increases and the branch radius decreases. In fact, in the fractal model, $n_k \sim n_s^{N-k}$, $r_{s,k} \sim n_s^{(N-k)/2}$, and $r_k \sim n_s^{\bar{a}/[2(N-k)]}$ (West *et al* 1999), so that, $r_{s,k} \sim r_k^{1/\bar{a}}$ and $n_k \sim r_k^{2/\bar{a}}$. Note that we left the tapering exponent free in order to test whether its value, as estimated from observed scaling relations nears the theoretical value, $\bar{a} = 1/6$. It follows that the number n_k of conduits in a branch of order k per unit conductive area, $n_k \pi r_k^2$, decreases with the conduit radius at each node as $\sim r_k^{-2}$, consistent with observations (McCulloh *et al* 2010). This result is independent of the tapering exponent, because conduits in West *et al*'s model furcate precisely as branches. Moreover, the -2 'packing' exponent would not change by assuming that the conductive fraction of the cross-sectional area is constant, and that conduits fill all the available space (Savage *et al* 2010).

The assumption of invariant conduit number does not, however, hold in general. Conduits do, in fact, furcate at branching nodes to maintain conductive area, while at the same time taper with increasing branching order (McCulloh *et al* 2003, Savage *et al* 2010). When conduit furcation is considered, as shown in figure 8.2(B), not only conduit radius, but also conduit number and their relation to branching order must be re-calculated. These parameters can be predicted based on various criteria. Here, we present two approaches to address this problem: Murray's hypothesis of energy consumption minimization and the more recent hypothesis that plant-scale conductance is maximized, while minimizing cavitation risk (Savage *et al* 2010).

8.4.1 Minimizing energy consumption

Murray's law (Murray 1926a, 1926b) is based on the hypothesis that power dissipation and maintenance costs are minimized (Sherman 1981, McCulloh *et al* 2003). To illustrate the consequences of this hypothesis, let us consider a uniform conduit radius at each branching order, with n_k conduits per branch and n_s^k branches of rank k as before. The flow rate through all conduits of radius r_k is determined using the Hagen–Poiseuille law and scales as, $E_k \sim n_s^k n_k r_k^4 l_k^{-1} \Delta \psi_k$, where $\Delta \psi_k$ is the water potential difference driving the flow. The power required to maintain E_k is, thus, $E_k \Delta \psi_k$, where $\Delta \psi_k \sim E_k n_s^{-k} n_k^{-1} r_k^{-4} l_k$. Combining the equations, the power for flow maintenance (per unit conduit length) is expressed as,

$$P_{E,k} \sim E_k^2 n_s^{-k} n_k^{-1} r_k^{-4}. \tag{8.16}$$

Conduits require an investment to be built and maintained, corresponding to an energy consumption $P_{M,k}$. Such an investment can be parameterized as a maintenance rate times the conduit cross-sectional area, i.e., volume per unit length, so that,

$$P_{M,k} \sim n_s^k n_k r_k^2. \tag{8.17}$$

Combining equations (8.16) and (8.17) and setting a specific flow rate E_k, the radius minimizing the combined power contributions is computed by setting, $\partial(P_{E,k} + P_{M,k})/\partial r_k = 0$, which yields, $r_k \sim E_k^{1/3} n_k^{-1/3} n_s^{-k/3}$, or

$$n_k n_s^k r_k^3 = \text{constant}. \tag{8.18}$$

Therefore, this approach predicts that the sum of the conduit radii cubed should be conserved across branch ranks. The same result can be derived by simultaneously minimizing conduit volume and resistance to flow at the branch level (McCulloh *et al* 2003). The same optimization procedure can also include the role of the scaling, $l_k \sim r_k^{2/(3\bar{a})}$. This approach has not yet been attempted, but could provide further constraints between optimal r_k and the taper exponent $\bar{a}$.

8.4.2 Maximizing hydraulic conductance

One may also argue that tapering and conduit size should be selected for minimizing resistance at the plant level, as done in the estimation of the tapering exponent $\bar{a}$ (West *et al* 1999). However, when considering conduit furcation, a more complex model than described above is required. Toward this goal, Savage *et al* (2010) proposed to separate the structural (subscript s) and hydraulic components of the network, defining a furcation number for xylem conduits as, $n = n_s^p$, where $0 \leqslant p \leqslant 1$ (recall that in the previous derivations $n = 1$, i.e., $p = 0$). With this approach, the number of xylem conduits may increase distally faster than the increase in branch number; see figure 8.1. Based on the same assumptions regarding the architecture of the structural branch network, and on the premises regarding conduit furcation, the first and last expressions of equation (8.4) are substituted by (see figure 8.2),

$$n_k = n_N n_s^{(N-k)(1-p)},$$
$$r_k = r_N n_s^{p(N-k)/2}. \tag{8.19}$$

Therefore, larger values of p promote larger conduits at the base of the tree, which might in turn increase the risk of cavitation, raising again the question as to how efficiency and safety can be balanced in the optimal transport network. The issue is addressed by calculating the plant-scale hydraulic resistance (and, thus, conductance). At each level, the overall resistance is given by the inverse of the sum of the conductances of each conduit, $g(j, k)$, since conduits of a given order are in parallel. Thus, the plant-scale resistance of all flow paths from rank $k = 0$ to $k = N$ is determined as the sum of the resistances at each level k, as branches across orders are in series,

$$g_P^{-1} = \sum_{k=0}^{N} \left(\sum_{j=1}^{n_k n_s^k} g(j, k) \right)^{-1} = \left(\frac{Z_N}{n_N n_s^N} \right) \frac{1 - \left[\left(\frac{l_t}{l_N} \right) \left(n_s^{1/3} - 1 \right) \right]^{1-3p}}{n_s^{1/3-p}} \approx \left(\frac{1}{n_N n_s^N} \right) \left(\frac{Z_N}{1 - n_s^{1/3-p}} \right), \tag{8.20}$$
$$\rightarrow g_P \approx n_N n_s^N \left(1 - n_s^{1/3-p} \right) Z_N^{-1},$$

which is valid for $p > 1/3$. Note that calculating the plant conductance based on equation (8.20), normalized by the number of terminal units, is analogous to using equation (8.6) with $\bar{a} = p/2$. As before, the optimal trade-off between efficient water transport—with g_P increasing monotonically with p—and safety from cavitation, which increases monotonically with p, is obtained at the threshold $p = 1/3$, which is used in the following derivations.

The predictions of the model can now be compared to other approaches and data. First, the conductive area of a branch of order k is given by the product of conduit lumen area and number of branches at that level, $A_{S,k} = n_k r_k^2 \sim n_s^{N-k}$. Therefore, the total conductive area at level k is given by,

$$A_S = A_{S,k} n_s^k \sim n_s^N \tag{8.21}$$

that is, A_S depends only on the number of terminal units and the radius of their conduits. A corollary of equation (8.21) is that, $A_S \sim E \sim M^{3/4}$, and, $A_{s,k} \sim A_{s,t}$, which should be compared with the exponent in equation (8.15) with a value less than 1. Therefore, the model predicts that conduit area is conserved across branching levels—a fundamentally different prediction compared to Murray's model, where the sum of the conduit radii cubed is conserved—see equation (8.18). Based on equation (8.4) regarding the structural architecture, and equations (8.19) regarding conduit architecture, a set of scaling relations linking stem size, mass, and conduit radius at the tree base is obtained,

Table 8.2. Scaling exponents in allometric relations $y(x)$ linking parameters y in each row to parameters x in each column, obtained by combining equations (8.9), (8.11), (8.19), and (8.22), and assuming $a = 1$ and $p = 1/3$ (after Savage *et al* 2010). Differences with predictions by West *et al* (1999) are highlighted in bold.

$y \downarrow x \rightarrow$	E	M	h, l_0	r_0	$r_{s,0}$	$A_{S,0}$
E	1	3/4	3	**6**	2	**1**
M	4/3	1	4	**8**	8/3	**4/3**
h, l_0	1/3	1/4	1	2	2/3	**1/3**
r_0	**1/6**	**1/8**	**1/2**	1	**1/3**	**1/6**
$r_{s,0}$	**1/2**	3/8	3/2	3	1	**1/2**
$A_{S,0}$	**1**	**3/4**	**3**	**6**	**2**	1

$$r_0 \sim M^{3p/2(3+a)} \sim M^{1/8},$$
$$A_{s,0} \sim M^{3/(3+a)} \sim M^{3/4}. \tag{8.22}$$

Other scaling exponents are listed in table 8.2. Figure 8.5 shows the data for conduit radii at the tree base and tapering ratios r_0/r_{N-1} as a function of stem size (radius and length) across various plant functional types. In general, the predictions of the model proposed by Savage *et al* (2010) appear more in line with the observed scaling relations. In particular, the scaling exponents were overestimated when neglecting conduit furcation, hence suggesting that optimal tapering can only be achieved by furcating conduits.

8.5 Scaling of above-ground and below-ground characteristic sizes

In support of the results that will be presented in chapter 9, it is important to determine the relation between the root system size and plant height. The size of the root system is characterized by its radial extent L_R. To link L_R to plant height, we start from the isometric scaling of root and stem mass (Niklas 2007),

$$M_R = \beta_{RS} M_S, \tag{8.23}$$

where the proportionality coefficient is, $\beta_{RS} \simeq 0.423$. Using the relations in table 8.1, it is possible to show that stem length scales as $M_S^{1/4}$, but stem length also scales linearly with the height. Following the same rationale adopted for the above-ground organs, one may argue that the same scaling relations among root structure and mass hold, so that, $L_R \sim M_R^{1/4}$ with the same scaling coefficient as for the stems (Kempes *et al* 2011). Using these relations and equation (8.23), the root radial extent is determined:

$$L_R = \beta_{RS}^{1/4} h \approx 0.8h. \tag{8.24}$$

An approximately linear relation between L_R and h is apparent in trees from a range of ecosystems, but only in individuals taller than about a meter; see figure 8.6. Younger trees tend to grow deeper roots first, and then extend them laterally (Phillips *et al* 2015). Lacking detailed scaling relations for herbaceous species (and

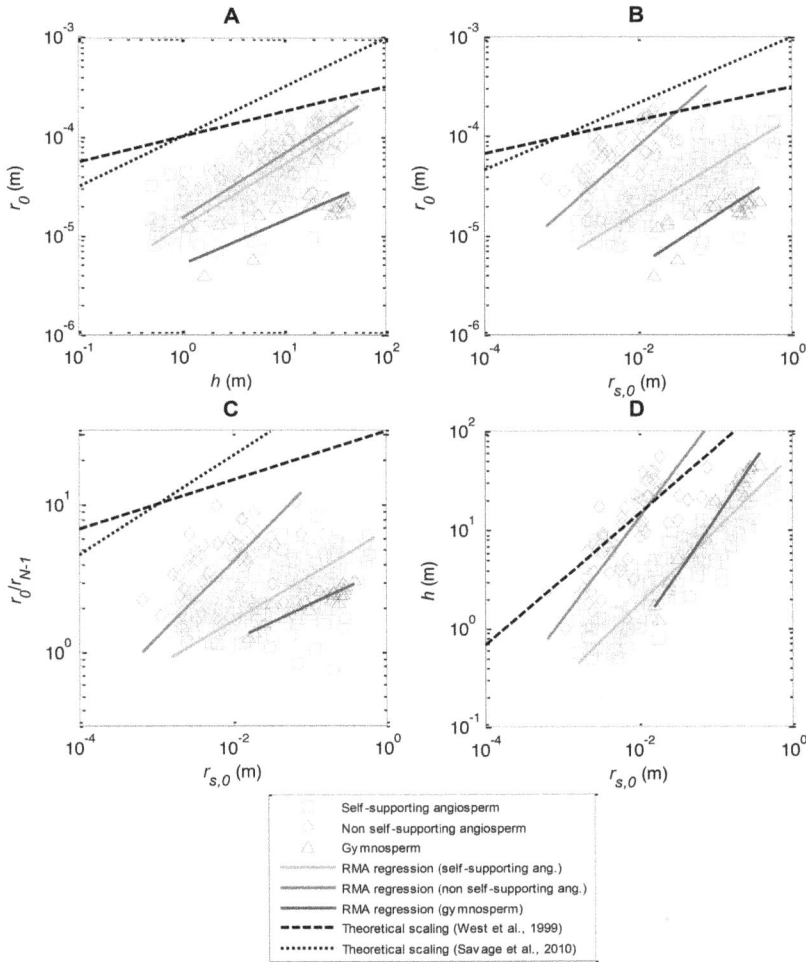

Figure 8.5. Scaling relations between conduit radius and plant size. (A) Scaling of conduit radius at the plant base r_0 and plant height h; (B) scaling of r_0 and stem radius $r_{s,0}$; (C) scaling of widening ratio—ratio of r_0 and conduit radius in terminal branches, r_{N-1} and $r_{s,0}$, and (D) scaling of h and $r_{s,0}$. In each panel, data for different plant functional types are shown. The data are from Anfodillo *et al* (2006) and Olson *et al* (2014). Solid colour lines are reduced major axis regressions of the data; dashed and dotted black lines indicate the slopes predicted by allometric theory (tables 8.1 and 8.2), plotted for reference (not fitted), and regression statistics are reported in table 8.3.

crops in particular), we assume that equation (8.24) holds in general, and briefly discuss the shortcomings of the assumption in the following. Only one study could be found that compared root radial extent to canopy geometry—specifically canopy volume—in both woody and herbaceous species in arid ecosystems (Schenk and Jackson 2002). In this study, L_R scales as canopy volume to the powers 0.442, 0.413, and 0.365 in, respectively, woody species, forbs, and grasses. In the calculation, the canopy was assumed to have an ellipsoidal shape, so that its volume is proportional

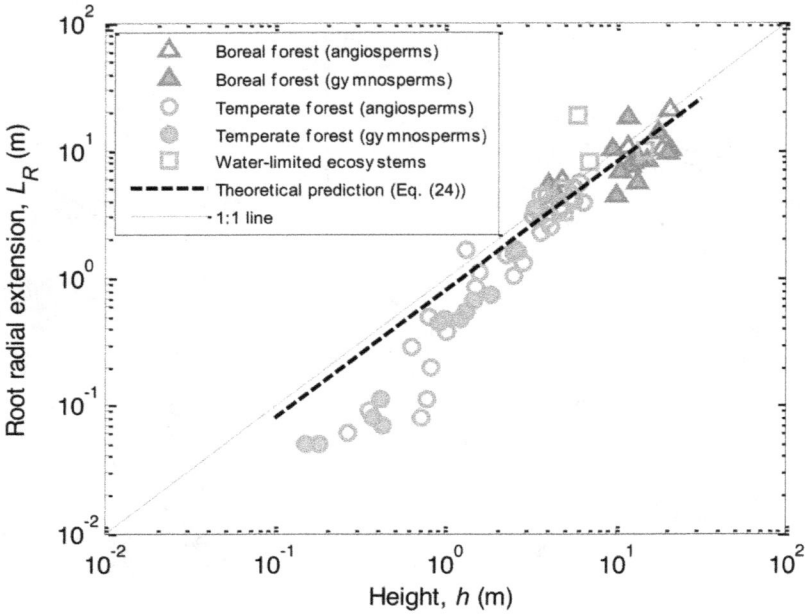

Figure 8.6. Comparison of the theoretical relation between root radial extent and tree height with data for gymnosperms and angiosperms for various ecosystems (Stone and Kalisz 1991, Kalliokoski *et al* 2008, Phillips *et al* 2014, 2015).

to the product of the square of canopy width and height. If we further assume that width and height are also proportional, canopy volume scales as h^3. As a result L_R is related to h by,

$$L_R \sim \begin{cases} h^{1.326} & \text{woody species,} \\ h^{1.239} & \text{forbs,} \\ h^{1.095} & \text{grasses.} \end{cases} \qquad (8.25)$$

The above exponents were estimated empirically for species growing in arid and semiarid environments, where all plant functional types, and woody species in particular, heavily invest in their root systems (Schenk and Jackson 2002), possibly resulting in a scaling exponent larger than the theoretical value of one. While we cannot confirm that the isometric scaling between L_R and h is universal, equation (8.25) does suggest that at most the scaling exponent may be slightly larger than one, perhaps exhibiting higher values in more arid conditions.

8.6 Scaling of size and age

The allometric relations derived in the previous sections can be used to predict plant growth trajectories (Enquist *et al* 2007). For this purpose, the metabolic rate—or, equivalently, equation (8.2) or (8.11)—is expressed as growth rate as,

$$\frac{dM}{dt} = \beta_G M^{3/4},$$ (8.26)

where β_G, expressed in, for example, $kg^{1/4} \, y^{-1}$, accounts for the all the size-independent proportionality constants that link network properties, allocation patterns, and biochemical efficiencies to the metabolic rate. Values of β_G are estimated by theoretical considerations, or by fitting the data, with the two approaches yielding comparable results, and vary between 0.24 (gymnosperms) and 0.43 $kg^{1/4} \, y^{-1}$ (angiosperms) (Enquist *et al* 2007). Recalling that, $h = \beta_h M^{1/4}$, the plant height increase rate is shown to be size-independent,

$$\frac{dh}{dt} = \frac{dh}{dM}\frac{dM}{dt} = \frac{1}{4}\beta_h \beta_G.$$ (8.27)

Using, $\beta_h \approx 2.6$ m $kg^{-1/4}$ (Niklas and Enquist 2001), equation (8.27) can be integrated to derive the height-age relations, yielding, $h \approx 0.28\tau$ for angiosperms and $h \approx 0.16\tau$ for gymnosperms at time τ. Therefore, the analysis identifies characteristic velocities h/τ on the order of a few decimetres per year, i.e., on the order of 0.01 $\mu m \, s^{-1}$.

8.7 Ecosystem scale

The plant-scale relation between metabolism (transpiration) and plant mass can be scaled up to the ecosystem level by considering a homogeneous stand with individuals of mean biomass M growing under resource-limited, i.e., water-limited, conditions at a maximum plant density N_{max}. Using equation (8.11), the total transpiration rate per unit area, E_{tot} is, thus, linked to N_{max} by

$$E_{tot} = N_{max} E \sim N_{max} M^{3/4}.$$ (8.28)

Because of resource supply E_{tot}, constraining stand growth is mainly dependent on external (e.g., climatic) conditions and, therefore, it follows that (Enquist *et al* 1998)

$$E_{tot} \sim M^0 \rightarrow N_{max} \sim M^{-3/4}.$$ (8.29)

The above scaling of the number of individuals versus biomass matches plot-scale observations of thinning rates (Enquist *et al* 1998). More recent analyses confirmed that ecosystem productivity varies across wide climatic gradients, predictably as a function of plant size and age, which mediate the climatic effects (Michaletz *et al* 2014).

The relations expressed by equation (8.29) neglect, however, the positive feedbacks of plant cover on resource supply. Vegetation in arid ecosystems improves infiltration rate, which is proportional to $M^{0.43}$ (Thompson *et al* 2010), and over large areas may provide moisture to sustain convective rainfall (Rodriguez-Iturbe *et al* 1991, Katul *et al* 2012). Using the relation between infiltration rate and biomass as a proxy for the $E_{tot} - M$ relation, together with equation (8.29), leads to a novel prediction for arid ecosystems: plant density is expected to scale as $M^{-1/2}$. In other words, where there is a positive feedback between vegetation and resource supply,

Table 8.3. Regression statistics for figures 8.3, 8.4, and 8.5. Reduced major axis regressions and 95% confidence intervals (CI) for the slope were calculated using the Standardised Major Axis Tests and Routines (SMATR) package (Warton *et al* 2006). Bold numbers indicate when the CI bracket the theoretical allometric slopes. Theoretical slopes refer to the model by West *et al* (1999). The values in brackets refer to the model by Savage *et al* (2010). The statistics for $E_{\max}$ are for figure 8.3; for E for figure 8.4; and for the rest are for figure 8.5.

Scaling relation	Theoretical slope	Data group	Estimated slope	CI of the slope
$E_{\max}(A_S)$	$6/7 = 0.857$	$h > 10$ m	0.968	(0.867, 3.38)
$E_{\max}(h)$	3	$h > 10$ m	**2.88**	**(2.53, 1.082)**
$E(A_S)$	$6/7 = 0.857$	Angiosperms	**0.873**	**(0.813, 0.936)**
$E(A_S)$	$6/7 = 0.857$	Gymnosperms	**0.993**	**(0.779, 1.27)**
$E(r_{s,0})$	2	Angiosperms	1.63	(1.48, 1.76)
$E(r_{s,0})$	2	Gymnosperms	1.39	(1.04, 1.84)
$E(M)$	$3/4 = 0.75$	Angiosperms	0.624	(0.567, 0.686)
$E(M)$	$3/4 = 0.75$	Gymnosperms	**0.585**	**(0.448, 0.765)**
$r_0(h)$	$1/4 \ (1/2)$	Self-supp. angios.	0.628	(0.570, 0.692)
		Self-supp. gymnos.	**0.442**	**(0.315, 0.619)**
		Non self-supp.	0.660	(0.566, 0.768)
$r_0(r_{s,0})$	$1/6 \ (1/3)$	Self-Supp. angios.	0.473	(0.429, 0.521)
		Self-supp. angios.	0.500	(0.365, 0.685)
		Non self.supp.	0.678	(0.547, 0.842)
$r_0/r_N(r_{s,0})$	$1/6 \ (1/3)$	Self-supp. angios.	**0.310**	**(0.275, 0.349)**
		Self-supp. angios.	**0.243**	**(0.206, 0.287)**
		Non self-supp.	0.523	(0.420, 0.650)
$h(r_{s,0})$	$2/3 = 0.667$	Self-supp. angios.	0.754	(0.714, 0.795)
		Self-supp. angios.	1.13	(1.04, 1.24)
		Non self-supp.	1.03	(0.883, 1.20)

relatively more plants for a given mean biomass can be sustained, when compared to an ecosystem without such a feedback. These predictions are valuable in evaluating the risk of ecosystem transitions from vegetated to bare conditions. While this theory holds for idealized communities with uniform stem radius (and age), extensions have been proposed to communities characterized by a distribution of radii. These extensions are beyond our scope in this book. The interested reader is referred to previous publications on this topic (West *et al* 2009, Simini *et al* 2010, Michaletz *et al* 2014).

The regression statistics for figures 8.3, 8.4, and 8.5 are compiled in table 8.3.

References

Anfodillo T, Carraro V, Carrer M, Fior C and Rossi S 2006 Convergent tapering of xylem conduits in different woody species *New Phytol.* **169** 279

Banavar J R, Maritan A and Rinaldo A 1999 Size and form in efficient transportation networks *Nature* **399** 130

Bejan A, Lorente S and Lee J 2008 Unifying constructal theory of tree roots, canopies and forests *J. Theor. Biol.* **254** 529

Chang X X, Zhao W Z, Zhang Z H and Su Y Z 2006 Sap flow and tree conductance of shelter-belt in and region of China *Agric. For. Meteorol.* **138** 132

Cowan I 1986 Economics of carbon fixation in higher plants *On the Economy of Plant Form and Function* ed t J Givnish (Cambridge: Cambridge University Press) pp 133–70

Edwards W R N 1986 Precision weighing lysimetry for trees, using a simplified tared-balance design *Tree Physiol.* **1** 127

Enquist B J, Brown J H and West G B 1998 Allometric scaling of plant energetics and population density *Nature* **395** 163

Enquist B J, Kerkhoff A J, Stark S C, Swenson N G, McCarthy M C and Price C A 2007 A general integrative model for scaling plant growth, carbon flux, and functional trait spectra *Nature* **449** 218

Feldpausch T R *et al* 2011 Height-diameter allometry of tropical forest trees *Biogeosciences* **8** 1081

Givnish T J, Wong S C, Stuart-Williams H, Holloway-Phillips M and Farquhar G D 2014 Determinants of maximum tree height in Eucalyptus species along a rainfall gradient in Victoria, Australia *Ecology* **95** 2991

Granier A, Loustau D and Breda N 2000 A generic model of forest canopy conductance dependent on climate, soil water availability and leaf area index *Ann. For. Sci.* **57** 755

Kalliokoski T, Nygren P and Sievanen R 2008 Coarse root architecture of three boreal tree species growing in mixed stands *Silva Fenn.* **42** 189

Katul G G, Oren R, Manzoni S, Higgins C and Parlange M B 2012 Evapotranspiration: a process driving mass transport and energy exchange in the soil-plant-atmosphere-climate system *Rev. Geophys.* **50** RG3002/2012

Kempes C P, West G B, Crowell K and Girvan M 2011 Predicting maximum tree heights and other traits from allometric scaling and resource limitations *PLoS One* **6** e20551

Kozlowski J and Konarzewski M 2004 Is West, Brown and Enquist's model of allometric scaling mathematically correct and biologically relevant? *Funct. Ecol.* **18** 283

Manzoni S, Vico G, Katul G, Palmroth S, Jackson R B and Porporato A 2013a Hydraulic limits on maximum plant transpiration and the origin of the safety-efficiency tradeoff *New Phytol.* **198** 169

Manzoni S, Vico G, Katul G, Palmroth S and Porporato A 2014 Optimal plant water-use strategies under stochastic rainfall *Water Resour. Res.* **50** 5379

Manzoni S, Vico G, Palmroth S, Porporato A and Katul G 2013b Optimization of stomatal conductance for maximum carbon gain under dynamic soil moisture *Adv. Water Resour.* **62** 90

McCulloh K, Sperry J S, Lachenbruch B, Meinzer F C, Reich P B and Voelker S 2010 Moving water well: comparing hydraulic efficiency in twigs and trunks of coniferous, ring-porous, and diffuse-porous saplings from temperate and tropical forests *New Phytol.* **186** 439

McCulloh K A, Sperry J S and Adler F R 2003 Water transport in plants obeys Murray's law *Nature* **421** 939

Meinzer F C, Bond B J, Warren J M and Woodruff D R 2005 Does water transport scale universally with tree size? *Funct. Ecol.* **19** 558

Mencuccini M 2002 Hydraulic constraints in the functional scaling of trees *Tree Physiol.* **22** 553

Michaletz S T, Cheng D, Kerkhoff A J and Enquist B J 2014 Convergence of terrestrial plant production across global climate gradients *Nature* **512** 39

Muller-Landau H C, Condit R S, Chave J, Thomas S C *et al* 2006 Testing metabolic ecology theory for allometric scaling of tree size, growth and mortality in tropical forests *Ecol. Lett.* **9** 575

Murray C D 1926a The physiological principle of minimum work: I. The vascular system and the cost of blood volume *Proc. Natl Acad. Sci. USA* **12** 207

Murray C D 1926b The physiological principle of minimum work: II. Oxygen exchange in capillaries *Proc. Natl Acad. Sci. USA* **12** 299

Niklas K J 2007 Maximum plant height and the biophysical factors that limit it *Tree Physiol.* **27** 433

Niklas K J and Enquist B J 2001 Invariant scaling relationships for interspecific plant biomass production rates and body size *Proc. Natl Acad. Sci. USA* **98** 2922

Olson M E, Anfodillo T, Rosell J A, Petit G, Crivellaro A, Isnard S, Leon-Gomez C, Alvarado-Cardenas L O and Castorena M 2014 Universal hydraulics of the flowering plants: vessel diameter scales with stem length across angiosperm lineages, habits and climates *Ecol. Lett.* **17** 988

Pataki D E and Oren R 2003 Species differences in stomatal control of water loss at the canopy scale in a mature bottomland deciduous forest *Adv. Water Resour.* **26** 1267

Phillips C J, Marden M and Suzanne L M 2014 Observations of root growth of young poplar and willow planting types *N.Z. J. For. Sci.* **44** 15

Phillips C J, Marden M and Suzanne L M 2015 Observations of 'coarse' root development in young trees of nine exotic species from a New Zealand plot trial *N.Z. J. For. Sci.* **45** 1

Rodriguez-Iturbe I, Entekhabi D and Bras R L 1991 Nonlinear dynamics of soil-moisture at climate scales. 1. Stochastic analysis *Water Resour. Res.* **27** 1899

Rodriguez-Iturbe I and Rinaldo A 1997 *Fractal River Basins. Chance and Self-Organization* (Cambridge: Cambridge University Press)

Savage V M, Bentley L P, Enquist B J, Sperry J S, Smith D D, Reich P B and von Allmen E I 2010 Hydraulic trade-offs and space filling enable better predictions of vascular structure and function in plants *Proc. Natl Acad. Sci. USA* **107** 22722

Schenk H J and Jackson R B 2002 Rooting depths, lateral root spreads and below-ground/above-ground allometries of plants in water-limited ecosystems *J. Ecol.* **90** 480

Schiller G and Cohen Y 1995 Water regime of a pine forest under a mediterranean climate *Agric. For. Meteorol.* **74** 181

Sherman T F 1981 On connecting large vessels to small–the meaning of Murray Law *J. Gen. Physiol.* **78** 431

Shinozaki K, Yoda K, Hozumi K and Kira T 1964 A quantitative analysis of plant form–The pipe model theory. I. Basic analyses *Jpn. J. Ecol.* **14** 97

Simini F, Anfodillo T, Carrer M, Banavar J R and Maritan A 2010 Self-similarity and scaling in forest communities *Proc. Natl Acad. Sci. USA* **107** 7658

Stone E L and Kalisz P J 1991 On the maximum extent of tree roots *For. Ecol. Manage.* **46** 59

Tang J W, Bolstad P V, Ewers B E, Desai A R, Davis K J and Carey E V 2006 Sap flux-upscaled canopy transpiration, stomatal conductance, and water use efficiency in an old growth forest in the Great Lakes region of the United States *J. Geophys. Res.-Biogeosci.* **111** G02009

Thompson S E, Harman C J, Heine P and Katul G G 2010 Vegetation-infiltration relationships across climatic and soil type gradients *J. Geophys. Res.-Biogeosci.* **115** G02023

Warton D I, Wright I J, Falster D S and Westoby M 2006 Bivariate line-fitting methods for allometry *Biol. Rev.* **81** 259

West G B, Brown J H and Enquist B J 1997 A general model for the origin of allometric scaling laws in biology *Science* **276** 122

West G B, Brown J H and Enquist B J 1999 A general model for the structure and allometry of plant vascular systems *Nature* **400** 664

West G B, Enquist B J and Brown J H 2009 A general quantitative theory of forest structure and dynamics *Proc. Natl Acad. Sci. USA* **106** 7040

IOP Publishing

Networks on Networks (Second Edition)
Role of connectivity in physics of geobiology and geochemistry
Allen G Hunt and Muhammad Sahimi

Chapter 9

Edaphic constraints: role of soil in vegetation growth

9.1 Background

Vegetation growth and, ultimately, ecosystem productivity are limited by a combination of biotic (internal) and abiotic (external) factors. Biotic limitations on plant growth include predominantly tradeoffs among eco-physiological traits, genetics, and competition. In addition to disturbance events, abiotic limitations to plant growth arise from four main factors: temperature, light, water, and nutrient availability (Vitousek and Howarth 1991, Boisvenue and Running 2006). The latter two are intrinsically related to edaphic conditions and dynamics, and interact with the other limiting factors for at least a fraction of any growing season. While such limitations may give rise to co-limitations or co-adaptations, the fundamental purpose in this chapter is, first, to identify some underlying order in how the complex system that we call soil constrains terrestrial biota. We present the overarching hypothesis that flow and transport within soil provide limits on the development of both the soil and the macro-organisms that depend on it for their growth, particularly vascular plants. We demonstrated independent confirmation of the solute transport theory in chapter 5, both insofar as it predicts solute arrival time distributions, and the scaling of chemical reaction rates over enormous time intervals. In this chapter we utilize the results directly to study the soil production function and the soil depth as a function of time. Secondly, we address the specific influence that the non-Gaussian solute transport characteristics can have on the growth of roots in the soil and, hence, of entire plants, and construct a theoretical organism size-age relation based on such characteristics.

The focus is on how edaphic constraints limit plant growth. The solute transport limitations are applied to the most important nutrients limiting growth, nitrogen (N) and phosphorous (P). The diminishing rate of solute transport with increasing time from solute source injection point represents a major limitation for effective plant growth. The hierarchical root system developed by plants, perhaps for other reasons, such as growth efficiency, stability, or areal coverage, also creates paths to solutes that greatly shorten the typical time of transport when compared with that of the transport of the same nutrients through the medium. One may, of course, argue that this particular assertion may need to be re-evaluated, in the case in which concentration gradients are so strong that ordinary diffusion, i.e., one with a constant diffusion coefficient, is not applicable. Root growth to nutrients is influenced by the solute transport paths within the soil, whereas allocation of carbon to the roots relates their growth and ultimately to entire plant growth. Therefore, root growth rates can become the limiting link in the plant growth (Leuzinger *et al* 2013). The key difference between the soil medium and a root network is the absence (or near absence) of loops within the root system. The scaling function derived from this consideration is nevertheless nonlinear in time, leading to a diminishing growth rate over long times, but the reduction is much less limiting than the relationship for abiotic transport. Although the relationship is verified over many orders of magnitude of length and time scales, and for vascular plants as well as for fungi, the bulk of the relevant data is for woody plants, especially forest trees.

It is important to understand the growth of forest trees, both because of their role in carbon sequestration and for their use in wood products. Thus, there is a wide range of research done in both managed and natural forests. In fact, the relevance of nutrient limitations is evident in both plantations (Binkley *et al* 1992); as stated in the reports by the United States National Academy of Sciences (1979, 1984),

Fast-growing tree species are the backbone of forest plantations and agroforestry systems covering millions of hectares throughout the humid tropics. Nitrogen availability commonly limits the productivity of these systems, leading to great interest in the development of systems that include nitrogen-fixing trees,

and in natural systems (Lynch 1995),

Water and nutrient availability limit plant growth in all but a very few natural ecosystems. They limit yield in most agricultural ecosystems. Sub-optimal availability of N and P is nearly universal.

However, intensive management of plantations to ameliorate such nutrient shortages effectively eliminates nutrient supply heterogeneity and, thus, renders most of the research on plantation tree growth irrelevant to the testing of edaphic constraints in natural vegetation. Nevertheless, growth rates of trees on plantations are relevant in the context of intensively managed systems, where delivery of nutrients via advection is predominant.

Focusing, first, on natural vegetation growth, we quote Lynch (1995) to help understand the need for the research in this book: 'Although there are no accurate quantitative estimates (!) of the extent and importance of edaphic (soil) constraints to plant productivity, they must certainly be very substantial. The importance of root architecture in plant productivity stems from the fact that many soil resources are unevenly distributed, or are subject to localized depletion, so that the spatial deployment of the root system will in large measure determine the ability of a plant to exploit those resources.'

The uneven distribution of such resources requires the scaling laws for solute transport based on microscopic variability in solute concentrations. Lynch (1995) also acknowledges the necessity of modeling in view of the limitations on observation:

The reason we know relatively little about root architecture is that it is difficult to observe, quantify, and interpret. Roots grow in soil, an opaque medium from which they cannot be extricated or readily observed without introducing artifacts, destroying the native root architecture, or precluding subsequent analysis of the same individual. Root systems themselves are exceedingly complex structures, typically being composed of thousands of individual root axes that vary developmentally, physiologically, and morphologically. At present there are no satisfactory analytical frameworks or quantitative tools to describe or summarize this complexity when it is characterized.

It will be shown that the rationale for the research reported in this book dovetails nicely with the basic needs and challenges perceived by Lynch.

In water- or nutrient-limited ecosystems, root growth and search for soil resources is particularly important. However, even if roots are not a limiting factor per se, they certainly are co-limiting. In fact, evolution tends to select for coordinated resource acquisition structures that minimize the negative effect of limiting resources (Rastetter and Shaver 1992). The isometric scaling of root and leaf growth rates is a good example of the coordination, leading to co-limitation (Niklas and Enquist 2002).

We start with the main hypotheses and their rationales. They are based on theoretical descriptions of water and solute transport in soil, the optimal flow paths through such heterogeneous physical networks, and what plants can do to alter, or exploit, the paths. We focus on Lynch's comment, 'The importance of root architecture in plant productivity therefore derives directly from the need to exploit a spatially heterogeneous environment.' Thus, being able to quantify the solute and water flow paths within the environment itself is a necessary condition for being able to assess how plants can evolve strategies to overcome edaphic limitations. Additional innovations discussed by Lynch (1995) and Fitter and Stickland (1992) are small, though important, modifications, that bean root structures address immobile nutrient paucity (P, for example) on the short time scale, but herringbone structures do so on longer time scales; that the root system fractal dimension increases with age, and that some species are able to change the fractal dimensions of their root system in order to access nutrients. These are all variants of the fundamental topology, which is singly, instead of multiply, connected. The advantage is the distinction between spatial and temporal tortuosity, produced by the multiply-connected paths, as discussed in Lee (1999); the latter, relevant for abiotic transport, are far more restrictive.

Three scaling relationships have been proposed (Hunt 2017) as providing limits to, (i) the depth of soil; (ii) sizes of the belowground structures of natural vascular plants (and fungi), and (iii) of intensively managed plants, each as a function of time, and summarized in table 9.1. Each relationship is tied to the fundamental time scale of 1 s and length scale of 1 μm, whose quotient gives the fundamental velocity of 1 μm s^{-1} occurring at the smallest scales. In the case of soil depth, the scaling law relates such microscopic spatial scales to processes that take place over as long a period as approximately 100 million years, but with vascular plants only to about 100 000 years, the longest (and the largest) lived plant known, and with annual crops

Table 9.1. The proposed power-law relations that link size and time in the processes of soil formation, root growth (in natural and intensively managed conditions), and water flow in the subsurface. For all the processes, the characteristic scales x_0 and t_0 are $1\,\mu$m and 1 s, respectively. Plant height is a proxy for of root radial extent, while by plant age we mean the age at the time of measuring the height. $D_{bb} \simeq 1.87$ and $D_{op} \simeq 1.21$ are, respectively, the fractal dimensions of the backbone and optimal paths.

Process	Spatial scale x	Time scale t	Scaling relation	Limiting factor
Soil formation	Soil thickness	Soil age	$x/x_0 = (t/t_0)^{1/D_{bb}}$	Solute transport in 3D
Natural root growth	Plant height	Plant age	$x/x_0 = (t/t_0)^{1/D_{op}}$	Solute transport in 2D
Managed root growth	Plant height	Plant age	$x/x_0 = (t/t_0)^1$	Water transport
Subsurface water flow	Distance traveled	Travel time	$x/x_0 = (t/t_0)^1$	Water transport

to scales limited to one growing season. While bi-annual, or perennial, crops are known, they have not been considered here, except for tree plantations. The velocity of $1\,\mu$m s^{-1} represents a typical gravity-induced water flow velocity (Blöschl and Sivapalan 1995) in unconsolidated porous media at Earth's surface, while $1\,\mu$m was chosen as a fundamental spatial scale (a typical pore size for fine silt), relevant to physical, chemical, and biological processes alike. In the later work related to the water balance and soil formation, the authors recognized that, in most cases, the near-surface infiltration flow rates are more accurately limited by precipitation and evapotranspiration, while a typical silt particle size, more nearly in the middle of that range of particle sizes, is also a better choice for a fundamental length scale. P—ET has a maximum of about 10 m yr^{-1}, with maximum (P—ET)/ϕ = 25 m yr^{-1} (rather than the implied value for $1\,\mu$m s^{-1} of 32 m yr^{-1}), slowing the estimated predictions for maximum growth rates slightly. However, in cases with sufficiently low values of near-surface hydraulic conductivity (e.g., bedrock surfaces), the factor P—ET can be replaced by the more limiting result from the hydraulic conductivity

Upon closer comparison with the data, it appears that, for biological processes also, a better choice for a fundamental spatial scaling length might have been more nearly 10 μm, which is the typical cross-section dimension of xylem conduits; see figure 7.2. Due to their elongated structure (shown in figure 7.2), and large driving force (about ten times higher than the gravitational forces typically acting in soils), flow velocities in xylem vessels are on the order of 10^2–10^4 μm s^{-1} (Lambers *et al* 1998)—much faster than typical velocities in the subsurface. But, altering the characteristic pore diameter alone would produce a constant numerical discrepancy of only about 50% in the absolute magnitudes of the predictions relevant to biota, which is mostly indistinguishable on the scales shown. Considering also the faster time scale, it becomes apparent that water transport in plants is much more efficient than in soils, requiring completely different fundamental scales. Our argument here, as will be made clearer in the following, does not deal with transport in plants, but rather with plant growth belowground and how belowground structures facilitate the acquisition of slowly-moving resources in the soil medium.

The three proposed scaling relationships relate to solute transport in soils. All three preserve the proportionality of solute to fluid flow velocities and, thus, address both water and nutrient limitations simultaneously. The theory of solute transport was presented in chapter 5. It was shown there that a wide range of experimental data support the proposed result that surface reaction rates in porous media are proportional to solute fluxes, as calculated based on conservative solute transport, and that these are also proportional to the fluid flow velocity. The chief input into the dependence of abiotic solute fluxes is the fractal structure of the percolation backbone, which has been analyzed, at least approximately, by a simple scaling formulation. The result that weathering reaction rates are proportional to solute fluxes (Hunt *et al* 2015) begs the simplifying assumption that weathering depths are the solute transport distances. The scaling formulation for solute transport sets the stage for the other hypotheses, which relate to the adaptability of plants, and the intervention of humans.

The value of the arguments lies in their simultaneous relevance to three distinct processes that all trace back to the same fundamental scales of space and time in soil, namely, approximately the size of a single pore, and the time it takes water to flow through it. The length scales addressed vary over 10 orders of magnitude, and time scales over 15 orders of magnitude, but the relevance of the pore scale prevails even at such immense spatial and temporal scales. This implies that the classification of subsurface length scales and their relevant processes is of secondary importance to the recognition of the role of the pore-scale processes, and that the specific classification of the dominant type of heterogeneity present at any scale is secondary to the fact that there is heterogeneity at every scale, and that percolation theory quantifies the appropriate impact of heterogeneity of most kinds at most scales.

9.2 Fundamental predictions based on percolation scaling

We now describe the fundamental predictions based on the scaling relations given in table 9.1. At the time of the original publication, the focus on using the hydraulic conductivity to relate the network scale to the fundamental time scale was misplaced. In fact, as subsequent research showed, these two scales are related by the actual flow rates, which are more often limited by the climate variables, precipitation and evapotranspiration, though in many cases, the hydraulic conductivity of the soil or rock may provide additional limitations.

9.2.1 Soil formation

Solute transport through three-dimensional (3D) strongly heterogeneous porous media should reasonably follow the scaling law, originally suggested by Sahimi (1987) and Sahimi and Imdakm, and modified by Lee *et al* (1999) and Sahimi (2012), $x = x_0(t/t_0)^{1/D_{bb}}$, where $D_{bb} \approx 1.87$ is the fractal dimension of the percolation backbone in random percolation in 3D (see chapter 3). Random percolation is appropriate for saturated conditions, as well as for saturating conditions. Note that the tortuous length of the shortest path connecting two points is described analogously by the fractal dimension D_{min}, while the length of an optimal path through a strongly disordered medium is governed by the fractal dimension D_{op}. Generally speaking, $D_{bb} > D_{op} > D_{min}$, and the temporal tortuosity is the largest of the three. The most accurate values of these three exponents were computed by Sheppard *et al* (1999), and are used here.

The fundamental length and time scales, $x_0 = 1$ μm and $t_0 = 1$ s, were already given above. The origins of the soil-depth relationship are in the topology of the percolation backbone, which is consistent with a multiply-connected fluid flow path and, consequently, a 'temporal' tortuosity much greater than the spatial tortuosity of the individual paths. The fundamental underlying hypothesis is that over the full range of spatial scales, soil formation by chemical weathering is solute transport-limited. Consequently, when theory is compared with the data, the soil 'depth' is typically taken to be the bottom depth of the Bw horizon. The B horizon is the subsoil layer that generally changes the most because of soil-forming processes, and w refers to weathering. Using such a relationship, with x_0 taken as 1 μm, implies that the relevant chemical heterogeneity is on the order of 1 μm. The actual value may be somewhat larger. Since the silicate weathering reaction involves dissolved CO_2, such minerals as plagioclase or other feldspars, biotite, augite, and hornblende, as well as reaction products (Berner 1992, White and Brantley 2003), the chemical hetero-geneity that is important should be assumed to be a property of the constituent with the smallest scale of variability. This scale could be as small as a single pore, presuming that the positions of the individual grains of the porous medium are not highly correlated. Choice of a length scale equal to that of a pore justifies a fundamental heterogeneity scale in the microns.

The soil production function is defined as the increase in soil depth per unit time (Heimsath *et al* 1997, Dietrich and Perron 2006), and is denoted by R_m. Assuming that the soil depth evolution $x(t)$ is known, R_m is obtained by differentiating the power law for $x(t)$, and then substituting $t(x)$ into the derivative, so as to generate the rate of change of depth as a function of depth,

$$R_m = dx/dt = \frac{d}{dt}\left[x_0\left(\frac{t}{t_0}\right)^{1/D_{bb}} \right] = \left(\frac{x_0}{D_{bb}t_0}\right)\left(\frac{x}{x_0}\right)^{1-D_{bb}} \sim x^{-0.87}. \qquad (9.1)$$

Equation (9.1) predicts that soil formation rates decrease as the its depth increases, in qualitative agreement with other formulations based on empirical exponential func-tions (see, e.g., Heimsath *et al* 1997). R_m is also proportional to x_0/t_0, which is the flow velocity. Used together with an erosion term, equation (9.1), the soil production function, provides the simplest of landscape evolution models, in that the details of the surface transport in soil, including its divergence and convergence, are omitted.

9.2.2 Natural plants

Growth of natural plants is proposed to be limited by the ability of roots to grow towards sources of nutrients and moisture. Lynch (1995) states that, 'The exploi-tation of soil resources through root activity may consume more than half of the available photosynthate in mature plants (Fogel 1985),' as also reflected by isometric growth rates of aboveground and belowground organs across species (Niklas and Enquist 2002). Such an enormous resource allocation must produce commensurate rewards in terms of access to water and nutrients. Such allocation to root structures is hypothesized to be optimal, if it has been selected for in a wide range of species (Franklin *et al* 2012). For example, rooting depth has been shown to optimize (i.e., maximize) water use for a given carbon investment belowground (Guswa 2010). At a microscopic level and focusing on resource acquisition as a proxy of plant fitness, the optimal strategy is for root growth to follow the optimal flow paths of the medium itself on account of their high solute velocities and (detectable) solute gradients, as well as for the minimization of energy expenditure. According to the definition of optimal paths in percolation theory, roots that follow an optimal path are fractal objects. Quoting Lynch (1995) again, 'It is reasonable to hope that fractal geometry may provide quantitative summaries and functional insights into root architecture that have eluded researchers using Euclidean geometry (Berntson *et al* 1995).' As discussed in chapter 2, other works have in fact confirmed that root architecture is fractal and compared root networks to fractal aboveground branch networks (Tatsumi *et al* 1989, Salas *et al* 2004, Kempes *et al* 2011). In the present case, the edaphic input, coupled with the response of the medium to imposed root potentials, allows universal aspects of transport in random media (pore networks) to guide the plant development towards a fractal architecture. Flow along optimal paths that intersect nutrient sources induced by the root potential will also tend to signal the best local connection to grow along for nutrient and water access.

The adaptation of roots to develop a hierarchical topology makes the solute transport paths simply connected, eliminating the distinction between spatial and temporal tortuosity along percolation paths (Lee *et al* 1999). We hypothesize that roots minimize energy expenditure and grow to optimize nutrient delivery; that is, roots follow optimal paths (chapter 3). As a consequence, the exponent in the travel time–distance relation for nutrient transport is expected to be D_{op}, rather than D_{bb}. The change in fractal dimension reduces the scaling exponents from 1.87 to 1.43 (in 3D) or from 1.64 to 1.21 (in 2D), leading to a huge reduction in the transport time at larger spatial scales. Roots grow towards nutrients at the rate of nutrient transport on the optimal path, hence rendering a much wider range of spatial scales accessible within useful time scales. Even though the building of the architecture of the root system involves a holistic optimization over the entire organism (Barlow 1993, Franklin *et al* 2012), the tortuosity of the root paths so constrained will be set by percolation theory.

The hypothesis regarding plant growth is based on the following understanding. While diffusion controls the adsorption across the root soil boundary (root hairs), as well as the ability to detect the optimal direction to grow in, the hydraulic conductivity and the pore diameters of the soils control root growth rates at the root tip, and root, and root hair sizes, respectively. Even though the rates of root tip growth may remain relatively constant, the tortuosity of the root path will increase with its length, as is a general characteristic of fractal objects (Mandelbrot 1982). This tortuosity is a product of the changing directions of the root growth over time, in response to the existing soil heterogeneity.

The optimal paths for flow, as described by percolation theory, yield the largest flow velocities and, consequently, the largest solute fluxes, promoting the detection of such pathways by the plant roots. The tortuosity of the roots is, therefore, controlled by that of the optimal paths, which also control the flow properties of the porous medium. In other words, it is hypothesized that roots grow at a rate comparable to solute velocities along an optimal pore pathway. Growing slower, along a more tortuous path, would provide a smaller advantage in terms of nutrient acquisition, whereas growing faster would require larger carbon and energy investments with diminishing returns.

Two alternate possibilities can be thought of regarding the overall system topology. One is that relevant root development for nutrient extraction is predominantly within the surface layer, where nutrients are concentrated. That would be consistent with choosing the exponent, $D_{op} \simeq 1.21$ for 2D systems. The alternate hypothesis is that at smaller length scales it might be appropriate to choose the 3D value of the optimal paths exponent D_{op}, with a crossover to a 2D exponent at scales larger than decimeters. The second possibility is so far not supported by the available data. The alternate hypothesis was not, however, *a priori* excluded, and might yet prove relevant in some cases. In any case, with a 2D geometry, we propose

$$ x = x_0 \left(\frac{t}{t_0} \right)^{1/D_{op}}, \tag{9.2} $$

where x represents the root radial extent, and t the time required to achieve such an extent. The choice of the 2D value of $D_{op} \simeq 1.21$ is consistent with the limitations of growth within 2D 'skin' of Earth, while D_{op} describes the tortuosity of a path through the medium that is selected to minimize the dissipation for the flowing fluid. In other words, root structures are assumed to be prevalently 2D, as would occur in most soils where the majority of roots are concentrated near the surface (Jackson *et al* 1996) and extend laterally much more than in depth; see table 9.1. In dry ecosystems, however, rooting depth is comparable to or larger than root spread over most plant functional types (Schenk and Jackson 2002). Furthermore, for small plant sizes (also an important range of length scales) one could reasonably postulate that the nutrient sources are accessed three-dimensionally. But consider (Lynch 1995), 'Spatial heterogeneity of nutrients is evident in the weathered Oxisols and Ultisols of the humid tropics, which are warm and moist throughout the soil profile but in which available P, Ca, and other mineral nutrients are concentrated in a thin surface layer. The Spodosols common to temperate coniferous forests have striking horizonation, with a black layer of acidic leaf litter over a white layer of leached sand, followed by an orange layer of clay and nutrient accumulation, all within 1 m or less of the soil surface.' These statements imply that the tortuosity exponent (fractal dimension) of the paths constrained to 2D is likely to be appropriate, and is compatible with the specific scaling results for tree growth in temperate and tropical forests (see below).

It must be emphasized that edaphic constraints, as described by equation (9.2), do not facilitate growth, where it is not genetically supported, or where other limitations are at play, such as light, temperature, external disturbances, and herbivores or pathogens. Thus, such edaphic constraints should be seen as limits to growth of the dominant individuals, or larger species, where conditions are otherwise optimal. It follows that edaphic constraints need not be relevant to growth of smaller organisms filling particular ecological niches, or of any species under other limitations. For testing the relevance of these constraints, we need to access information about the largest organisms with the fastest growth rates, since for such organisms the theoretical size-age relation of equation (9.2) should have predictive power. In contrast, we expect that the growth rates under sub-optimal conditions will be upper-bounded by the theoretical relation.

To summarize, our simple scaling relationship, equation (9.2), contains information on heterogeneity scale, x_0, flow rates, t_0, and the architecture of roots in the soil, their fractal dimension. Both x_0 and t_0 depend on soil type and are very sensitive to changes in soil moisture, hence introducing a site dependence in the scaling relation. Thus, we predict a scaling relationship that has its origins in the delivery of heterogeneously distributed nutrients to the plant that depends on water flow rate through the soils, and the solute transport over the tortuous preferred paths as primarily constrained by the root architecture. The constraints are formulated solely by nutrient (mainly N and P) delivery, though the proportionality of the solute transport to the flow rate also brings in the water flow rate. Therefore, the hypothesized equivalence of root growth rate and solute velocity along an optimal path links the two limitations to biological aspects in the context of ecological optimality (table 9.2).

9.2.3 Intensively managed plants

In the case of irrigated and fertilized crops and tree plantations, we conjecture that the growth rate is limited by the nutrient delivery to the root system, which is proposed to be linear in time for the reasons given below, not by the ability of the root system to grow to the sources of nutrients. Thus, the pressure on the root system, and the organism as a whole, to find the optimal paths to the nutrients is eliminated. Why should this be the case? The most important feature of fertilization in agriculture is the uniform, or homogeneous, delivery of nutrients to the crops, eliminating the heterogeneity in nutrient distribution present in natural soils. Moreover, soil tillage breaks down aggregates and levels micro-topographic features, thus effectively homogenizing nutrient sources. One advantage in this is that the flux of nutrients towards the plant remains constant, whether or not its roots grow. Thus, local heterogeneities in N and P, postulated by Lynch to be critical to the root architecture, do not develop as a result of nutrient depletion. A second advantage is that root tips need not change their path directions in response to local heterogeneities in nutrients. This change from natural vegetation may reduce the range of spatial scales over which the root architecture is tortuous. A constant rate of root tip growth would then produce a maximum overall extension at any time, commensurate with a linear temporal increase in organism size, and the maximum ability to absorb nutrients simultaneously.

Table 9.2. Summary of L_R/Z_R, the ratio of root radial extent L_R and depth Z_R, for various types of climates and plant functional. Numbers indicate the median values of the reported or calculated ratios in each category. DSGS represents deserts, scrublands, grasslands and savannas with <1000 mm rainfall. Median across climatic conditions (MACC) is for each plant functional type, while APC stands for after planting cuttings.

Ecosystem	Plant functional type	L_R/Z_R	Source	Notes
DSGS	Trees	3.33	Schenk and Jackson (2002)	MACC
DSGS	Shrubs	0.91		MACC
DSGS	Semi-shrubs	0.5		MACC
	Perennial grasses	0.34		
	Perennial forbs	0.28		
	Annuals	0.3		
	Succulents	5.63		
Temperate forests	Riparian species	7.86	Phillips *et al* (2014)	Age: 1 year APC
	Angiosperms	7.53	Phillips *et al* (2015)	Age: 2–3 years APC
	Gymnosperms	5.54		
	Angiosperms	4.0	Stone and Kalisz (1991)	Age: > 10 years
	Gymnosperms	4.14		
Boreal forests	Angiosperms	7.5	Kalliokoski *et al* (2008); Stone and Kalisz (1991)	Age: > 10 years
	Gymnosperms	6.25		

Thus, as the above discussion indicates, the goal of agriculture is—or should be—making the nutrient delivery simultaneous with that of water, with the crop root growth rate hypothesized to follow a scaling relationship identical to the flow of water, i.e., $x = x_0(t/t_0)$. Synchronizing water and solute delivery can probably be accomplished by various techniques, which are: (i) hydroponics, (ii) watering with uniform nutrient concentration; (iii) applying fertilizer uniformly in a plane perpendicular to the water flow; and (iv) massive application of fertilizer. Deleterious effects on water quality, particularly in stream headwaters, may be reduced by using strategies (ii) or (iii), rather than (iv). In strategy (iii) one should assume that the appropriate scaling relationship is the same as the one for natural vegetation growth, but that the appropriate fundamental length scale is equal to the transverse length scale of the homogenization of the nutrient delivery, i.e., along the horizontal for crops (Rajaram and Gelhar 1993, Hunt *et al* 2015). Thus, effectively, the solute transport distance varies linearly with time up to the length scale that defines the variability of solute concentration transverse to the water flow. This homogeneous distance can vary from centimeters to meters, depending on planting density, and would, therefore, speed up the delivery of nutrients by two to three orders of magnitude when compared with transport in soil, but only about one order of magnitude relative to plants. Indeed, a one order of magnitude increase in the growth rates is approximately the contrast observed between crops and natural woody vegetation. Therefore, the typical vertical water flow rate in soils is suggested to provide a maximum growth rate in intensively managed crops. Water and nutrients are again co-limiting factors on growth since, in the optimal case provided by intense agricultural management, solute fluxes have the same time dependence as water flux.

9.2.4 Testing the scaling relations

The validity of the three scaling relationships is investigated in figure 9.1, where we employ plant height h as a proxy for root lateral extent—x in the relations above— and plant age t as characteristic size and time scale. Our analysis rests on the assumption that height and root radial extent L_R are proportional, or related, by power laws with exponent close to one. The advantage of the approach is that plant height h is commonly measured, whereas data on root extent are scarce. Its disadvantage is that the relation between height and root extent is not universal (section 8.5), so that it is difficult to generalize the fundamental assumption of the

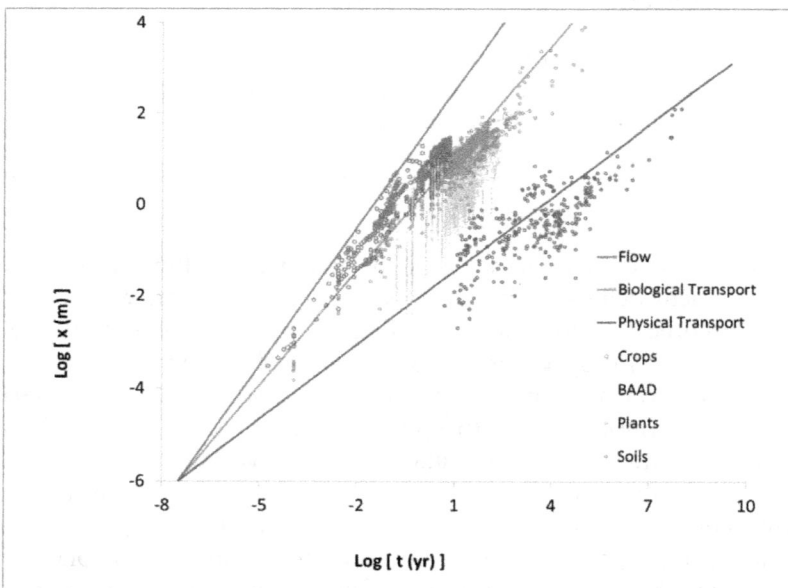

Figure 9.1. Test of the three scaling relationships between size and age in 325 pairs of data for soil depthl (brown; section 9.3), 910 pairs for the linear dimension of natural vegetation and fungi (green; sections 9.4.2 and 9.4.4), 1041 pairs, including *Eucalyptus* and other plantations, for crops (magenta; section 9.4.5), and 5538 (lighter green) pairs from the BAAD dataset (Falster *et al* 2015, section 9.4.3). The common origin at 1 μm and 1 s relates to typical fluid velocities and pore sizes of a fine-textured soil. The different slopes for the green and brown lines arise from the distinct topologies of solute transport through soils and roots, while the magenta line represents a scale-independent fluid velocity (table 9.1).

approach. In trees growing in moist conditions, however, which are more likely nutrient limited, allometric theory predicts an isometric scaling with approximately unitary coefficient, $L_R \approx 0.8h$ [equation (8.24)], and data from diverse ecosystems and plant functional types support this result; see figure 8.5. The relation between root spread and height in herbaceous crops is more uncertain, but there are not enough data to conclude that isometric scaling is not reasonable.

In order to collect the data for figure 9.1, a very wide range of sources was consulted, which will be given later, when the individual scaling relationships are considered in greater detail. It should, however, be mentioned here that the data at very long time intervals correspond to clonal underground growth, or to extensive fungi, whereas the data at very short time intervals correspond to root tip growth. Thus, in neither extreme limits is it necessary to relate above- to below-ground growth. Data points from 'Fast Plants and Fungi' may overlap with 'Woody Plants,' due to the finite size of the data points and the fact that the biomass and allometry database (BAAD) were not edited to select only the slower growing plants. The BAAD data are, however, overwhelmed by the Fast Plants data in the vicinity of the scaling prediction of equation (9.2), since we have accessed a large number of studies not in BAAD that document this region of the scaling plot. Tree height data are upper bounded by hydraulic and edaphic limitations (section 9.5.2). As indicated in section 9.4.5, some tree plantation data were better put into the intensively managed plant category, in view of the associated fertilization. These plantation trees and crops grow noticeably faster than the scaling prediction of equation (9.2)— consistent with the hypothesis of water flow limitation. Finally, figure 9.1 suggests that the absolute minimum growth rate is very near the solute transport rate for abiotic media, while the data for 'Woody Plants' is rather generally not found in regions occupied by soil depths. The two coincidences might support the idea that the principle adaptive advantage of plants is in the strategy of root development.

The tendency for many data points to approach the proposed limiting curves suggests that the hypotheses are indeed valid. Of course, further investigations, perhaps guided by the hypothesis proposed here, is desirable and has, in fact, been carried out recently (Hunt et al 2020, 2022). The distinction between intensively managed and natural vegetation is not necessarily clear, while the heights of many cultivated species have continued to increase over the last 20 years with the development of new cultivars, and more attention to individual plants (informal global competitions). On the other hand, the needs of industrial agriculture no longer support the development of cultivars with a maximum growth rate. Therefore, individual examples of such plants continue to get larger, while the typical sizes in mass cultivation diminish.

Considerable spread exists in each data category shown in figure 9.1. To explain such variability, it is important to recall that our predicted scaling functions are valid for the most common, nominally saturated hydraulic conductivity in soil, and when nearly saturated conditions are present most of the time. In practice, soil moisture varies dramatically through time (Rodriguez-Iturbe and Porporato 2004) and, thus, the hydraulic conductivity takes on a wide range of values due to the occurrence of dry periods characterized by low conductivity (Blöschl and Sivapalan 1995, Dingman 2002).

Later work that relates plant growth rates to transpiration suggests, in fact, an alternative formulation of the hypothesis based on the hydraulic conductivity discussed here, namely, that the typical water content of soils adjusts itself to generate a hydraulic conductivity that is similar in magnitude to the annual infiltration flux. Due to the underestimation of the characteristic time t_0 in dry soils and/or dry climate, transport distances in systems that are frequently unsaturated are overestimated, introducing a roughly two-order of magnitude variability into the predicted soil depth. Moreover, temperature effects are neglected, but we know that both weathering and biological reaction rates increase nearly exponentially with temperature, so that variability in the data is likely caused by this effect when moving along wide climatic gradients. Regarding plants, height-age data for species from water-limited ecosystems are expected to lie below the theoretical $h(t)$ curve for the same reason. According to Blöschl and Sivapalan (1995), most K values are within an order of magnitude of the one we assumed. Much larger variability does, however, exist. Whether variability in K ultimately explains most of the variability in the actual results is a topic for additional research. In any case, in the more specific comparisons below, we indicate an uncertainty in predicted soil depth consistent with the spread in potential soil K values.

A secondary conclusion that emerges from study of the data is that the maximum size reached by fungi at very long time scales appears to be not quite as great as for vascular plants (roughly a factor two less), perhaps related to the slightly slower pore-scale growth rate of fungal hyphae compared with root tips (about half an order of magnitude; see Watt *et al* 2006). While this issue, even at the smaller scales is not fully resolved, the inference is that it is a necessary condition for the survival of vascular plants that root extension is more rapid, in view of the apparent ability of each to detect and respond to the other over distances up to about 1 mm. Since the chemical signals are thought to propagate by diffusion, though at distances near 1 mm, advection should be the dominant means of solute transport. Moreover, careful examination of the data for the growth of natural vegetation indicates several series where trees reach maximum heights, at about 25, 50, and 100 m, for reasons other than the proposed basis of the power law. Low leaf water potential at the top of tall trees does not allow leaf function, hence limiting the height reached by trees (Koch *et al* 2004). Moreover, there might be genetic reasons, or tree height may be limited by other causes, namely, wind, lightning, and snow loads. Although reports of trees up to 120–130 m, which had been logged exist, and are likely reliable, on the present logarithmic scale, the distinction is minimal. A more important difficulty is that many of the largest trees mentioned (see, e.g., Waring and Franklin 1979) are paired with dates typical for the life of the tree, not with the date at which a tree reached a given height. This tends to overestimate the ages of the largest trees and, thus, produce a deviation to slower growth rates, or even constant heights, at longer times.

9.3 Soil data

Let us describe soil data that are essential for testing the theory that we have proposed.

9.3.1 Soil formation data

The data for soils used by Hunt (2017) were originally reported by Alexandrovskiy (2007), Borman (1995), Bockheim and Tarnocai (1998), Colman and Pierce (1986), Evans and Hartemink (2014), Goodman *et al* (2001), Heimsath *et al* (2001a), Jacobson *et al* (2002), McFadden and Weldon (1987), Pillans (1997), Schülli-Maurer *et al* (2007), Stevens (1968), Van den Bygaart and Protz (1995), and White *et al* (1996). Further datasets at both early and later times were accessed in the meantime, including those reported by Frouz *et al* (2008), Li *et al* (2007), Smale *et al* (1997), Mavris *et al* (2010), Trustrum and De Rose (1988), Almond *et al* (2007), Lichter (1998), and Dethier (1988), as well as a set of studies on ancient tropical 'soils' called *laterite*. The additional studies were chiefly accessed in order to extend the range of the time scales for soil formation investigated, both to shorter time periods (in the years to decades after landslides, burial by sand, exposure due to glacial retreat, or on abandoned mine tailings), as well as to longer periods, the laterites. Previously accessed information included recent exposures due to treethrow and glacial retreat. Note that we have always used data for soil 'depth' when given; when depths to various horizons were presented, we chose Bw preferentially, B if no distinction was made, and A, if the B horizon was not mentioned [the A horizon is the surface horizon of a mineral soil], or if it had not developed. Bw is, for example, the oxidation depth, while B generally is the depth to chemical changes without addition of components. In this context, for the study of Trustrum and De Rose (1988), we chose to use the depth that excluded the effects of added 'rafts.' We show eight of these soils that individually correspond reasonably well with our predictions in figure 9.2.

The soils studied by Trustrum and De Rose (1988), as well as by Smale (1997), are from humid temperate climates in New Zealand, with year round moist and temperate conditions. The soil water content in the Baltic soils is also high, whereas the Australian, McCall Idaho, and the Cajon Pass soils are from much more arid regions. Thus, our analysis implies the relevance of a larger numerical prefactor for the humid New Zealand and Baltic regions than for the intermountain West and Australia. This suggests the relevance of a larger hydraulic conductivity, presumably due to much more frequent conditions near saturation. Jenny (1941) noted long ago that soil formation was strongly dependent on water content. Since we use the hydraulic conductivity under saturated conditions to define the fundamental time scale, we expect that soils that are unsaturated most of the time will have much less time to develop (smaller value of t than the actual time elapsed), or a larger value of t_0 (smaller hydraulic conductivity from less than saturated conditions). Consequently, one should also expect that the depth of a carbonate soil horizon should increase with precipitation, as indeed observed (Jenny 1941). However, the precise form of this relationship would depend on whether t or t_0 is the factor where the effect of increased precipitation is most strongly felt. Since the calcic horizon depth dependence on precipitation appears to be superlinear, rather than sublinear, the best choice appears to be in the factor t_0, inversely proportional to K, since K is roughly proportional to a large power of the moisture content. If, however,

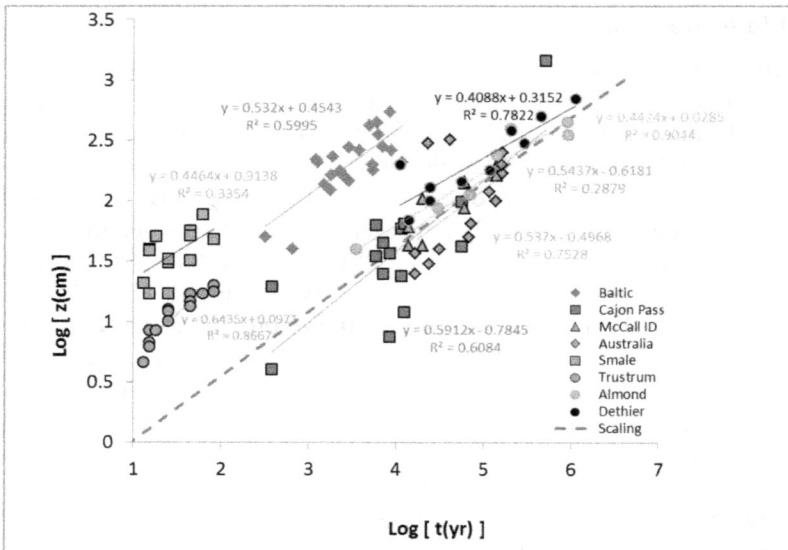

Figure 9.2. Soil depth–age relations for eight soils with soil development functions similar to the predicted distance for solute transport. Here, the mean exponent is 0.52 ± 0.08, which compares well with the predicted value of 0.53 based on the percolation model. Omitting a single data point in the Dethier (1988) database at early times with exceptional depth would increase the slope in that experiment to 0.505 and, thus, the mean exponent to 0.53, and decrease the standard deviation from 0.08 to 0.07.

desaturation conditions are predominant, then the form of the soil depth–age relationship should change, as is noted for the case of weathering rinds on surface rocks (Hunt 2017). The separate case of weathering rinds is considered below.

The results for laterite thicknesses (Chowdhury *et al* 1965, Chowdhury 1986) in the West Indian Ghats and Australia (Gardner 1957) extend data on the depth of a chemically weathered layer to tens and even hundreds of meters, and the time scales of formation to at least tens of millions of years, and perhaps 100 million (Gardner 1957) (figure 9.3). Laterite is thought to be a product of chemical weathering of iron-rich saprolites (highly fractured bedrocks, such as basalts) in tropical-to-subtropical climates with high total precipitation, but prolonged dry periods during the year. It is noteworthy that such saprolite hydraulic conductivities and weathering rates are typically presumed to be approximately the same as that of unconsolidated deposits (Dixon *et al* 2009). Primary laterites are believed to have formed by weathering in place; secondary laterites are presumably younger, and are believed to have been transported from their place of origin (Chowdhury 1986). The depths of the Indian laterites are up to 30 m (Madhya Pradesh) and their ages are believed to be Oligocene (30–40 million years or Ma) or younger, but could be as old as 65 Ma, the time of the formation of the Deccan Traps by flood basalts. Thus, an age of about 40 ± 25 Ma is reasonable. Note that, on the logarithmic scale, the uncertainty is relatively minor. The important point is that the date are not wrong by orders of magnitude. Ghosh and Guchhait (2015) add as a constraint the necessity of a

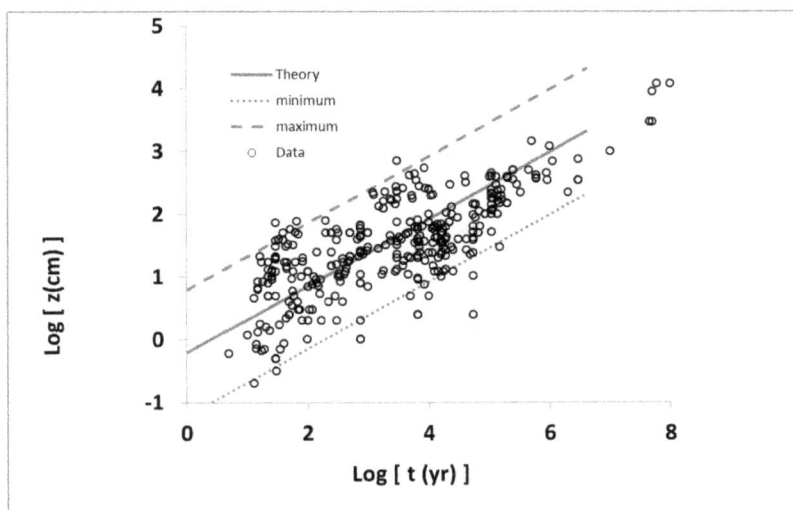

Figure 9.3. The depths of all 24 soils as functions of time, and compared with the scaling function that incorporates an order of magnitude faster or slower predictions related to the two orders of magnitude spread in subsurface flow rates, centered about 1 μm s^{-1}. Several soils in Alpine regions of rapid tectonic uplift (not shown in figure 9.2, but included here) exhibit, however, little correlation between age and depth. The prediction is consistent with a loss of surface material in hundreds of meters per million years, which is less than the mean rate of denudation in the Himalayas, for example, which is 1000 m Ma^{-1}.

monsoon climate, which set on after the end of the Mesozoic, in agreement with the geological constraints noted by Chowdhury (1986).

The Australian laterites are as deep as 120 m, and may be as ancient as Middle Mesozoic, but are thought to be closer to 65 Ma, or even younger. The stability of Australia, both tectonically and climatologically, over even a larger age range than for India and other provenances, is considered likely (Gardner 1957). Thus, approximate ages of 40 Ma (to the Indian laterites) and 100 ± 50 Ma (to the Australian laterites) years were assigned to these deposits, although the uncertainty may seem to be unacceptably large. Nevertheless, an uncertainty by a factor of 2 in the age (early Cenezoic to middle Mesozoic) translates to only 40% uncertainty in depth, in view of the approximately square root dependence of depth on age. Note that even a factor of 2 in time represents less than half an order of magnitude. But application of a scaling relationship over 8 orders of magnitude variations in length scale (microns to 100 m) can lead to 1.5 orders of magnitude error in time, if the scaling exponent is in error by only 10%. In any case, laterite depth variability of between 10 m and 120 m over an age range of 5 to 150 Ma is quite compatible with the general trend of the data of shallower soils, and with the scaling relationship proposed.

Differentiating $x(t)$ with respect to time and making some rearrangements, the soil production as a function of depth is obtained; see equation (9.1). The predicted soil depth does not yet contain the effects of soil erosion, though it is not a difficult matter to introduce it. The soil production function as a function of depth is independent of erosion, however, and our prediction is compared with the data

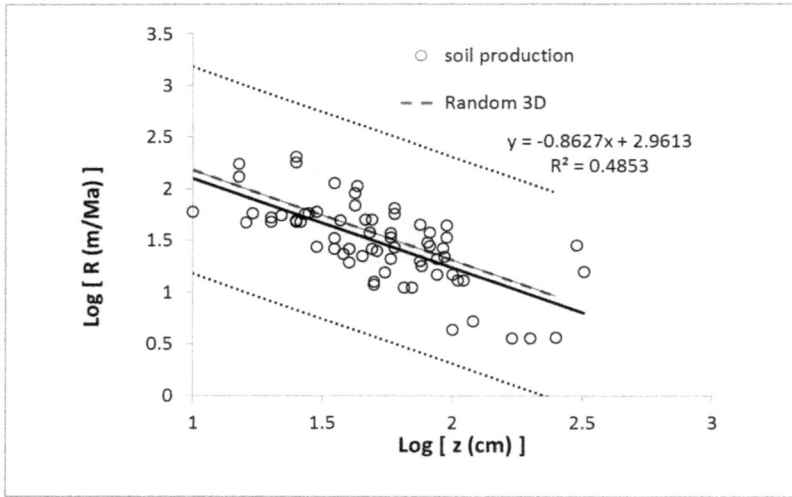

Figure 9.4. Comparison of the predicted soil production function with data from six studies of Heimsath and co-workers. Note that the exponent obtained from the regression is -0.86, which differs by 1% from -0.87 predicted by the percolation model, while the numerical prefactor differs by 8%. Thus, the entire dataset, considered of high relevance to the study of geomorphology, is in very close agreement with the predictions of the percolation model for soil depth as a function of time, as well with the scaling relationship. The dotted lines represent the variability introduced by the range of typical flow velocities found in soils (e.g., Blöschl and Sivapalan 1995). The value $R^2 = 1.0$ for the theoretical result merely reflects its lack of experimental uncertainty.

reported in six papers of Heimsath and co-workers (1997, 1999, 2001a, 2001b, 2005) and Burke *et al* (2007), and shown in figure 9.4.

9.3.2 Weathering rind data

Weathering rinds are chemically-altered outer surfaces of rocks that are exposed to the elements. Such surfaces tend to have lower density, less material strength, and, in the case of basalt, pronounced color differences compared with the parent rock. Their higher porosity endows them with a higher hydraulic conductivity as well, but the necessity to penetrate the unweathered rock, as well as to remove any products of the weathering reaction require that water flow into the unweathered pore space, with its very low hydraulic conductivity. Rocks lying on the surface, or at shallow depths in soils, are also subject to climatic effects on water fluxes, with limits provided by precipitation, but it is easy to see that when parallel paths of much greater hydraulic conductivity exist, the contrast between soil and unweathered rock hydraulic conductivity values will add an additional constraint to weathering rind fluxes. Basalt hydraulic conductivities are about six orders of magnitude smaller than unconsolidated media (Freeze and Cherry 1979). If one were to make the prediction that weathering rinds should develop under the same saturation conditions as soils, then the approximately square root time dependence of the rind thickness (analogous to the soil depth) would change the six order of magnitude K

contrast to a three order of magnitude contrast in weathering rind thicknesses. This is actually a very good initial approximation; after 100 000 years or so, soil depths are measured in meters, whereas weathering rind thicknesses tend to be measured in millimeters or centimeters. However, the surface, or near-surface, exposure of weathered rock surfaces measured means that typical conditions are unsaturated, and a somewhat more careful analysis is appropriate. The data for weathering rinds are not plotted simultaneously with crops, natural vegetation, and soils, because the fundamental time scale for water to traverse a single pore is so much longer. Thus, these data would not diverge from the same point on a bilogarithmic plot of space and time. The reason is that they form mostly under unsaturated conditions on rocks. Under such conditions, the relevant hydraulic conductivity is many orders of magnitude slower.

For weathering rinds the proportionality, $r \sim t^{1/D_{bb}}$ for the rind thickness r can also be rewritten as an equation, if it is possible to assign length and time scales, r_0 and t_0, which are effectively demanded by unit considerations,

$$r = r_0 \left(\frac{t}{t_0} \right)^{1/D_{bb}}, \tag{9.3}$$

The ratio r_0/t_0 is a pore-scale fluid flow velocity, v_0, which (using mass conservation) under the influence of gravity can be written as, $v_0 = K/\phi$, with ϕ being the porosity. For unfractured basalts, the typical value of K for saturated conditions is (Freeze and Cherry 1979) slightly larger than 10^{-10} cm s^{-1}, but the smallest saturated value is 10^{-12} cm s^{-1}. The same value is, thus, also barely larger than the largest possible hydraulic conductivity, for which the material is certain to be unsaturated. Basalt grain sizes are constrained by definition to lie between roughly nanometers (associated with a glassy texture) and millimeters (larger grained materials that are called gabbro), and typically require magnification, such as from a microscope, to be visible. A geometric mean value between these limits is 1 μm. Since, as will be seen, most weathering rinds form under unsaturated conditions, with much smaller K values than under saturated conditions, a reasonable single value to apply for K is the smallest saturated value, 10^{-12} cm s^{-1}. Given a common porosity of 5%, a fundamental (mineral heterogeneity) length scale of one micron, and the exponent $0.69 = 1/D_{bb} = 1/1.46$, appropriate to 3D desaturating conditions (and a common value observed), one can make a concrete prediction for the rind thickness as a function of time,

$$r = 10^{-6} \left(\frac{t}{5 \times 10^6} \right)^{0.69}, \tag{9.4}$$

with r expressed in meters and t in seconds. Note that, $t_0 = 5 \times 10^6$ s is about 59 days. Thus, this result cannot be valid at time scales less than 59 days, the time required for fluid to move a grain separation. The comparison of weathering rinds of various materials demonstrates the importance of the hydraulic conductivity in determining weathering rind thickness. The predictions of this simple prediction produces a slope that is generally consistent with the data from fifteen different studies (see figure 9.5). The Bohemia dataset, reported by Cernohouz and

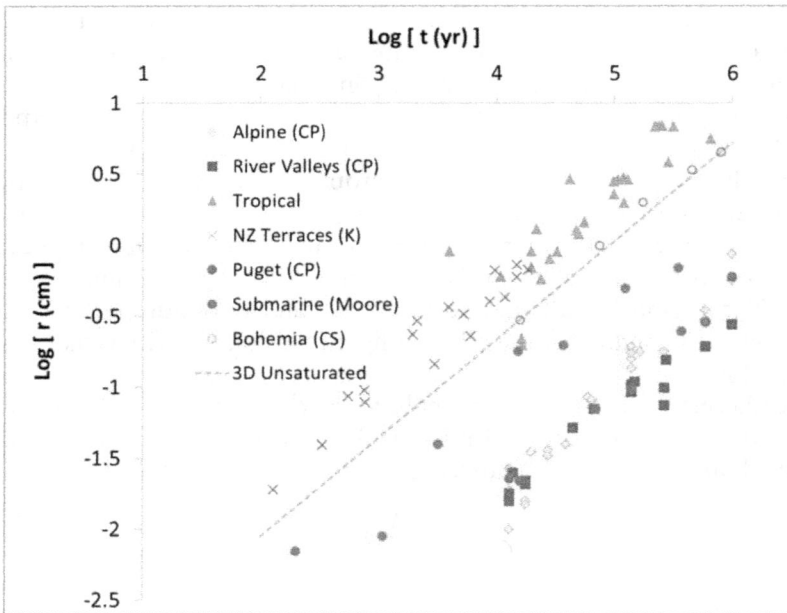

Figure 9.5. Weathering rind thickness, as a function of time, and its comparison with the predictions of the same scaling law and fundamental grain size, developed for soil thickness, but with a different t_0, appropriate for the much smaller hydraulic conductivity of unsaturated basalts, and a value of D_{bb} appropriate for 3D flow under desaturating conditions. Many clasts (rocks) taken for weathering rind analysis were subaerial (on the surface), some were in the shallow subsurface. The majority were likely unsaturated virtually the entire time, justifying the use of the percolation exponent for unsaturated conditions.

Solc (1966), lies almost exactly along the line; the rinds reported by Sak *et al* (2004), Navarre-Sitchler *et al* (2007), Ma *et al* (2012), Pelt *et al* (2008), and Oguchi and Matsukura (1999) (called collectively, 'tropical,') are a factor two or so greater, as are the sandstone rinds reported by Knuepfer (1988), while the Colman and Pierce (1981, 1986) data are approximately a factor 10 smaller than predicted and conform to prediction using the 2D unsaturated exponent, 1.217, which produces a time dependence to the 0.82 power, also commonly observed (Hunt, 2015).

When analyzed separately, the weathering rinds of Moore (1966) yielded an exponent $0.61 = 1/1.643$ [with 1.643 being the fractal dimension D_{bb} of the backbone of 2D percolation clusters], appropriate for 2D saturated conditions, which is what one might expect from highly fractured submarine pillow basalts. Finally, it is worth comparing the scaling results for soils and weathering rinds together, but in the absence of vegetation data, as in figure 9.6.

9.4 Vascular plant data

For natural plants, there are sufficient sources to obtain over 900 data pairs over length scales varying from fractions of millimeters to 10 km, in order to test the scaling hypothesis, equation (9.2). Aside from tree plantations, only about 150 data

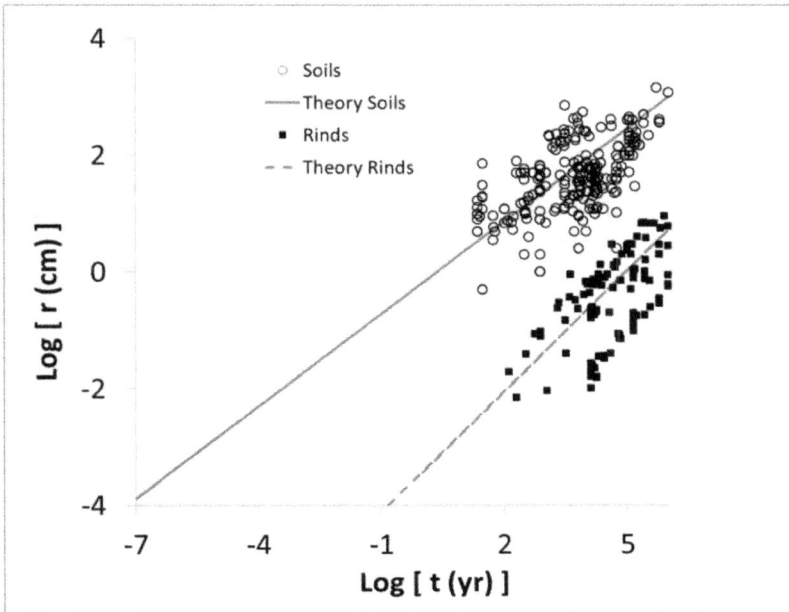

Figure 9.6. Comparison of soil depths with weathering rind thicknesses, incorporating the differences in the fundamental time scale and the relevant fractal dimension of the percolation backbone. Note the approximately six order of time slowing in the initial rate of formation of weathering rind, which is due to the combined effects of lower water contents (unsaturated conditions) and the approximately four orders of magnitude smaller hydraulic conductivity of basalt, when compared with typical unconsolidated media that form, or are, soils.

pairs of intensively managed plant growth (crops) are available, but including tree plantations, approximately 1000 are accessed. On the assumption that dimensions of the root system underground are roughly proportional to the height of the canopy above ground, for a range of plant sizes, from centimeters up to about 100 m, the maximum height of the canopy was used to test equation (9.2), including crops. Due to hydraulic and structural limitations (Koch *et al* 2004, Niklas 2007) and atmospheric phenomena, such as wind, or human activities, such as logging, it is rare for any trees to reach heights significantly greater than 100 m. Much larger organisms do exist, however, and the extent of these individuals is revealed in the subsurface. For the analysis of the subsurface organisms, the step relating subsurface and above-ground growth can be omitted.

A strategy for the following analysis was developed *a posteriori*, in particular after it was determined which data sources were available. Since the scaling functions are proposed to be limits to vascular plant growth provided by the soil, maximum growth rates in each vegetation category, natural or intensively managed, are sought. The distinction between the two categories was determined based on the existence, lack thereof, of irrigation and fertilization, which alter the natural water and nutrient delivery. In the case of natural plants, we looked for: (i) studies or compilations of rapidly growing species; (ii) studies of seedlings responding to canopy disturbances; (iii) studies that referred to tree heights of dominant individuals; and (iv) species that are first in succession after fires. Since many of the rapidly growing tree species are also adapted to succeeding fires, there is some overlap in the strategies. Additionally, since a number of the rapidly growing tree species are also used in tree plantations, it was necessary to distinguish carefully which growth patterns are naturally supported and that require intensive human intervention.

In the case of intensively managed plants, the highest growing are difficult to document, except in the case of tree plantations. In fact, most studies address biomass, or productivity, and in most cases crops are not optimized for height growth, but rather for productivity, uniformity, seed quality, and time to market. Thus, the highest managed plants for time scales of less than a year are often described in the context of contests for the largest individual of a given species. For tree plantations, however, where an important component of the productivity is tree height, this variable is more often documented. At smaller length scales, some data address root growth directly, and inferences relating above- and below-ground growth are also unnecessary. However, the root growth data cannot be assigned either to intensively managed or to natural vegetation, because experimental conditions could be altered in ways that could accommodate the distinctions between crop and natural vegetation, even for a given species. Thus, root growth is discussed in a separate subsection, before we address separately the managed and natural vegetation.

9.4.1 Root growth

Table 9.3 presents data at very short time scales that directly provide root tip extension rates. A range of plants are included in individual studies, which often seek to quantify the distinctions between optimal and sub-optimal conditions, hence providing some examples that belong in the managed plant category, and others that belong in natural vegetation. It was decided to refer only to crop species that were actually tested under ideal growing conditions as crops (intensively managed). Otherwise, the results were placed with the natural plants. Rates of root tip extension are measured on scales of days (Watt *et al* 2006; data summarized in

Table 9.3. Root tip extension rates under varying conditions (modified from Watt *et al* 2006).

Genus	Conditions	Root extension rate (μm s^{-1})
Triticum	Moist, loose soil at 15 °C	0.2–0.4
	Moist, hard soil at 15 °C	0.1
Zea	Moist, loose soil at 29 °C	0.8–0.9
	Dry, loose soil at 29 °C	0.3
Gossypium	Hydroponics, low salt	0.3
	Hydroponics, high salt	0.03
Hyphae of root pathogens	Various conditions	0.004–0.2

table 9.3) through scales of hours (Sugimoto *et al* 2000, Walter *et al* 2002, Wu *et al* 2007, Cordoba-Pedegrosa *et al* 1996, Muzuroglu and Gerkil 2002, Sharp *et al* 1994, Saab *et al* 1990) down to minutes (Evans 1976). In some cases, the rates were stated to remain approximately constant over 24 h periods (Walter *et al* 2002). The data reported by Walter *et al* (2002) were for Zea mays, and were put among the crops. The data reported by Sugimoto *et al* (2000) were for Arabidopsis, and were assigned to natural vegetation. The data of Wu *et al* (2007) were for *Nicotiana attenuata*, a wild tobacco, and were also assigned to the natural vegetation. Watt *et al*'s data (2006) included both Arabidopsis and crop species. When the conditions of temperature, medium, and moisture were ideal, the particular examples were considered as crops; when additional salt, particularly hard soils, or low temperatures were used, the species were treated as natural vegetation, along with Arabidopsis. In this study, variations in root growth rates over a factor nearly 20 were measured, depending on the conditions.

9.4.2 Rapidly growing trees under (mostly) natural conditions

The most rapidly growing tree species that we can be locate are *Eucalyptus regnans* (and *saligna*, as well as some other *eucalypts*), *Populus deltoides*, Sequoia (*Sempervirens, Sequoiadendron giganteum*, and *Metasequoia*), *Pseudotsuga menziesii, Liriodendron tulipifera*, and a wide range of tropical dipterocarps. We start with the gymnosperms, namely, the three sequoia species, *Sequoiadendron giganteum, Metasequoia glyptostroboides*, and *Sequoia sempervirens*, though they do grow to different heights, and yet appear to have very similar maximum growth rates. Data for *S. sempervirens* that were found were reported by Burns and Honkala (1990), Kuser *et al* (1995), Berrill and O'Hara (2014), and Waring and O'Hara (2007). In 1948 a number of seeds of Metasequoia were brought to the United States and Europe and planted at a variety of arboretums, hotels, and other locations. Kuser (1982, 1983, 1998) tracked the growth of many of these individuals through time, but only the fastest-growing examples are included. Other data points were obtained from a study of the Ohio State University (Chatfield *et al* 2006), three reported by Kuser (1998), and a few from the internet site, Monumental Trees. A number of data pairs for *Sequoiadendron* were located (table 9.4 and figure 9.7). Once it is established that the three sequoia species grow at essentially the same rate, they can be included later as a single tree type (figure 9.5).

Data for the fast-growing angiosperms were reported in studies on eucalypts, dipterocarps and eastern cottonwoods (*Populus deltoides*) (Johnson and Burkart 1976). The eastern cottonwoods measured were in regions along the Mississippi River, which had recently been flooded, removing existing natural cottonwood stands. In one case (one age), the new stand was thinned, but, otherwise, there was no management. Aside from tree plantations, the fastest-growing individual angiosperms appear largely to be in recently burned areas (Tng *et al* 2012), or in gaps in tropical forest canopies formed by various reasons (see below). The following studies addressed growth rates of eucalypts seedlings (see figure 9.8): Wood *et al* (2010), Van Der Meer *et al* (1999, 2007), and Whitmore and Brown

Table 9.4. Data sources for individual *Sequoiadendron* tree heights and ages (see also figure 9.7). 'Monumental own' refers to http://www.monumentaltrees.com/en/trees/giantsequoia/growing_your_own/, while 'Monumental elsewhere' means Monumental trees that were planted away from their places of origin: http://www.monumentaltrees.com/en/trees/giantsequoia/elsewhere/. Encyclopedia refers to http://www. statemaster.com/encyclopedia/Sequoiadendron#Europe. 'Big Stump' was logged in 1883 and surveyed in 1968, though the references are from later imes later (Mcdonald 1992, Stohlgren 1992).

Age (years)	Height (m)	Source
100	60	McDonald (1992), Gasser (1992)
50	35	Weatherspoon (1990)
32	20	Schmid and Schmid (2012)
13	8	Heald (1986)
60	35	Heald (1986)
50	31	York *et al* (2006)
15	8.7	Stephenson (1992)
5	4	Gasser (1992)
8	6	Gasser (1992)
85	60	Big Stump
400	73	Weatherspoon (1990)
4	3	Heald (1986)
14	10.6	Heald (1986)
14	11.5	Heald (1986)
17	22	Encyclopedia
4	7	York *et al* (2013)
5	8	York *et al* (2013)
6	9	York *et al* (2013)
7	10	York *et al* (2013)
11	11.5	York *et al* (2013)
15	10.5	York *et al* (2013)
19	12.5	York *et al* (2013)
24	13.5	York *et al* (2013)
4	6	York *et al* (2013)
5	7	York *et al* (2013)
6	7	York *et al* (2013)
7	8	York *et al* (2013)
11	10	York *et al* (2013)
15	10	York *et al* (2013)
19	11	York *et al* (2013)
24	12	York *et al* (2013)
50	40	Monumental own
10	10	Monumental own
42	35	Monumental elsewhere
75	51	Monumental elsewhere
50	27	York *et al* (2002)
60	34	York *et al* (2002)

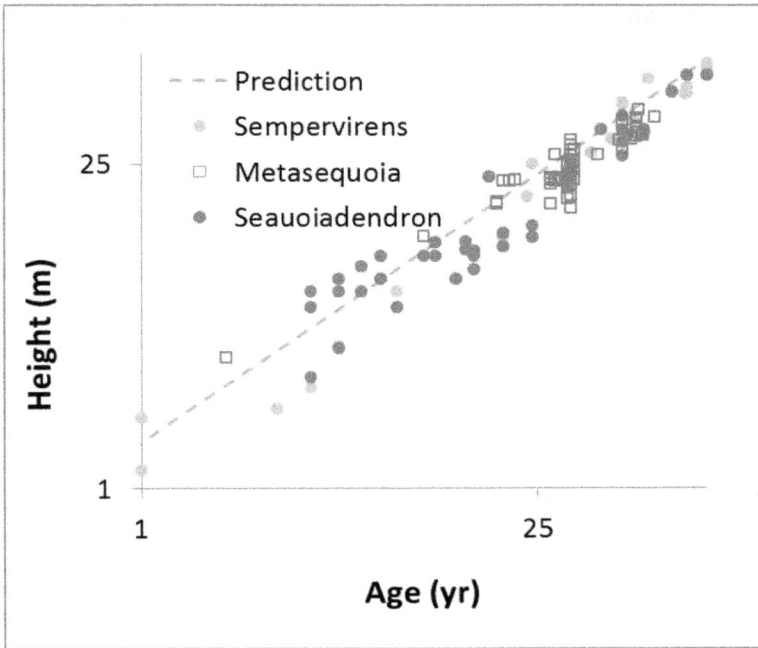

Figure 9.7. Heights of fast-growing individual trees for three sequoia species as a function of time (data sources are given in table 9.4).

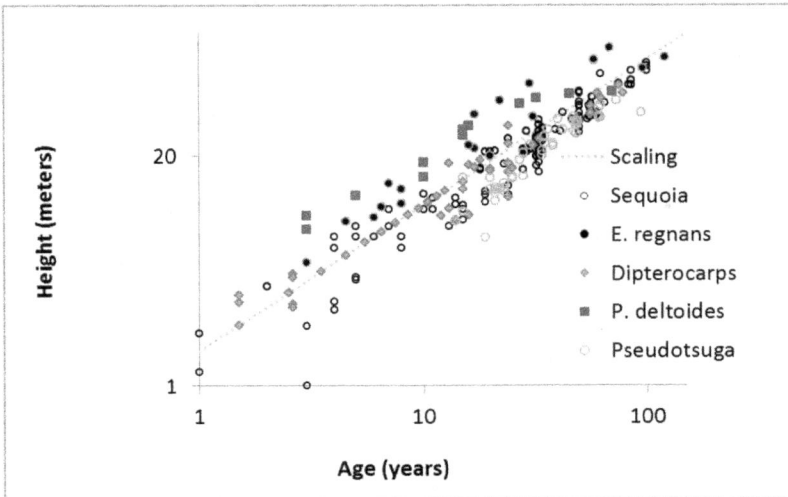

Figure 9.8. A scaling plot of tree height versus tree age for some of the world's fastest growing trees. The sequoias include all three species together shown in figure 9.7.

(1996). For dipterocarps, data were obtained from Bungard *et al* (2002) and Davies (2001). Other studies included ages of large numbers of mostly dipterocarps in a mature rainforest site (Worbes *et al* 2003), from which only the 10 fastest individual trees were chosen. Understory trees were largely excluded thereby, as trees under the canopy tend to grow much more slowly, whereas even the trees much older than 100 years did not exceed 60 m and, thus, also had slower growth rates. Clark and Clark (2001) described the life histories of four dipterocarps; the one that was chosen here was the only one made it through the 13 year study without severe damage from falling branches. Nicolle (2011) gives the largest tree in Europe as *Eucalyptus regnans*, 72 m at 120 yr. The dipterocarps, eucalypts and eastern cottonwoods are shown together with the sequoias and P*seudotsuga menziesii* (Fontes *et al* 2003) in figure 9.8.

Rapidly growing trees were the subjects of a number of other studies on their height as a function of time. The data from a number of such studies are included in figure 9.9, as well as two model predictions (Biging 1985, Lappi and Bailey 1988). Since only six data entries were reported by Gonzales *et al* (2005) and the bulk of their data could not be digitalized, these points were included in figure 9.1, but not here, or in figures 9.10 and 9.11. Except for the *P. nigra* data, which follow distinct curves in two distinct geographic regions, the remaining data in figure 9.9 could be

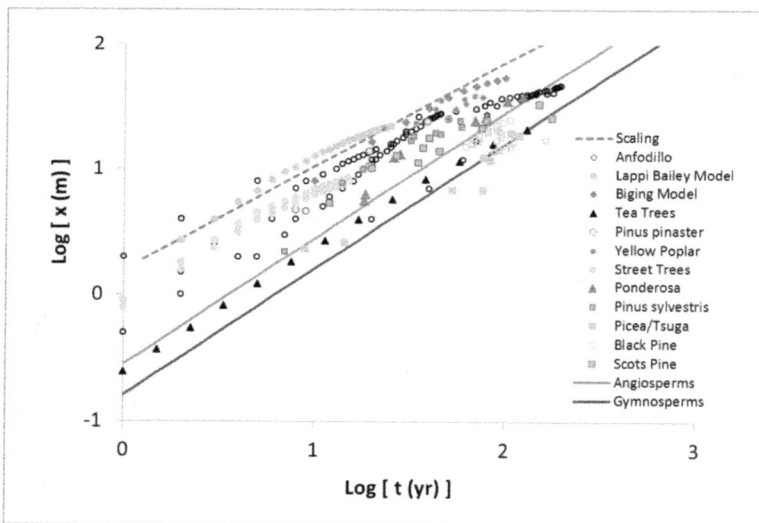

Figure 9.9. Scaling plot of a number of additional trees whose growth was mainly slightly slower than the predicted upper limit. The data are from McNab (1989) (*Liriodendron tulipifera*), Sala and Hoch (2008) (*Pinus ponderosa*) Seymour and Kenefic (2002) (*Tsuga canadensis* and *Picea rubens*), Martin-Benito *et al* (2008) (*Pinus nigra*), Stoffberg *et al* (2008) (three street tree species in Tshwane, South Africa), Magnani *et al* (2000) (*Pinus sylvestris*), Smale *et al* (1997) (*Kunzea ericoides*), and Anfodillo *et al* (2006) (*Fraxinus excelsior, Larix decidua,* and *Picea abies*). In addition, a *Pinus elliottii* plantation summary (Lappi and Bailey 1988) and the site index model of Biging (1985) are included. The lines labeled 'Angiosperms' and 'Gymnosperms' represent the predictions of allometric scaling (section 8.6), which yield predictions similar to equation (9.2) over the rather limited range of length scales represented here.

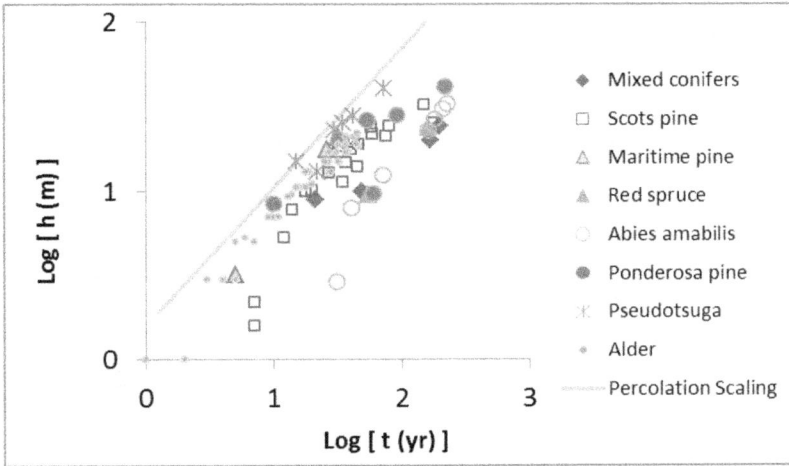

Figure 9.10. Scaling plot of (mostly) rapidly growing trees. The sources are 'Mixed conifers' (Franklin *et al* 2003), 'Scots pine' (Mäkela and Vanninen 1998, Mencuccini and Grace 1996), 'Maritime pine,' (Delzon *et al* 2004), 'Red spruce' (Maguire *et al* 1998), *Abies amabilis* (Buchmann *et al* 1998), 'Ponderosa pine' (Domec and Gartner 2003), and 'Alder' (Deal 1997). Additional Pseudotsuga data were found from Phillips *et al* (2002), Eis (1974), and Long and Turner (1975).

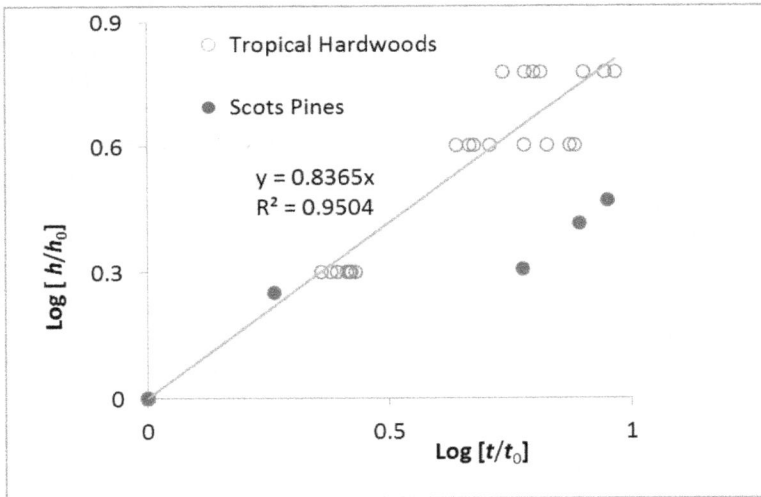

Figure 9.11. Normalized neotropical hardwood heights (O'Brien *et al* 1995) and *P. sylvestris* heights (Martinez-Vilalta *et al* 2007) as functions of tree age. The neotropical hardwoods are shown here in a separate figure, since the different species had rather widely different fundamental growth rates. In this separate analysis, in which each tree type is normalized to its initial height (in order to reduce the effects of scatter) a slope of 0.84 is obtained. Such an analysis is partially *a posteriori* justified by the general compatibility of the data with a power law, for which such a renormalization is allowed. The *P. sylvestris* growth curve leveled off after several decades, likely due to competition with other individuals for nutrients. After thinning, at around 200 years, the growth rate nearly resumed its initial trend. The slight decline in growth rate with age (a power roughly 0.9) was attributed to nutrient limitations, noting that other investigators had found larger trees of the same age in the same species.

reasonably interpreted as power laws. However, in those cases where the data levelled off, with a final range exhibiting virtually no increase in height with age (see, especially Anfodillo *et al* (2006)), the extracted power was obtained by omitting the data at the longest times. Such leveling off also occurs in the sequoia species, as will be seen explicitly later, and more generally across angiosperms (Muller-Landau *et al* 2006). Data from two studies are shown in a normalized representation in figure 9.11. While the normalization to initial heights described in the caption to figure 9.11 were likely appropriate for the various tropical hardwood species (O'Brien *et al* 1995), the data was not used elsewhere. Most other data, restricted mostly to one (or a few) tree species under the same conditions, do not exhibit greater scatter at any particular time scale, and a summary of the exponents extracted for these studies from simple bilogarithmic plots is given in table 9.5. While there was considerable variability in the individual powers, the overall tendency was to be generally consistent with the proposed scaling relationship [an exponent of 0.83 from equation (9.2)]. The mean and standard deviation of the exponents obtained are 0.82 ± 0.07 for gymnosperms and 0.81 ± 0.08 for angiosperms. Therefore, the results do not appear to depend on tree type. Significant differences in the exponent obtained for tropical and temperate species were not found either. In figure 9.9, the theoretical isometric relation between height and age from allometric theory (section 8.6) is also shown for angiosperms and gymnosperms separately. Although the allometric scaling predictions appear to have a slope generally consistent with data, they underestimate the height of the tallest trees.

Additional datasets for tall trees were also used to extend the range of observations, on the ground that the tallest trees also have a sustained fast growth rate. If slow growing, the probability of damage and reduction of height would increase. For some very tall species ('giants'), tree sizes and ages used were for the largest known examples of their species. Where ages (or heights) were given with uncertainties, the algebraic mean values were used. When ages for a given specimen were not given, the greatest age and tallest tree measurements were used. Waring and Franklin (1979) use a similar convention, namely, the greatest height and typical age. Note that this can result in a considerable overestimate of tree ages. The tallest coast redwood (*Sequoia sempervirens*)[1], known as 'Hyperion,' is thought to be about 600 years old, rather than 2000, but since its real age is not known, an alternate strategy had to be used. The same general remarks can also be made about the Sitka spruce (*Picea sitchensis*)[2], the Douglas fir (*Pseudotsuga menziesii*)[3], and Mountain ash (*Eucalyptus regnans*)[4]. The growth scaling data for the above individual studies, together with the neotropical hardwoods and Scots pines, are shown in figure 9.11.

Significant additional data for giant trees that are included in figure 9.1 (but not in figure 9.11) were reported by Waring and Franklin (1979), including Silver fir (*Abies amabilis*), noble fir (*Abies procera*), Port Orford cedar (*Chamaecyparis lawsoniana*), Alaska yellow cedar (*Chamaecyparis nootkatenis*), western larch (*Larix occidentalis*),

---

[1] http://www.monumentaltrees.com/en/trees/coastredwood/tallest_tree_in_the_world/#tallesttree

[2] http://www.visitoldgrowth.com/sites/OR-Spruce.htm

[3] http://online.sfsu.edu/bholzman/courses/Fall00Projects/Douglas-fir.html

[4] http://anpsa.org.au/eregn.html, and http://www.monumentaltrees.com/en/trees/mountainash/records/

Table 9.5. The ordinary least-square exponents and the R^2 values of the fits, obtained from the various studies of tree heights, as a function of time. AEFGT stands for 'almost Europe's fastest growing trees;' 175D-12CD means '175 dominants, 12 co-dominants;' FGTNA represents 'fastest growing tree in North America;' TMDTH is the abbreviation for 'tested model of dominant tree heights;' SSDE implies 'steady-state only in the driest conditions;' MCP represents 'mostly in canopy gaps;' MLS means 'medium to large species;' and O represents 'observations.'

Tree species	Source	Functional type	Slope	R^2	Comment
> 30 species, 50 individuals	Anfodillo et al (2006)	Angiosperm & gymnosperm	0.86	0.85	AEFGT
Mixed conifers (187 trees)	Biging (1985)	Gymnosperms	0.82	0.98	175D-12CD
Populus deltoides	Johnson and Burkhardt (1976)	Angiosperm	0.67	0.98	FGRNA
Pinus elliottii	Lappi and Bailey (1988)	Gymnosperms	0.82	0.99	TMDTH
Kunzea ericoides[1]	Smale et al (1997)	Angiosperm	0.90	N/A	SSDE
Eucalyptus regnans	Various studies	Angiosperm	0.77	0.93	O
Sequoiadendron giganteum	Various studies	Gymnosperm	0.71	0.98	MCG
Eight neotropical hardwoods	O'Brien et al (1995)	Angiosperm	0.84	0.95	O
Pinus sylvestris	Magnani et al (2000)	Gymnosperm	0.74	0.95	O
Tsuga Canadensis, Picea rubens	Seymour and Kenefic (2002)	Gymnosperm	0.81	0.91	O
CE[2], SL[3], SP[4] Searsia	Stoffberg et al (2008)	Angiosperms	0.81	0.98	MLS
Pinus ponderosa	Sala and Hoch (2008)	Gymnosperm	0.92	0.95	O
Sequoia sempervirens	Various studies	Gymnosperm	0.87	0.97	O
Alnus rubra	Deal (1997)	Angiosperm	0.74	0.93	O
Dipterocarps	Various studies	Angiosperm	0.74	0.95	MCG
Pseudotsuga menziesiii	Several studies[5]	Gymnosperm	0.91	0.88	O
Abies amabilis	Buchman (1998)	Gymnosperm	0.82	0.997	
Scots Pine	Mencuccini (1996)	Gymnosperm	0.82	0.83	O
Red alder	Deal (1997)	Angiosperm	0.86	0.96	O

[1]Tea trees; [2]Combretum erythrophyllum; [3]Searsia lancea; [4]Searsia pendulina; [5]Eis (1974); Long and Turner (1975); Phillips et al (2002); Fontes et al (2003).

incense cedar (*Libocedrus decurrens*), Engelmann spruce (*Picea engelmanii*), Sitka spruce (*Picea sitchensis*), sugar pine (*Pinus lambertiana*), western white pine (*Pinus monticola*), ponderosa pine (*Pinus ponderosa*), Douglas fir (*Pseudotsuga menziesii*), coast redwood (*Sequoia sempervirens*), western red cedar (*Thuja plicata*), western hemlock (*Tsuga heterophylla*), and mountain hemlock (*Tsuga mertensiana*). These species were chosen on the basis of their great contributions to the remarkable biomass of the northwestern conifer forests. Another option for determining tree growth is to measure the age and diameter at multiple heights along the trunk of freshly logged trees (Winter et al 2002). As in other datasets, the agreement with the predictions of equation (9.2) is excellent for approximately 60 years, whereas growth rates in older trees decline compared to the predicted scaling; see figure 9.12.

Changing growth rates of eucalypt species over shorter periods of time are given by Walsh *et al* (2008), and can be compared with our predicted growth in figure 9.13. Here, the variability in the growth rates from site to site offers the possibility of investigating variability in the parameters of equation (9.2). Clearly, a more mechanistic approach would be required to rigorously test how equation (9.2)

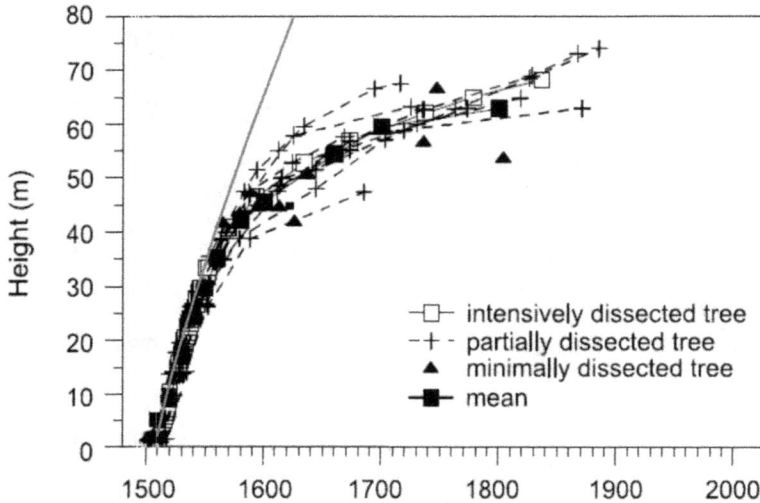

Figure 9.12. Deduced growth history of *Pseudotsuga menziesii* in the Pacific Northwest (Winter *et al* 2002), compared with the scaling predictions of equation (9.2). Felled trees from four different stands in different areas were dated and measured.

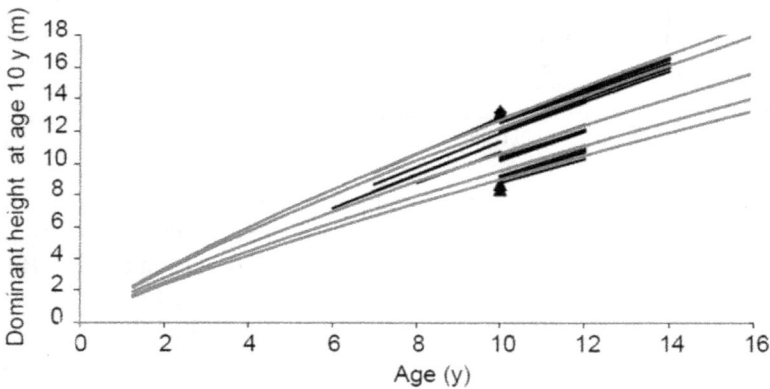

Figure 9.13. Growth curves of *Eucalyptus* (in black) and its comparison with the scaling predictions. The middle curve is equation (9.2). The other curves were predicted by equation (9.2) by multiplying by the numerical factors 1.2, 1.15, 0.9, and 0.85, respectively.

describes height-age relations along climatic gradients. Such an approach would need to account for the variability in soil moisture and, thus, in hydraulic conductance.

9.4.3 'Average' woody species under natural conditions

The datasets on the fastest-growing individuals were complemented with the more recent BAAD data for woody plants (Falster *et al* 2015), which include allometric data for plantation species (mostly tropical), as well as from plants growing under natural conditions. Some plantation data have been classified under the intensively managed category due to the large external nutrient supply. The data points from plants grown in natural conditions (5539 out of 6650 data points) are compared with the scaling prediction in figure 9.14. The figure also includes a 'Hydraulic Limit,' and a 'Scaling Minimum.' The hydraulic limit, as proposed by Koch *et al* (2004), is about 122–130 m. The minimum is motivated by the study of Givnish *et al* (2014) who determined that the single angiosperm species, *Eucalyptus regnans*, could span 'almost the entire range' of angiosperm mean maximum heights along a rainfall gradient. These heights ranged from 4.4 m to 87.2 m, and the correlation with the ratio of precipitation to pan evaporation was nearly linear with an R^2 value of 0.88. Thus, the same scaling function was used, but multiplied by the ratio of 4.4/87.2 in order to determine a minimum upper bound on a growth rate due to edaphic

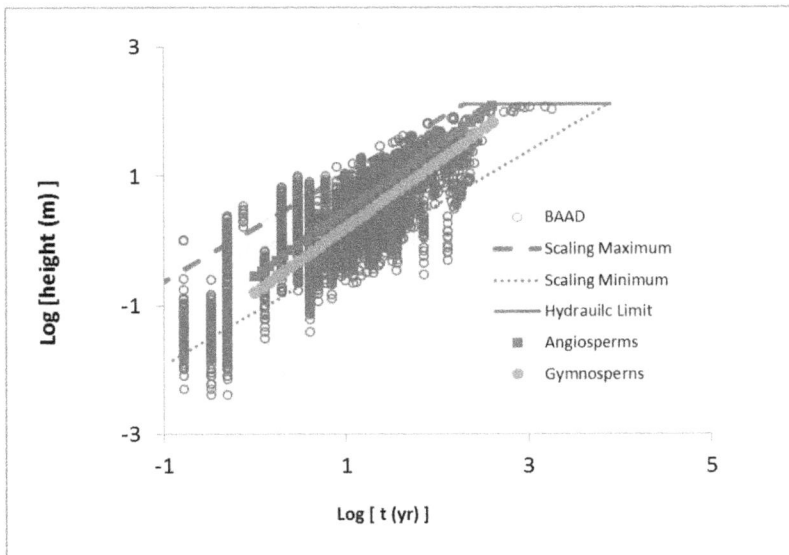

Figure 9.14. Comparison of the BAAD dataset, omitting data from intensively managed plantations, mostly of *Eucalyptus* species, with the prediction of the scaling relation, equation (7.2), and a proposed hydraulic limit. The description of the rationale for a lower scaling bound is based on the discussion of *E. regnans* presented by Givnish *et al* (2014); see the text for details. The lines labeled 'Angiosperms' and 'Gymnosperms' represent the predictions of allometric scaling (section 8.6), which yields the predictions in the middle of the dataset, but which do not match reported growth scaling at shorter and longer times.

constraints, which might be valid at least for *E. regnans*. It turned out that this estimate is also more or less consistent with the overall data from BAAD, as might be concluded based on the assertion of Givnish *et al* (2014) regarding its suitability as a representative for angiosperms generally. The ability of this species to 'explore' the entire range of angiosperm variability in terms of a single variable, the moisture content, is conferred by its exceptionally large hydraulic conduction properties (section 9.5.2). Figure 9.14 also suggests that for periods of up to about 6–8 months, the observed maximum in the data still exceeds the scaling prediction by a small amount. This may be a result of the existence of optimal growing conditions over a growing season, which cannot be matched for the entire year, although it may also be an artifact of incomplete data or theory.

9.4.4 Other data for natural vascular organisms

Given the close correspondence of the observed maximum growth rates in natural trees with the predictions, it remains desirable to extend the range of time scales over which equation (9.4) can be tested. For larger length scales, owing to hydraulic limitations, this can only be accomplished by looking underground, where it is not necessary to consider the additional step of correlating above- and below-ground growth. The largest individual vascular plants and fungi include aspen clones (*Populus tremuloides*; see Grant (1993)), *Armillaria fungi* (of two different genera; see Smith *et al* (1992))[5], and seagrass clones (*Posidonia oceanica*)[6]. The data for these organisms represent the horizontal extents, rather than their height, as considered before. Note that there is some confusion about the ages of the fungi. Thus, in view of the wide ranges given, the geometric mean age was chosen. Exceedingly fast-growing kelp was excluded since it is not vascular and accesses nutrients directly from the seawater through diffusion (although this can also be a complicating factor in the case of *Posidonia oceanica*). A summary of clone data was also given by Arnaud-Haond *et al* (2012). Additional references to enormous invasive bamboo colonies exist (Suzuki and Nakagoshi 2011), with ages of only a few hundred years, but without evidence that they are composed of single organisms. In any case, data at such large length and time scales is still rather sparse, compared with documentation of tree heights. At shorter length scales, natural grasses (Leithead *et al* 1971) and woody vines, such as wisteria[7] can be considered. The grasses were problematic, however. As a protocol, vertical growth in some plants has been excluded, since, for example, shoots of bamboo grow from crowns that store energy (and nutrients) for years. In such cases, vertical growth rates do not clearly represent horizontal expansion rates. But the growth of the majority of referenced grasses is rhizomatic in character too, with single year vertical growth based on storage from previous years.

---

[5] http://www.sciam.com/article.cfm?id=strange-but-true-largest-organism-is-fungus
[6] http://www.ibtimes.com/oldest-living-thing-earth-posidonia-oceanica-discovered-be-over-100000-years-old-407412
[7] http://waynesword.palomar.edu/ww0601.htm

9.4.5 Cultivated vascular plants: herbaceous crops and tree plantations

The data for the growth of some fast-growing crops, including peas, beans, corn, sunflower, amaranth, and hemp, were collected. For sunflower and corn, the data refer to typical growth curves in time, but to the maximum height given for the species. The somewhat more slowly growing wheat is included as well. Better controlled studies were available on rice (Tilly *et al* 2014), though rice is not an especially fast growing crop. Some specific cases of world record breaking individual plants, such as tomatoes, amaranth, hemp, corn, and sunflower, are given as well, when their ages are known or fairly well constrained (by, for example, a growing season for outdoor individuals). For crops, all the data referred to above-ground measurements. In any case, these data include corn[8], wheat[9], peas[10], beans[11], sunflowers[12] http://phe.rockefeller.edu/Bi-Logistic/, tomatoes (65 ft in about 12 months)[13], hemp[14] http://www.hemp-technologies.com/page83/page83.html, and amaranth[15] http://forums.gardenweb.com/forums/load/giants/msg1210392811910. html?49. Some of the data were accessed from public internet sites and may, thus, be less reliable than from peer-reviewed studies. However, fewer controlled studies address the question of how fast vegetation, cultivated or not, can grow. When bi-logistic curves were available, they were scaled to the largest individual of the cultivar; otherwise, the data for the largest specimens of a particular cultivar were restricted to single points. The results are compared with equation (9.6) in figure 9.15. Note that the largest corn plant was grown almost 70 years ago; modern agriculture focuses on other goals, such as reducing time to market, maximizing yield, and generating uniformity in stand height. The slowest growing crop was rice, which barely exceeded the predicted natural plant growth curve. The data for crop species heights are shown in figure 9.15.

Some plantation data pairs from the BAAD database were also classified among the managed plants. Tropical plantations are known to suffer from nutrient limitations, which is overcome with fertilization, and mixing Eucalyptus species with nitrogen-fixing species (National Academy of Sciences 1979, 1984). It is, therefore, not surprising that intensively managed plantations can lead to Eucalyptus growth rates of up to $5\,\mathrm{m}\,\mathrm{yr}^{-1}$ for up to 7 years, which is three to five times the maximum growth rate of trees in natural environments. Because fertilization generally precludes the relevance of edaphic constraints on nutrient availability from the soil, some Eucalyptus plantation data were simply classified as managed plants (Battaglia *et al* 1998, Nouvellon *et al* 2010, Epron *et al* 2012), as well as the

---

[8] http://thegazette.com/2009/09/30/iowas-tall-corn-contest   http://threethingsverydullindeed.blogspot.com/ 2010_06_01_archive.html

[9] http://www.hgca.com/media/185687/g39-the-wheat-growth-guide.pdf

[10] http://ucanr.org/sites/asi/db/covercrops.cfm?crop_id=17

[11] http://www.selah.k12.wa.us/soar/sciproj2003/AllisonE_DATA.pdf

[12] http://www.guinnessworldrecords.com/records-1000/tallest-sunflower/

[13] http://www.openwriting.com/archives/2007/09/the_biggest_tom_1.php

[14] http://www.japanhemp.org/uncleweed/agriculture.htm

[15] http://www.pandpseed.com/Merchant2/merchant.mvc? Screen=PROD&Store_Code=pandpseed&Product_Code=GAMAR&Category_Code=GiantAmaranth

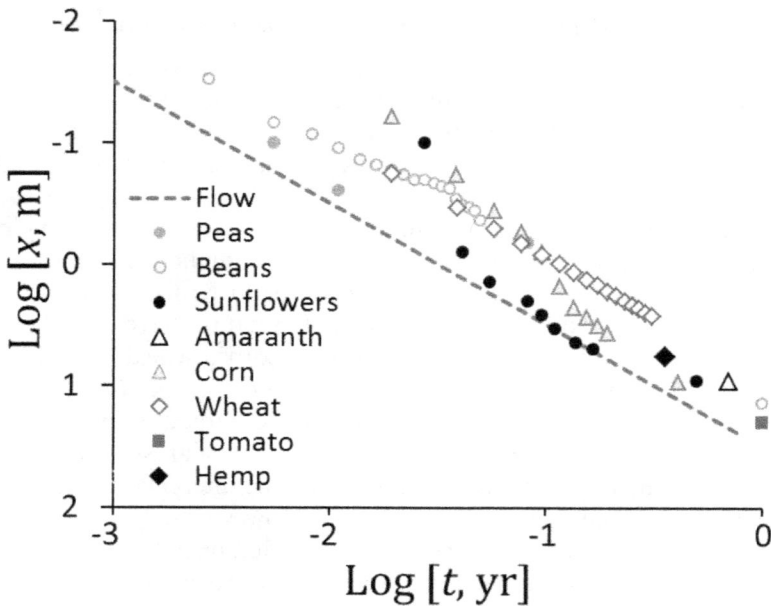

Figure 9.15. Data for the height of various cultivated species as a function of time.

Acacia study, and an intensively managed *Pinus elliottii* plantation from the southeastern USA. Height growth data from well fertilized tropical Eucalyptus plantations in independent studies (Barnard *et al* 2003, DeBell *et al* 1987) were compared with the plantation data from BAAD, and found to be substantially equivalent (not shown), thus supporting our choice to include plantation data from BAAD among the intensively managed plant category.

9.5 Generalizations and implications

Two aspects of plant growth are worth emphasizing in connection to the proposed scaling relation. First, can one explains the observed diminishing growth rate with age (section 9.5.1)? Second, can global patterns in tree height be interpreted using the scaling relation (section 9.5.2)? Finally, section 9.5.3 provides a more general discussion.

9.5.1 Diminishing growth rate with age

In figure 9.16 we compare our predicted growth rate, obtained as the numerical derivative of equation (9.2), with Ryan and Yoder's (1997) summary and the data reported by Givnish *et al* (2014), as well as with the growth rates summarized for the Sequoia, the time derivative of the Lappi and Bailey (1988) model for *P. elliottii*, Metasequoia growth rates from Kuser (1982, 1998), and the growth rate of Armillaria reported by Smith *et al* (1992). Since the growth rate for Armillaria is given after 1500 years, it is important to show the first 150 years separately on such a linear graph, with the second panel covering the entire range of the data. Note that

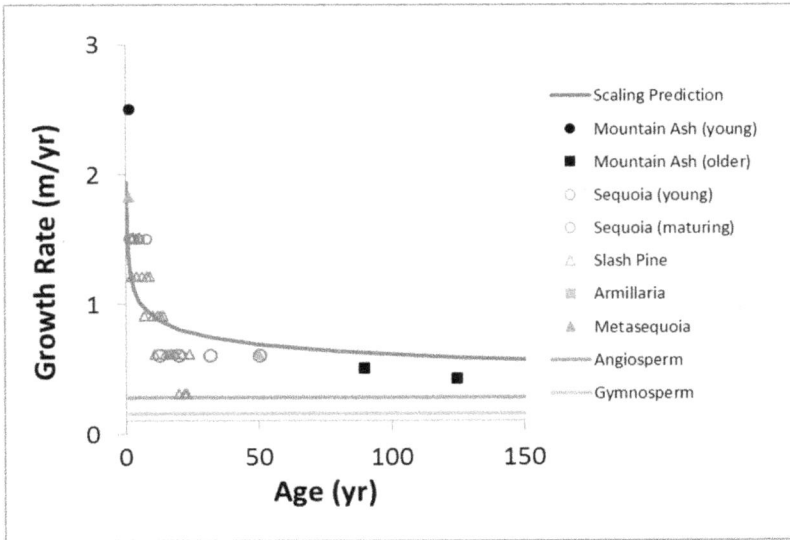

Figure 9.16. Comparison of the predictions of the growth rate by the scaling equation with individual data points from both *Sequoia* and *Eucalyptus regnans*. Data sources for the Sequoia are described in the text, whereas those for the *Eucalyptus regnans* are given by Ryan and Yoder (1997) and Givnish *et al* (2014). The growth rates for *Metasequoia glyptostroboides* were reported by from Kuser (1982, 1998), while those for Armillaria were reported by Smith *et al* (1992). The data for *Pinus elliottii* data were obtained by a time-derivative of the model of Lappi and Bailey (1988). 'Gymnosperm' and 'Angiosperm' refer to the summaries from allometric scaling (section 8.6).

the initial point of the predicted growth curve corresponds approximately to a growing season at 0.8 yr. These rather limited data are generally in accord with the prediction. Considering that the prediction uses no adjustable parameters, the agreement seems quite good, indeed. For comparison, the two growth curves from allometric scaling (section 8.6) are shown as well. While the distinction between allometric scaling and equation (9.2) is difficult to see on a log–log graph of height compared with time, such as in figure 9.9, the two predictions are very different, as seen in figure 9.16, which represents, on a linear graph, the time derivative of the height as a function of time. The distinction between a constant growth rate and a negative power of time is even greater at shorter time scales, and the rates of root tip growth of up to 70 mm day^{-1} (Watt *et al* 2006), that are also in accord with equation (9.2), is worked out to be 25 m yr^{-1}, which exceeds the maximum yearly height growth rates by one or more orders of magnitude. It should be noted that the allometric relations serve here as a reference, because, (1) they are derived for mature trees only, and (2) they represent an 'average' rather than the particularly fast-growing individuals depicted in figure 9.13. Such data are also in general accord with a *Euclayptus regnans* shoot growth rate measured by Cremer (1975) of 6.2 m yr^{-1} at 30 days. In particular, multiplying the highest growth rate of *Eucalyptus regnans* at one year (3 m yr^{-1}) (Ryan and Yoder 1997) by $12^{0.21}$ to determine the commensurate growth rate at 30 days, and one finds a value 5.1 m yr^{-1}. Then, we take $30^{0.21}$ times 6 m yr^{-1} to find the growth rate at one day to be 12 m yr^{-1}, only a factor of two different from the fastest growing root tips of Watt *et al* (2006). Thus, the proposed scaling law explains why growth rates slow down with age—one of the open questions posed by Ryan and Yoder (1997), while variations in relevant water fluxes, i.e., transpiration, explain spatial variations in intra-species growth rates (Hunt *et al* 2020).

9.5.2 Characteristics of the world's tallest trees

Consider the following assertion of Ryan and Yoder (1997): 'A mechanism that can explain differences in maximum tree height at different locations and patterns in height growth with age has eluded ecologists and plant physiologists.' Following this statement, several theories were developed that explained such a pattern based on hydraulic limitations to leaf functioning and biomechanical stability (Koch *et al* 2004, Niklas 2007, Kempes *et al* 2011). This value is certainly in accord with the data acquired for figure 9.11, as well as for the data of Waring and Franklin (1979). Continuing, Ryan and Yoder state, 'A young mountain ash (*Eucalyptus regnans*) growing east of Melbourne may grow 2–3 m per year in height. By 90 years of age, height growth has slowed to 50 cm per year. By 150 years, height growth has virtually stopped, although the tree may live for another century or more.' Later evidence suggests that growth can continue at a rate of 20 cm yr^{-1} in 300 years old *E. regnans* (Givnish *et al* 2014). Cessation of growth at greater ages, but continuous diameter increment, is common and likely genetic in nature where environmental limitations (water in particular) are not important (King 1990). Therefore, it has been argued that, 'When height growth ceases to offer a competitive advantage through avoidance of shading, then (genetically programmed) resource allocation will be adjusted to enhance tree survival and reproduction, not necessarily wood production' (Becker *et al* 2000) (figure 9.17).

What characteristics do the world's tallest trees have in common? Three such features stand out, which are rapid growth, long lifetimes, and geographic location. The combination of very similar rapid growth rates, but mortality distinctions, appears to explain a contrast in tree heights in between tropical and temperate rainforests. While the upper limit on tree height in a given ecosystem is moderately influenced by climatic conditions (Moles *et al* 2009), the occurrence of the tallest individuals globally is restricted to limited geographic locations. Quoting Givnish *et al* (2014), 'It is remarkable that 9 of the 11 tallest tree species in the world are found in tall (i.e., wet) sclerophyll forests and adjacent temperate rainforests, including five conifers in the Pacific Northwest of the United States and four Eucalyptus species in Tasmania and Victoria in Australia (Tng *et al* 2012). Only two occur in tropical rainforests, despite the warmer, rainier, and more humid conditions prevailing there.' Tng *et al* (2012) considered a slightly less restrictive subset of the world's trees, 'Although any definition of gigantism is necessarily arbitrary, a practical threshold of 70 m maximum height […] delimits ca. 50 species, representing < 0.005% of an estimated total of 100 000 tree species,' and added southeast Asia and Indonesia to the relevant area. If a threshold of 88 m is chosen, the two tropical species Givnish *et al* (2014) consider, *Shorea faguetiana* and *Araucaria hunsteinii*, as

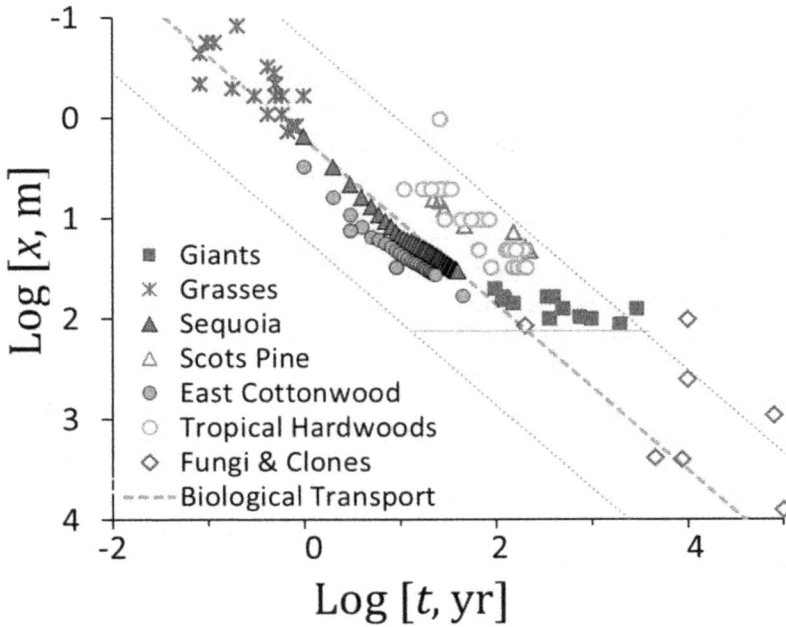

Figure 9.17. Observed scaling of size and age of a number of individual vascular plants and fungi, and their comparison with the predicted scaling relationship. The dotted lines provide an estimate of the uncertainty in the scaling prediction that stems from the variability in groundwater flow velocities. The orange horizontal line is a proposed hydraulic limit on tree height of 125 m (Koch *et al* 2004).

well as their geographic regions, Borneo and New Guinea, are included, respectively. The geographic region that supports the largest tree growth is restricted almost as severely as the number of species.

What is the purpose of the rapid growth to considerable heights, and what factors enable it? Tng *et al* (2012) proposed that the giant *Eucalyptus* species are unique fire-adapted trees, a statement that, however, applies equally to *Sequoiadendron giganteum* and reasonably to *Pseudotsuga menziesii* as well. Consider the overall characteristics of the environment after fires in such sclerophyll and temperate rainforests that include (Tng *et al* 2012): (i) high, near-surface nutrient concentration; (ii) abundant light (in conjunction with species adaptation to low shade tolerance); and (iii) variable, but, in places, high water content. Along a rainfall gradient from the sclerophyll to the rainforest, the moisture content will increase, while the fire frequency diminishes. Rapid growth is an adaptation to a competition for light. Shade intolerant eucalypts (van der Meer *et al* 1999, 2007) and sequoias (York *et al* 2002, 2013, York 2006) are known to grow much more rapidly in gaps than under the canopy, and natural seed germination may require fire (see, e.g., York *et al* cited above). In the tropics, greater light can be provided by canopy gaps (Whitmore and Brown 1996). The rapid growth of dipterocarps in gaps shows that light is likely as important for rapid dipterocarp growth (Whitmore and Brown 1996, Davies 2001, Bungard *et al* 2002) as it is for eucalypts and sequoias.

Additional factors required for rapid growth are warm temperatures, high soil water contents (Givnish *et al* 2014), and high hydraulic conductivity of the tree (Fan *et al* 2012). One way of developing a large hydraulic conductivity is based on a large xylem diameter. *Populus tremuloides* growth rates in regions not seriously impacted by freezing correlate strongly with xylem diameter (Schreiber *et al* 2011, Hacke *et al* 2006). *Populus deltoides*, one of the two fastest growing trees located in our literature review, have xylem conduits 30–50 μ m in diameter, near the upper limit of common xylem diameters. These grow fastest in bottomland forests, along rivers bank and lakes, on well-drained soils characterized by sandy or loamy texture (http://plants.usda.gov). As noted by Freeze and Cherry (1979), the geometric mean hydraulic conductivity of silty sand is 10 μm s^{-1}, one order of magnitude larger than geometric mean of all soils. Thus, the large xylem of *P. deltoides* can exploit soils with larger pore diameters and higher water contents. The correlation of efficient xylem and height can be explained using the lumped plant hydraulic model introduced earlier in this book. Assuming that leaf water potential is maintained stable, and neglecting catastrophic cavitation, the rate of transpiration per unit sapwood area is shown to scale as $E/A_S \sim K_S/h$ [equation (8.10)]. Thus, to maintain sapwood specific transpiration at increasing heights, the xylem conductivity must increase proportionally, as confirmed empirically (Gleason *et al* 2012, Fan *et al* 2012). Petit *et al* (2010) measured the vertical profiles of the vessel dimensions and density of *E. regnans* trees of varying heights and found that xylem vessels in *E. regnans* exhibit strong tapering to compensate for the hydraulic limitations caused by increased tree height (see also chapter 8). They concluded that, relative to other fast-growing trees, *E. regnans* has evolved a xylem design that ensures a high hydraulic efficiency, with a mean hydraulic diameter exceeding 200 μm. An efficient xylem, however, is also prone to hydraulic failure (section 8.4), thereby posing limits to *E. regnans* height. This tradeoff might help explain why *E. regnans* is environmentally limited and its maximum height varies by almost a factor 20 over regions where the precipitation to pan evaporation ratio varies by essentially the same amount (Givnish *et al* 2014).

Differences in height within a species across sites can, therefore, be explained by soil moisture limitations. Continuing with the discussion of Ryan and Yoder (1997), 'Why do trees grow to different heights in different places? The mechanism that determines maximum tree height varies among sites. For example, in the front range of the Colorado Rocky Mountains a seed from a 25 m tall ponderosa pine (*Pinus ponderosa*) may fall into a rocky crevice and never grow more than 1–2 m. On the eastern slope of the Cascade Mountains in Oregon ponderosa pine soar to 50 m. 30 km farther east, in a drier climate, the same species struggles to attain 10 m.' Similarly, Givnish *et al* (2014) note a virtually linear relationship between rainfall and maximum tree height. These observations are indeed generally consistent with the scaling form of our function, view of the rapid increase in soil hydraulic conductivity with increasing moisture content, providing a smaller characteristic time scale, t_0. Studies that attribute growth variation to terrain effects may also reflect surface moisture variability, because moisture availability is higher in convergent terrain. McNab (1989) found that trees in concave terrain were typically about 15 m higher than in convex terrain. But such a variation is also consistent with the generally higher water contents found near the soil surface in concave, rather than convex, topographies, and the concomitant higher hydraulic conductivity.

We can also understand the distinction between tropical and temperate tree heights. If the same scaling function is used for all trees (as suggested by figure 9.8 to be relevant at least for dipterocarps, sequoias, and douglas firs, though *E. regnans* does seem to grow more rapidly), then, the tendency for the maximum tropical tree heights to be smaller than in temperate forests is likely a result of much higher tropical tree mortality (1.7% of canopy/yr; see Stephenson and van Mantgem 2005) compared with temperate rainforests (0.1%–0.4% of canopy/yr) (Givnish *et al* 2014). The former statistic predicts a typical age of about 60 years, whereas the latter predicts a typical age of 250–1000 years. In fact, Becker *et al* (2000) state, 'On Barro Colorado Island in Panama, for example, where average life expectancy of canopy trees is about 100 years, 60% of tree deaths were caused by snapped trunks, 17% by uprooting.' A tree following our predicted growth curve will reach about 50 m in height at about 60 years, a little over 70 m in 100 years (though Sequoiadendron heights are closer to 60 m), but would reach 125 m after 200 years, implying that maximum heights of trees in temperate forests are set by mechanisms other than mortality. Thus, one would expect that even the emergent trees in tropical forests would typically have heights between 50 m and 60 m, particularly since the higher mortality rates in tropical forests might encourage a co-limitation based on adaptation. It may be of additional importance that, at least among tropical species, rapid growth is typically associated with higher mortality (Wright *et al* 2010), a result that has no equivalent in the temperate forests.

For general confirmation of the lower tropical tree heights, consider the following results. Clark *et al* (2004) used remote sensing to measure the height of 59 'old-growth emergent' trees in a tropical rainforest, and found a maximum tree height of 56 m, while Lefsky *et al* (2005) found a maximum tree height of 55.7 m in their roughly 7000 m^2 study area in Santarem, Brazil, about half of which was old-growth forest. Kenzo *et al* (2006) also found that three of their five studied species had maximum emergent heights near 50 m: *Dryobalanops aromatic* (49.4 m), *Shorea beccariana* (52.5 m), and *Dipterocarpus globosus* (46 m), while the other species were smaller. Clark and Clark (2001) found maximum emergent tree heights in 5 species in La Selva Costa Rica to be as follows, *Lecythis ampla* (51 m), *Hymenolobium mesoamericanum* (58 m), *Dipteryx panamensis* (54 m), *Balizia elegans*, (55 m), and *Hyeronima alchorneoides* (50 m). Quoting Tng *et al* (2012), 'The tallest measured tropical angiosperm (*Shorea faguetiana; Dipterocarpaceae*) stands at 88.1 m (R Dial, personal communication) but among dipterocarps such heights are the exception, with emergent dipterocarps typically <60 m tall (Wyatt-Smith 1964, Cao and Zhang 1997, Whitmore 1998).' Therefore, in accord with our scaling function and, as mentioned also by Givnish *et al* (2014), an important cause for shorter tree heights in tropical forests is the much higher mortality rate. Such factors are not included in our scaling arguments, which are instead based on edaphic constraints. As a result, the proposed height-age curves are expected to overestimate height for a given age in most individuals, but the fastest-growing ones, and to predict continuous growth beyond the limits imposed by mortality (wind, drought stress, and disease) and xylem hydraulics.

As Hunt *et al* (2020) show, specific cases of the variation in tree growth rates with slope aspect, slope curvature, soil physical properties, climate, and xylem diameter can all be more parsimoniously explained by relating the flow rate, x_0/t_0, to the plant transpiration rate, which may be affected equally by any of the following: climatic water fluxes, convergence of subsurface flow paths in positively curved topography (with higher soil moisture content), greater direct evaporation due to southerly slope exposures—in the northern hemisphere, soil hydraulic conductivity, which may be reduced by soil compression, or hydrophobicity, or xylem hydraulic conductivity.

9.5.3 Growth of fractal root networks

Organisms that live in the soil, including vascular plants, do not exist in isolation from the soil. For example, water required for evapotranspiration is obtained from the soil. If the resistance of root tips to flow is greatly different from that of the medium, inefficiency is introduced, either in the function of the roots (if they are too thin), or in their construction (if they are too thick). In fact, the range of pore sizes that contain water that can be extracted by plants, is from about 0.2 μm to about 30 μm (Watt 2006). Larger pores tend to lose water on time scales too short—less one day—to be particularly useful, whereas pores smaller than 0.2 μm hold their water too tightly for vascular plants to extract. In an approximately ideal medium with, a typical porosity of 0.4 in which pore sizes are between the two limits, the application of the Rieu and Sposito model (see chapter 4) gives rise to a fractal pore space with a fractal dimensionality of 2.89, which is rather similar to values considered typical of soils, such as 2.84 (Bittelli *et al* 1999). Applying critical path analysis to such a medium generates a critical pore size under perfectly saturated conditions of 20 μm, but under conditions of 90% saturation, which is the highest common soil saturation value, of only 13 μm.

A size of 13 μm is very close to the 10 μm value, given as the geometric mean diameter of xylem vessels. Fungal hyphae tend to be smaller than 5 μm (Bååth and Soderstrom 1979). 10 μm is also the geometric mean of the usual range of xylem conduit diameters, 1–100 μm (Watt *et al* 2006; see figure 6.2), although root diameters are more nearly a factor 10 larger, closer to 100 μm. Since the hydraulic conductivity is determined by the smallest pore conductance on a connected path, the hydraulic conductivity of the optimal paths in the soil is nearly that of the xylem conduits. The saturated hydraulic conductivity of a medium with all pores having a 10 μm diameter will be about 1 μm s^{-1} (calculated by the standard Kozeny–Carman equation, valid for a homogeneous porous medium; see Bernabe and Bruderer 1998), consistent with near saturation vertical flow rates of about 1 μm s^{-1}. Thus, a porous medium consisting of pores useful for plants that are distributed according to a power law has a fractal dimensionality similar to the typical natural medium, a critical pore diameter for flow under near optimal conditions that is roughly equal to the geometric mean diameter of xylem in vascular plants, and a hydraulic conductivity that closely matches the typical value for porous media, as well as the value we chose to generate a typical scaling function.

Considering the fact that root tips often change their direction of growth, it is reasonable to hypothesize that this behavior is part of a search for an optimal channel for growth, which is not narrower than about 10 μm, and that the size chosen is optimal for soils—or that soils and plants have evolved together to generate such a correspondence. The increasing energy (with age) needed by the plant in its search for nutrients, and the slowing time scale for success, is postulated to be the control factor on the slowing growth rate, as indicated by equation (9.2). Finally, such a search for the optimal paths would allow root tips to continue to grow at approximately the same rate. However, the Euclidean distance from the root tip to the base of the plant increases more slowly on account of the tortuosity of the paths, consistent with that of the optimal paths in percolation theory, and the hierarchical architecture of a fractal root system.

9.5.4 Consequences for allometric scaling

The earlier works on allometric theory neglected the implications of resource limitations. Muller-Landau *et al* (2006) tested the predictions of allometric scaling in view of the possibility that light availability could limit the impact of the scaling hypotheses. Their results indicated that a number of the scaling predictions had to be modified to account for resource limitations. Although they interpret the resource limitations as being due to light reduction, the inference here is for nutrient transport constraints imposed by soil. In particular, light constraints diminish with age of the individual, though they increase with stand age, whereas the edaphic constraints become stronger with the age of either individual or with that of the stand. More importantly for our particular study here, however, is the specific result that they obtained for the exponent describing the scaling of the height, h, as a function of diameter, D. This exponent is predicted by allometric scaling to be $2/3 = 0.667$. But, if actual height is proportional to a 'potential' height to the power of $1/1.21 = 0.83$, as suggested by the edaphic controls, then, $h \sim (D^{0.67})^{0.83} = D^{0.55}$. The meta-study of Muller-Landau *et al* (2006) disclosed three values for the exponent, 0.46, 0.59, and 0.64. The mean of the three values is 0.567, much more nearly in accord with the tortuous length than with a linear increase in height with time. One possible hypothesis would then be that an overall (tortuous, rather than Euclidean) root length might scale exactly as diameter to the two thirds power. Note that of the ten sites considered, Barro Colorado Nature Monument in Panama alone included individual measurements on 9042 trees of 223 different species (table 9.6). Later, Feldpausch *et al* (2011) investigated the scaling exponent for h as a function of D in a meta-study of nearly 40 000 trees, and reported values, with a mean of 0.53, which are much lower than predicted by allometric theory; see table 7.1.

If one relies on the mean values of the exponents more than on the variability, the implication is that it may be sufficient to revise the current understanding of allometric scaling in line with the results of this chapter. It is not, however, necessarily the case that the variability is random, and that only the mean values

Table 9.6. Scaling of tree height with diameter in tropical trees (Feldpausch *et al* 2011).

Region	Exponent
Pan-tropical	0.53
Asia	0.57
Central Africa	0.57
East Africa	0.65
West Africa	0.64
Brazilian Shield Amazonia	0.46
East-Central Amazonia	0.49
Guyana Shield Amazonia	0.50
West Amazonia	0.47
Australia	0.52
Allometric theory (table 9.1)	0.67
Modified allometric theory	0.55

are relevant. On one hand, we found that, in the case of crops, where neither edaphic constraints, nor presumably light limitations, played a role, the growth rate should be linear, and we derived a reasonable confirmation for this. Thus, if no limitations are relevant, the exponent 0.67 may be appropriate. On the other hand, when root architectures are not confined to two dimensions, but three, the appropriate tortuosity exponent for optimal paths is 1.43, and one has the quotient, $0.67/1.43 = 0.47$. The three values calculated, 0.67, 0.55, and 0.47, span the observed variability. However, one could predict a similar range of values of the height-diameter exponent by simply considering the observed variability for trees of the height-age exponent in table 9.5, which vary from 0.9 to 0.71 (leading to the range 0.6–0.48), if one wishes to avoid a theoretical interpretation.

If one assumes that the variation in the exponent from region to region is meaningful, then one would reach the conclusion that nutrient limitations are minimal in East and West Africa, but in Central Africa and in Australia, two-dimensional constraints are most important, and in most of Amazonia, three-dimensional constraints operate. We know, however, of no evidence that suggests the relevance of much deeper, nutrient laden, soils in Amazonia. In any case, the discrepancy between an exponent of 0.55 and the predicted exponent of 0.67, a difference of 17%, is significant.

A natural question at this point would be, why have such large discrepancies not been seen in other exponents? To explain this, we recall that a simplified starting point for allometric scaling is that the tree mass M scales as the product of the square of the diameter D and the height h. M is proportional h^4, but only to $D^{8/3}$, which yields both $M \sim D^2 h$ and $h \sim D^{2/3}$ (see the previous chapters for a step-by-step derivation for a fractal tree architecture, leading essentially to the same result). A consequence of these results is that, while the height may be linear in the time, equation (8.27), the diameter must increase as the 3/2 power of time. The mass then increases as t^4, and $dM/dt \propto t^3$ is proportional as well to $M^{3/4}$. More generally, one can write,

$$\frac{dM}{dt} = D^2\frac{dh}{dt} + 2Dh\frac{dD}{dt}.$$

(9.5)

Equation (9.2) is consistent with,

$$\frac{dh}{dt} \sim t^{(1/D_{op})-1}.$$

(9.6)

The empirical correction inspired by the previous studies (Feldpausch *et al* 2011, Muller-Landau *et al* 2006), together with equation (9.2) suggest the change,

$$h \sim D^{2/(3D_{op})}.$$

(9.7)

Rearranging equation (9.6), differentiating with respect to t, and using equation (9.7) yields

$$\frac{dD}{dt} \sim h^{3D_{op}/2-1}\frac{dh}{dt} \sim t^{1/2} \rightarrow D \sim t^{3/2},$$

(9.8)

which is consistent with the allometric scaling exponent between tree diameter and height. Now, it is possible to use $M = D^2h$ to link mass and time, $M \sim t^{3+1/D_{op}}$, the time derivative of which yields,

$$\frac{dM}{dt} \sim M^{(2+1/D_{op})/(3+1/D_{op})} \approx M^{3/4}.$$

(9.9)

This expression yields exactly the metabolic equation, with a proportionality to $M^{3/4}$, in the case that $D_{op} = 1$. However, even with $D_{op} \simeq 1.21$, the exponent $3/4 = 0.75$ changes only to 0.74.

Summarizing this chapter, data for growth rates undoubtedly cross orders of magnitude. Plants have lifetimes that vary from weeks to 100 000 years or longer. But variability may hide simple laws governing maximum growth rates, and the wide range of time scales that can be investigated allows distinctions that are difficult at short time and length scales. The same scaling relationship links different scales and different individuals within and across species. It describes the slowing of growth rates with age in individuals, and within species for increasing stand age. In view of its relative universality, the scaling relationship helps understand the tendency for the largest trees to be in areas with lower mortality. It also helps to understand the partial correlation between fast-growing and large trees. It is at least generally compatible with the tendency for large trees to be found in certain, limited, geographic regions, where edaphic constraints may be spatially correlated.

Thus, the implications of what was presented in this chapter for forest growth are as follows. We focus on the lifespan of a single tree, in view of the conclusions of Binkley *et al* (2002): 'We conclude that part of the universal age-related decline in forest growth derives from competition-related changes in stand structure and the resource-use efficiencies of individual trees.' In an individual tree's lifetime, there may be an initial period of slower above-ground growth during which root allocation dominates, but this period is not our focus. Over a span of 20 to 200 years, trees tend to follow the scaling law we propose, and the length of time over which this law is followed depends primarily on the constraints to water transport in the xylem, genetics, and local conditions. The proposed power law is a result of soil constraints on nutrient delivery, not availability. Once the height at maturity is reached, as determined by resource availability and cost-benefit optimization, the height growth ceases, though other measures may continue to increase, such as diameter. Many other factors can influence tree growth sufficiently to produce deviations from the power law, seasonally or longer, such as carbon allocation to reproduction. But the edaphic scaling constraint provides an excellent guide to tree growth interpretation and complements traditional explanations. Mathematically, averages over power laws tend to generate power laws as well and, therefore, some of the characteristics of individual trees will carry over to stands, if the stand is aging. On the other hand, power-law growth characteristics will also tend to be reflected in power-law mortality rates, if the stand is at steady state.

References

Alexandrovskiy A L 2007 Rates of soil-forming processes in three main models of pedogenesis *Rev. Mex. Cienc. Geol.* **24** 1007

Almond P, Roering J and Hales T C 2007 Using soil residence time to delineate spatial and temporal patterns of transient landscape response *J. Geophys. Res.* **112** F03S17

Ando T and Takeuchi I 1973 Growth and production structure of *Acacia mollissima Willd Bull. Gov. For. Prod. Res. Inst.* **252** 149

Anfodillo T, Carraro V, Carrer M, Fior C and Rossi S 2006 Convergent tapering of xylem conduits in different woody species *New Phytol.* **169** 279

Arnaud-Haond S C, Duarte M, Diaz Almela E, Marba N, Sintes T and Serrao E A 2012 Implications of extreme lifespan in clonal organisms: millenary clones in meadows of the threatened seagrass *Posidonia oceanica PLoS One* **7** e30454

Bååth E and Soderstrom B 1979 The significance of hyphal diameter in calculation of fungalbiovolume *Oikos* **3** 11–14

Barlow P W 1993 The response of roots and root systems to their environment–an interpretation derived from an analysis of the hierarchical organization of plant life *Environ. Exp. Bot.* **33** 1

Barnard H R and Ryan M G 2003 A test of the hydraulic limitation hypothesis in fast-growing *Eucalyptus saligna Plant Cell Environ* **26** 1235

Battaglia M, M Cherry M, Badle C, Sands P J and Hingston A 1998 Prediction of leaf area index in eucalypt plantations: effects of water stress and temperature *Tree Physiol.* **18** 521

Becker P, Meinzer F C and Wullschleger S D 2000 Hydraulic limitation of tree height: a critique *Funct. Ecol.* **14** 4

Bernabé Y and Bruderer C 1998 Effect of the variance of pore size distribution on the transport properties of heterogeneous networks *J. Geophys. Res.* **103** 513

Berner R A 1992 Weathering, plants, and the long-term carbon-cycle *Geochim. Cosmochim. Acta* **56** 3225

Berntson G M, Lynch J P and Snapp S 1995 Fractal geometry and plant root systems: current perspectives and future applications *Fractals in Soil Science* ed P Baveye, J-Y Parlange and B A Stewart (Lewis Publishers) p 3225

Berrill J-P and O'Hara K L 2014 Estimating site productivity in the basal area or volume increment of the dominant species *Can. J. For. Resour.* **44** 92

Biging G S 1985 Improved estimates of site index curves using a varying-parameter model *For. Sci.* **31** 246

Binkley D, Dunkin K A, DeBell D and Ryan M G 1992 Production and nutrient cycling in mixed plantations of *Eucalyptus* and *Albizia* in Hawaii *For. Sci.* **38** 393

Binkley D, Stape J L, Ryan M G, Barnard H R and Frownes J 2002 Age-related decline in forestecosystem growth: an individual-tree, stand-structure hypothesis *Ecosytems* **5** 58–67

Bittelli M, Campbell G S and Flury M 1999 Characterization of particle-size distribution in soils with a fragmentation model *Sci. Soc. Am. J.* **63** 782

Blöschl G and Sivapalan M 1995 Scale issues in hydrological modelling: a review *Hydrol. Process.* **9** 251

Bockheim J G and Tarnocai C 1998 Nature, occurrence and origin of dry permafrost *PERMAFROST—Proc. of 7th Int. Conf. (Yellowknife, California) Collection Nordicana* **55** 57

Boisvenue C and Running S W 2006 Impacts of climate change on natural forest productivity-evidence since the middle of the 20th century *Glob. Chang. Biol.* **12** 862

Bond B J, Czarnomski N M, Cooper C, Day M E and Greenwood M S 2007 Developmental decline in height growth in Douglas-fir *Tree Physiol.* **27** 441

Borman P T, Spaltensein H, McClellan M H, Ugolini F C, Cromack K and Nay S M 1995 Rapid soil development after windthrow disturbance in pristine forests *J. Ecol.* **83** 747

Buchmann N, Hinckley T M and Ehleringer J R 1998 Carbon isotope dynamics inAbies amabilis stands in the Cascades *Can. J. For. Res.* **28** 808

Bungard R A, Zipperlen S A, Press M C and Scholes J D 2002 The influence of nutrients on growth and photosynthesis of seedlings of two rainforest dipterocarp species *Funct. Plant Biol.* **29** 505

Burke B C, Heimsath A M and White A F 2007 Coupling chemical weathering with soil production across soil-mantled landscapes *Earth Surf. Process. Landf.* **32** 853–73

Burns R M and Honkala B H 1990 *Silvics of North America Volume: 1. Conifers. Agriculture Handbook 654* (Washington, DC: US Department of Agriculture) p 675

Cao M and Zhang J 1997 Tree species diversity of tropical forest vegetation in Xishuangbanna, SW China *Biodivers. Conserv.* **6** 995

Cernohouz J and Solc I 1966 Use of sandstone wanes and weathered basaltic crust in absolute chronology *Nature* **212** 806

Chatfield J A, Chatfield A N and Cochran K D 2006 Dawn redwoods and Secrest Arboretum *Ornamental Plants; Annual Reports and Research Reviews 2005 Special Circular 197* ed J A Chatfield, E A Draper, D E Dyke, P J Bennett and J F Boggs (Ohio Agricultural Research and Development Center) p 166

Chowdhury M K R 1986 Concepts on the origin of Indian laterites in historical perspective *Proc. Indian Natl Sci. Acad.* **52** 1307

Chowdhury M K R, Venkatesh V, Anandalwar M A and Paul D K 1965 Recent concepts on the origin of Indian laterite *Geol. Surv. India* **31** 547

Clark D A and Clark D B 2001 Getting to the canopy: tree height growth in a neotropical rain forest *Ecology* **82** 1460

Clark M L, Clark D B and Roberts D A 2004 Small-footprint lidar estimation of sub-canopy elevation and tree height in a tropical rain forest landscape *Remote Sens. Environ.* **91** 68

Colman S M 1981 Rock-weathering rates as functions of time *Quat. Res.* **15** 250

Colman S M 1982a *Chemical weathering of basalts and basaltic andesites: evidence from weathering rinds* (U.S. Geological Survey 1246)

Colman S M 1982b Clay mineralogy of weathering rinds and possible implications concerning the sources of clay minerals in soil *Geology* **10** 370

Colman S M and Pierce K L 1981 Weathering rinds in andesitic and basaltic stones as a quaternary age indicator, western United States *U.S. Geological Survey Professional Paper 1210* (Washington, DC: U.S. Geological Survey) p 1

Colman S M and Pierce K L 1986 Glacial sequence near McCall, Idaho: weathering rinds, soil development, morphology, and other relative-age criteria *Quat. Res.* **25** 25–42

Cordoba-Pedegrosa MdC, Gonzales-Reyes J A, del Sgrarior Canadillas M, Navas P and Cordoba F 1996 Role of apoplastic and cell-was peroxidases on the stimulation of root elongation by ascorbate *Plant Physiol.* **112** 1119

Cremer K W 1975 Temperature and other climatic influences on shoot development and growth of *Eucalyptus regnans Aust. J. Bot.* **23** 27

Davies S L 2001 Tree mortality and growth in 11 sympatric macaranga species in Borneo *Ecology* **82** 920

Deal R L 1997 Understory plant diversity in riparian alder-conifer stands after logging in southeast Alaska *Pacific Northwest Research Station, Res. Note PNW-RN-523* (Portland, OR: US Dept. of Agriculture, Forest Service, Pacific Northwest Research Station)

DeBell D S, Whitesell C D and Crabbe T B 1987 *Benefits of* Eucalyptus–Albizia *mixtures vary by site on Hawaii island, U.S.D.A. Forest Service Research Paper PSW-187* (Berkeley, California: Pacific Southwest Forest and Range Experiment Station)

Delzon S M, Sartore R, Burlett R, Dewar R and Loustal D 2004 Hydraulic responses to height growth in maritime pine trees *Plant Cell Environ.* **27** 1077

Dethier D P 1988 The soil chronosequence along the Cowlitz River *Washington. U.S. Geological Survey* Report 1590F U.S. Geological Survey

Dietrich W E and Perron J T 2006 The search for a topographic signature of life *Nature* **439** 411

Dingman S L 2002 *Physical Hydrology* (Englewood Cliffs, NJ: Prentice-Hall)

Dixon J L, Heimsath A M and Amundson R 2009 The critical role of climate and saprolite weathering in landscape evolution *Earth Surf. Process. Landforms* **34** 1507

Domec J-C and Gartner B L 2003 Relationship between growth rates and xylem hydraulic characteristics in young, mature and old-growth ponderosa pine trees *Plant Cell Environ.* **26** 471

Eis S 1974 Root system morphology of western hemlock, western red cedan and Doublas fir *Can. J. For. Res.* **4** 28

Epron D, Laclau J P, Almeida J C R, Goncalves J L M, Ponton S, Sette C R, Delgado-Rojas J S, Bouillet J and Nouvellon Y 2012 Do changes in carbon allocation account for the growth response to potassium and sodium applications in tropical Eucalyptus plantations? *Tree Physiol.* **32** 667

Evans D M and Hartemink A E 2014 Terra Rossa catenas in Wisconsin, USA *Catena* **123** 148

Evans M E 1976 A new sensitive root auxanometer *Plant Physiol.* **58** 599

Falster D S *et al* 2015 BAAD, a biomass and allometry database for woody plants *Ecol. Arch.* **128** HO96–128

Fan Z-X, Hao G-Y, Slik J W F and Cao K-F 2012 Hydraulic conductivity traits predict growth rates and adult stature of 40 Asian tropical tree species better than wood density *J. Ecol.* **100** 732

Feldpausch T R L *et al* 2011 Height-diameter allometry of tropical forest trees *Biogeosciences* **8** 1081

Fitter A H and Stickland T R 1992 Fractal characterization of root system architecture *Funct. Ecol.* **6** 632

Fontes L, Tomé M, Coelho M B, Wright H, Luis J S and Savill P 2003 Modelling dominant height growth of Douglas-fir (*Pseudotsuga menziesii Mirb. Franco*) in Portugal *Forestry* **76** 509

Fogel R 1985 Roots as primary producers in below-ground ecosystems *Ecological Interactions in Soil: Plants Microbes and Animals* ed A H Fitter, D Atkinson, D J Read and M B Usher (Oxford: Blackwell Scientific Publications) 23 special publication No. 4, the British Ecological Society

Franklin S E, Hall R J and Smith L 2003 Discrimination of conifer height, age and crown closure classes using Landsat-5 TM imagery in the Canadian Northwest Territories, I *Int. J. Remote Sens.* **24** 1823

Franklin O, Johansson J, Dewar R C, Dieckmann U, McMurtrie R E, Brannstrom A and Dybzinski R 2012 Modeling carbon allocation in trees: a search for principles *Tree Physiol.* **32** 648

Freeze R A and Cherry J A 1979 *Groundwater* (Englewood Cliffs, NJ: Prentice-Hall)

Frouz J, Prachb K, Pizl V, Ha'ne'l L, Stary J, Tajovsky K, Maternad J, Balka V, Kalcka J and Rehounkova K 2008 Interactions between soil development, vegetation and soil fauna during spontaneous succession in post mining sites *Eur. J. Soil Biol.* **44** 109

Gardner D E 1957 Laterite in Australia, Record 1957/067 (Bureau of Mineral Resources, Geology and Geophysics, Canberra)

Gasser D P 1992 Young growth management of giant sequoia *Proc. of the Symp. on Giant Sequoias: Their Place in the Ecosystem and Society* p 120 (US Department of Agriculture, General Technical Report PSW-GTR-151, 1992)

Ghosh S and Guchhait S K 2015 Characterization and evolution of primary and secondary laterites in northwestern Bengal Basin, West Bengal, India *J. Palaeogeogr.* **4** 203

Givnish T J, Wong C and Stuart-Williams H 2014 Determinants of maximum tree height in *Eucalyptus* species along a rainfall gradient in Victoria, Australia *Ecology* **95** 2991

Gleason S M, Butler D W, Zieminska K, Warysak P and Westoby M 2012 Stem xylem conductivity is key to plant water status across Australian angiosperm species *Funct. Ecol.* **26** 343

Gonzales J G A, Gonzales A D R, Soalleiro R R and Anta M B 2005 Ecoregional site index models for *Pinus pinaster* in Galicia (northwestern Spain) *Ann. For. Sci.* **62** 115

Goodman A Y and Rodbell D T 2001 G.O. Seltzer, and B.G. Mark, Subdivision of glacial deposits in southeastern Peru based on pedogenic development and radiometric ages *Quat. Res.* **56** 31

Grant M C 1993 The trembling giant *Discover* **14** 82

Guswa A J 2010 Effect of plant uptake strategy on the water-optimal root depth *Water Resour. Res.* **46** W09601

Hacke U G, Sperry J S, Pockman W T, Davis S D and McCulloch K A 2001 Trends in wood density and structure are linked to prevention of xylem implosion by negative pressure *Oecologia* **126** 457

Hacke U G, Sperry J S, Wheeler J K and Castro L 2006 Scaling of angiosperm xylem structure with safety and efficiency *Tree Physiol.* **26** 689

Heald R C 1986   Management of giant sequoia at Blodgett Forest Research Station, Pacific Southwest Forest and Range Experiment Station, Forest Service (U.S. Department of Agriculture; Gen. Tech. Rep. PSW-95)

Heimsath A M, Dietrich W E, Nishiizumi K and Finkel R C 1997 The soil production function and landscape equilibrium *Nature* **388** 358

Heimsath A M, Dietrich W E, Nishiizumi K and Finkel R C 1999 Cosmogenic nuclides, topography, and the spatial variation of soil depth *Geomorphology* **27** 151

Heimsath A M, Chappell J, Dietrich W E, Nishiizumi K and Finkel R C 2001a Late Quaternary erosion in southeastern Australia: a field example using cosmogenic nuclides *Quat. Int.* **83–85** 169–85

Heimsath A M, Dietrich W E, Nishiizumi K and Finkel R C 2001b Stochastic processes of soil production and transport: erosion rates, topographic variation and cosmogenic nuclides in the Oregon coast range *Earth Surf. Process. Landforms* **26** 531

Heimsath A M, Furbish D J and Dietrich W E 2005 The illusion of diffusion: field evidence for depth dependent sediment transport *Geology* **33** 949

Heimsath A M, Fink D and Hancock G R 2009 The 'humped' soil production function: eroding Arnhem Land, Australia *Earth Surf. Process. Landforms* **34** 1674

Hunt A G 2015 Predicting rates of weathering rind formation *Vadose Zone J.* **14** 1–13

Hunt A G 2017 Spatio-temporal scaling of soil depth and vegetation growth: explicit predictions *Vadose Zone J.* **16** 1

Hunt A G, Faybishenko B A and Powell T L 2020 A new phenomenological model to describe root-soil interactions based on percolation theory *Ecol. Model.* **433** 109205

Hunt A G, Faybishenko B and Powell T L 2022 Test of model of equivalence of tree height growth and transpiration rates in percolation-based phenomenology for root soil interaction *Ecol. Model.* **465** 109853

Hunt A G, Ghanbarian-Alavijeh B, Skinner T E and Ewing R P 2015 Scaling of geochemical reaction rates via advective solute transport *Chaos* **25** 075403

Jackson R B, Canadell J, Ehleringer J R, Mooney H A, Sala O E and Schulze E D 1996 A global analysis of root distributions for terrestrial biomes *Oecologia* **108** 389

Jacobson A D, Blum J D, Chamberlain C P, Poage M A and Sloan V F 2002 Ca/Sr and Sr isotope systematics of a Himalayan glacial chronosequence: Carbonate versus silicate weathering rates as a function of landscape surface age *Geochim. Cosmochim. Acta* **66** 13

Jenny H 1941 *Factors of Soil Formation: A System of Quantitative Pedology* (New York: Dover)

Johnson R L and Burkhardt E 1976 Natural cottonwood stands—past management and implications for plantations *Proc. of Symp. on Eastern Cottonwood and Related Species* B A Thielges and S B Land p 20

Kalliokoski T, Nygren P and Sievanen R 2008 Coarse root architecture of three boreal tree species growing in mixed stands *Silva Fenn.* **42** 189

Kempes C P, West G B, Crowell K and Girvan M 2011 Predicting maximum tree heights and other traits from allometric scaling and resource limitations *PLoS One* **6** e20551

Kenzo T, Ichie T, Watanabe Y, Yoneda R, Ninomiya I and Koikes T 2006 Changes in photosynthesis and leaf characteristics with tree height in five dipterocarp species in a tropical rain forest *Tree Physiol.* **26** 865

King D A 1990 The adaptive significance of tree height *Am. Nat.* **135** 809

Knuepfer P L K 1988 Estimating ages of late Quaternary stream terraces from analysis of weathering rinds and soils *Geol. Soc. Amer. Bull.* **100** 1224

Koch G W, Sillett S C, Jennings G M and Davis S D 2004 The limits to tree height *Nature* **428** 851

Kuser J E 1982 *Metasequoia* keeps on growing *Arnoldia* **42** 130

Kuser J E 1983 *Metasequoia glyptostroboides* in urban forestry *Metria* **4** 20

Kuser J E 1998 *Metasequoia glyptostroboides*: Fifty years of growth in North America *Arnoldia* **58/59** 76

Kuser J E, Baily A, Francelet A, Libby W J, Martin J, Rydelius J, Schoenike R and Vagle N 1995 Early results of a rangewide provenance test of *Sequoia sempervirens Forest Genetic Resources No. 23*

Lambers H, Chapin F S and Pons T L 1998 *Plant Physiological Ecology* (Berlin: Springer)

Lappi J and Bailey R L 1988 A height prediction model with random stand and tree parameters: an alternative to traditional site index methods *For. Sci.* **34** 907

Lee Y, Andrade J S, Buldyrev S V, Dokholoyan N V, Havlin S, King P R, Paul G and Stanley H E 1999 Traveling time and traveling length in critical percolation clusters *Phys. Rev. E* **60** 3425

Lefsky M A, Harding D J, Keller M, Cohen W B, Carabajal C C, Espirito-Santo F D B, Hunter M O and de Oliveira R Jr 2005 Estimates of forest canopy height and aboveground biomass using ICESat *Geophys. Res. Lett.* **32** L22S02

Leithead H L, Yarlett L L and Shiflet T N 1971 100 Native forage grasses in 11 southern states *Agricultural Handbook No. 389, Soil Conservation Service* (Washington, DC: US Department of Agriculture)

Leuzinger S, Manusch C, Bugmann H and Wolf A 2013 A sink-limited growth model improves biomass estimation along boreal and alpine tree lines *Glob. Ecol. Biogeog.* **22** 924–32

le Maire G, Marsden C, Verhoef W, Ponzoni F J, Lo Seen D, Bégué A, Stape J and Nouvellon Y 2011 Leaf area index estimation with MODIS reflectance time series and model inversion during full rotations of Eucalyptus plantations *Remote Sens. Environ.* **115** 586

Li X R, He M Z, Duan Z H, Xiao H L and Jia X H 2007 Recovery of topsoil physicochemical properties in revegetated sites in the sand-burial ecosystems of the Tengger Desert, northern China *Geomorphology* **88** 54

Long J L and Turner J 1975 Aboveground biomass of understorey and overstorey in an age sequence of four douglas-fir stands *J. Appl. Ecol.* **12** 179

Lynch J 1995 Root architecture and plant productivity *Plant Physiol.* **109** 7

Ma Z, Bielenberg D G, Brown K M and Lynch J P 2001 Regulation of root hair density by phosphorus availability in *Arabidopsis thaliana Plant Cell Environ.* **24** 459

Ma L, Chabaux F, Pelt E, Granet M, Sak P B, Gaillardet J, Lebedeva M and Brantley S L 2012 Theeffect of curvature on weathering rind formation: Evidence from Uranium-series isotopes inbasaltic andesite weathering clasts in Guadeloupe, 3987 *Geochim. Cosmochimica Acta* **80** 92–107

Mäkela A and Vanninen P 1998 Impacts of size and competition on tree form and distribution of aboveground biomass in Scots pine *Can. J. For. Res.* **28** 216

Magnani F, Mencuccini M and Grace J 2000 Age-related decline in stand productivity: the role of structural acclimation under hydraulic constraints *Plant Cell Environ.* **23** 251

Maguire D A, Brissette J C and Gu L H 1998 Crown structure and growth efficiency of red spruce stands in Maine *Can. J. For. Res.* **28** 1233

Mandelbrot B B 1982 *The Fractal Geometry of Nature* (San Francisco, CA: W.H. Freeman)

Martin-Benito D, Gea-Izquierdo G, del Rio M and Canellas I 2008 Long-term trends in dominant-height growth of black pine using dynamic models *For. Ecol. Manage.* **256** 1230

Martínez-Vilalta J, Vanderklein D and Mencuccini M 2006 Tree height and age-related decline in growth in Scots pine (*Pinus sylvestris* L.) *Oecologia* **150** 529–44

Mavris C, Egli M, Pltze M, Blum J D, Mirabella A, Giacci D and Haeberli W 2010 Initial stages of weathering and soil formation in the Morteratsch proglacial area (Upper Engadine Switzerland) *Geoderma* **155** 359

McDonald J E 1992 The sequoia forest plan settlement agreement as it affects Sequoiadendron giganteum: a giant step in the right direrction *Proc. of the Symp. on Giant Sequoias: Their Place in the Ecosystem and Society* (U.S. Department of Agriculture, General Technical Report PSW-GTR-151, 1992) p 126

McFadden L D and Weldon R J 1987 Rates and processes of soil development on Quaternary terraces in Cajon Pass, California *Geol. Soc. Am. Bull.* **98** 280

McNab W H 1989 Terrain shape index: Quantifying effects of minor landforms on tree height *For. Sci.* **35** 91

Mencuccini M and Grace J 1996 Hydraulic conductance, light interception and needle nutrient concentration in Scots pine stands and their relations with net primary productivity *Tree Physiol.* **16** 459

Moles A T, Warton D I, Warman L, Swenson N G, Laffan S W, Zanne A E, Pitman A, Hemmings F A and Leishman M R 2009 Global patterns in plant height *J. Ecol.* **97** 923

Moore J G 1966 Rate of palagonitization of submarine basalt adjacent to Hawaii, U.S. Geol. Survey Professional Paper 550, 163

Muller-Landau H C *et al* 2006 Testing metabolic ecology theory for allometric scaling of tree size, growth and mortality in tropical forests *Ecol. Lett.* **9** 575

Muzuroglu O and Geckil H 2002 Effects of metals on seed germination, root elongation, and coleoptile and hypocotyl growth in *Triticum aestivum* and *Cucumis sativus Arch. Environ. Contam. Toxicol.* **43** 203

National Academy of Sciences 1979 *Tropical legumes: Resources for the future* (Washington, DC: National Academy Press)

National Academy of Sciences 1984 *Leucaena: Promising Forage and Tree Crop for the Tropics* (Washington, DC: National Academy Press)

Navarre-Sitchler A, Steefel C, Hausrath E and Brantley S 2007 Influence of porosity on basalt weathering rates from the clast to watershed scale *Geochim. Cosmochim. Acta* **71** A707

Navarre-Sitchler A, Steefel C J, Sak P B and Brantley S L 2011 A reactive-transport model for weathering rind formation on basalt *Geochim. Cosmochim. Acta* **75** 7644

Nicolle D 2011 Tallest trees in Europe found in Portugal and Spain; https://git-forestry-blog. blogspot.com/2011/05/tallest-trees-in-europe-found-in.html (accessed 18 August 2015)

Niklas K J 2007 Maximum plant height and the biophysical factors that limit it *Tree Physiol.* **27** 433

Niklas K J and Enquist B J 2002 On the vegetative biomass partitioning of seed plant leaves, stems, and roots *Am. Nat.* **159** 482

Nouvellon Y *et al* 2010 Within-stand and seasonal variations of specific leaf area in a clonal Eucalyptus plantation in the Republic of Congo *For. Ecol. Manage.* **259** 1796

O'Brien S T, Hubbell S P, Spiro P, Condit R and Foster R B 1995 Diameter, height, crown and age relationships in eight neo-tropical hardwood species *Ecology* **76** 1926

Oguchi C T and Y Matsukura Y 1999 Effect of porosity on the increase in weathering-rind thicknesses of basaltic andesite gravel *Eng. Geol.* **55** 77

Pelt E, Chabaux F, Innocent C, Navarre-Sitchler A K, Sak P B and Brantley S L 2008 Uranium-thorium chronometry of weathering rinds: Rock alteration rate and paleo-isotopic record of weathering fluids *Earth Plan. Sci. Lett.* **276** 98

Petit G, Pfautsch S, Anfodillo T and Adams M A 2010 The challenge of tree height in *Eucalyptus regnans*: when xylem tapering overcomes hydraulic resistance *New Phytol.* **187** 1146

Phillips N B, Bond J, McDowell N G and Ryan M G 2002 Canopy and hydraulic conductance in young, mature and old Douglas-fir trees *Tree Physiol.* **22** 205

Phillips C J, Marden M and Suzanne L M 2014 Observations of root growth of young poplar and willow planting types *N.Z. J. For. Sci.* **44** 15

Phillips C J, Marden M and Suzanne L M 2015 Observations of 'coarse' root development in young trees of nine exotic species from a New Zealand plot trial *N.Z. J. For. Sci.* **45** 1

Pillans B 1997 Soil development at a snail's pace: evidence from a 6 Ma soil chronosequence on basalt in north Queensland, Australia *Geoderma* **80** 117

Rajaram H and Gelhar L W 1993 Plume scale-dependent dispersion in heterogeneous aquifers: 2. Eulerian analysis and three-dimensional aquifers *Water Resour. Res.* **29** 3261

Rastetter E B and Shaver G R 1992 A model of multiple-element limitation for acclimating vegetation *Ecology* **73** 1157

Rodriguez-Iturbe I and Porporato A 2004 *Ecohydrology of Water-Controlled Ecosystems. Soil Moisture and Plant Dynamics* (Cambridge: Cambridge University Press)

Roth B E, Jokela E J, Martin T A, Huber D A and White T L 2007 Genotype environment interactions in selected loblolly and slash pine plantations in the Southeastern United States *For. Ecol. Manage.* **238** 175

Ryan M and Yoder B J 1997 Hydraulic limits to tree height and growth: What keeps trees from growing beyond a certain height? *Bioscience* **47** 235

Saab I N, Sharp R E, Pritchard J and Voetberg G S 1990 Increased endogenous absciisic acid maintains primary root growth and inhibits shoot growth of maize seedlings at low water potentials *Plant Physiol.* **93** 1329

Sahimi M 1987 Hydrodynamic dispersion near the percolation threshold: scaling and probability densities *J. Phys.* A **20** L1293

Sahimi M 2012 Dispersion in porous media, continuous-time random walks, and percolation *Phys. Rev.* E **85** 016316

Sahimi M and Imdakm A O 1988 The effect of morphological disorder on hydrodynamic dispersion in flow through porous media *J. Phys.* A **21** 3833

Sak P B, Fisher D M, Gardner T W, Murphy K and Brantley S L 2004 Rates of weathering rind formation on Costa Rican basalt *Geochim. Cosmochim. Acta* **68** 1453

Sala A and Hoch G 2008 Height growth declines in ponderosa pine are not due to carbon limitation *Plant Cell Environ.* **32** 22

Salas E, Ozier-Lafontaine H and Nygren P 2004 A fractal root model applied for estimating the root biomass and architecture in two tropical legume tree species *Ann. For. Sci.* **61** 337

Schenk H J and Jackson R B 2002 Rooting depths, lateral root spreads and below-ground/above-ground allometries of plants in water-limited ecosystems *J. Ecol.* **90** 480

Schmid R and Schmid M 2012 *Sequoiadendron giganteium (Cupressageae) at Lake Fulmor, Riverside County, California, Aliso* vol **30** (Claremont, CA: Rancho Santa Ana Botanic Garden) p 103

Schreiber S G, Hacke U W, Hamann A and Thomas B R 2011 Genetic variation of hydraulic and wood anatomical traits in hybrid poplar and trembling aspen *New Phytol.* **190** 150

Schülli-Maurer I, Sauer D, Stahr K, Sperstaad R and Sorensen R 2007 Soil formation in marine sediments and beach deposits of southern Norway: investigations of soil chronosequences in the Oslofjord region *Rev. Mex. Cienc. Geol.* **24** 237

Seymour R S and Kenefic L S 2002 Influence of age on growth efficiency of Tsuga canadensis andPicea rubens trees in mixed-species multiaged northern conifer stands *Can. J. Forest Res.* **32** 2032–42

Sharp R E, Wu Y, Voetberg G S, Saab I N and LeNoble M D 1994 Confirmation that abscisic acid accumulation is required for maize primary root elongation at low water potentials *J. Exp. Botany* **45** 1743

Sheppard A P, Knackstedt M A, Pinczewski W V and Sahimi M 1999 Invasion percolation: new algorithms and universality classes *J. Phys.* A **32** L521

Smale M C, McLeod M and Smale P N 1997 Vegetation and soil recovery on shallow landslide scars in tertiary hill country, East Cape region, New Zealand *N. Z. J. Ecol.* **21** 31

Smith M L, Bruhn J N and Anderson J L 1992 The fungus Armillaria bulbosa is among the largest and oldest living organisms *Nature* **356** 428

Stephenson N L 1992 Long-term dynamics of giant sequoia populations: Implications for managing a pioneer species *Proc. of the Symp. on Giant Sequoias: Their Place in the Ecosystem and Society* (U.S. Department of Agriculture, General Technical Report PSW-GTR-151) p 56

Stephenson N L and van Mantgem P J 2005 Forest turnover rates follow global patterns of productivity *Ecol. Lett.* **8** 524

Stevens P R 1968 A chronosequence of soils near the Franz Josef Glacier *Dissertation* (Lincoln College: Canterbury, New Zealand)

Stoffberg G H, van Rooyen M W, van der Linde M J and Groeneveld H T 2008 Predicting the growth in tree height and crown size of three street tree species in the City of Tshwane, South Africa *Urban For. Urban Green.* **7** 259

Stohlgren T J 1992 Resilience of a heavily logged grove of giant sequoia (*Sequoiadendron giganteum*) in Kings Canyon National Park, California *For. Ecol. Manage.* **54** 115

Stone E L and Kalisz P J 1991 On the maximum extent of tree roots *For. Ecol. Manage.* **46** 59

Sugimoto K, Williamson R E and Wasteneys G O 2000 New techniques enable comparative analysis of microtubule orientation, wall texture, and growth rate in intact roots of arabidopsis *Plant Physiol.* **124** 1493

Suzuki S and Nakagoshi N 2011 Sustainable management of Satoyama bamboo landscapes *Landscape Ecology in Asian Cultures, Part 1* ed S K Hong, J Wu, J E Kim and N Nakagoshi (Berlin: Springer)

Tatsumi J, Yamauch A and Kono Y 1989 Fractal analysis of plant root systems *Ann. Bot.* **64** 499

Tilly N, Hoffmeister D, Cao Q, Huang S, Lenz-Wiedeann V, Miao Y and Bareth G 2014 Multitemporal crop surface models: accurate plant height measurement and biomass estimation with terrestrial laser scanning in paddy rice *J. Appl. Remote Sens.* **8** 083671

Tng D Y P, Williamson G J, Jordan G J and Bowman D M J S 2012 Giant *eucalypts*—globally unique fire-adapted rain-forest trees? *New Phytol.* **196** 1001

Trustrum N A and de Rose R C 1988 Soil depth-age relationships of landslides on deforested hillsides, Taranaki, New Zealand *Geomorphology* **1** 143

VandenBygaart A J and Protz R 1995 Soil genesis on a chronosequence, Pinery Provincial Park, Ontario *Can. J. Soil Sci.* **75** 63

Van Der Meer P J, Dignan P and Saveneh A G 1999 Effect of gap size on seedling establishment, growth and survival at three years in mountain ash (*Eucalyptus regnans* F. Muell) forest in Victoria, Australia *For. Ecol. Manage.* **117** 33

Van Der Meer P J and Dignan P 2007 Regeneration after 8 years in artificial canopy gaps in Mountain Ash (*Eucalyptus regnans* F. Muell.) forest in south-eastern Australia *For. Ecol. Manage.* **244** 102

Vitousek P M and Howarth R W 1991 Nitrogen Limitation on land and in the sea–how can it occur *Biogeochemistry* **13** 87

Walsh P G, Barton C V M and Haywood A 2008 Growth and carbon sequestration rates at age ten years of some eucalypt species in the low to medium-rainfall areas of New South Wales, Australia *Aust. For.* **71** 70

Walter A, Spies H, Terjung S, Küsters R, Kirchgeßner N and Schurr U 2002 Spatio-temporal dynamics of expansion growth in roots: aturomatic quantification of diurnal course and temperature response by digital image sequence processing *J. Exp. Bot.* **53** 689

Waring R H and Franklin J F 1979 Evergreen forests of the Pacific Northwest *Science* **204** 1380

Waring K M and O'Hara K L 2007 Stand dynamics of coast redwook/tanoak Forests following tanoak decline *USDA Forest Service Gen. Tech. Rep. PSW-GTR-194* **475**

Watt M, Silk W K and Passioura J B 2006 Rates of root and organism growth, soil conditions and temporal and spatial development of the rhizosphere *Ann. Bot.* **97** 839

Weatherspoon C P 1990 *Sequoiadendron giganteum Lindl. (Buchholz)* (Washington, DC: U.S. Department of Agriculture) 675 Giant sequoia, in, R.M. Burns and B.H. Honkala (tech. cords.), *Silvics of North America Volume: 1. Conifers. Agriculture Handbook 654*

White A F and Brantley S L 2003 The effect of time on the weathering rates of silicate minerals: why do weathering rates differ in the lab and in the field? *Chem. Geol.* **202** 479

White A G, Blum A E, Schulz M S, Bullen T D, Harden J W and Peterson M L 1996 Chemical weathering rates of a soil chronosequence on granitic alluvium: I. Quantification of mineralogical and surface area changes and calculation of primary silicate reaction rates *Geochim. Cosmochim. Acta* **60** 2533

Whitmore T C 1998 *An Introduction to Tropical Rain Forests* 2nd edn (Oxford: Oxford University Press)

Whitmore T C and Brown N D 1996 Dipterocarp seedling growth in rain forest canopy gaps during six and a half years *Phil. Trans. Biol. Sci.* **351** 1195

Winter L E, Brubaker L B, Franklin J F, Miller E A and DeWitt D Q 2002 Initiation of an old-growth Douglas fir stand in the Pacific Northwest: a reconstruction from tree-ring records *Can. J. For. Res.* **32** 1039

Wood S W, Allen K J, Hua Q and Bowman D M J S 2010 Age and growth of a fire prone Tasmanian temperate old-growth forest stand dominated by *Eucalyptus* regnans, the world's tallest angiosperm *For. Ecol. Manage.* **260** 438

Worbes M, Staschel R, Roloff A and Junk W J 2003 Tree ring analysis reveals age structure, dynamics and wood production of a natural forest stand in Cameroon *Forest Ecol. Manage.* **173** 105-29

Wright S J, Kitajima K, Kraft N B, Recih P B, Zanne E *et al* 2010 Functional traits and the growth-mortality trade-off in tropical trees *Ecology* **91** 3664

Wu J, Kurten E L, Monshausen G, Hummel G M, Gilroy S and Baldwin I T 2007 NaRALF, a peptide signal essential for the regulation of root hair tip apoplastic pH in Nicotiana attenuata, is required for root hair development and plant growth in native soils *Plant J.* **52** 877

Wyatt-Smith J 1964 A preliminary vegetation map of Malaya with descriptions of the vegetation types *J. Trop. Geogr.* **18** 200

York R A 2006 Release potential of giant sequoia following heavy suppression: 20-year results *For. Ecol. Manage.* **234** 136

York R A, Battles J J and Heald R C 2002 Edge effects in mixed conifer group selection openings: tree height response to resource gradients *For. Ecol. Manage.* **116** 1

York R A, O'Hara K L and Battles J J 2013 Density effects on giant sequoia (*Sequoiadendron giganteum*) growth through 22 years: Implications for restoration and plantation management *West. J. Appl. For.* **28** 30

IOP Publishing

Networks on Networks (Second Edition)
Role of connectivity in physics of geobiology and geochemistry
Allen G Hunt and Muhammad Sahimi

Chapter 10

Geomorphological applications of percolation theory: river networks, and weathering and soil depths

10.1 Background

Study of the origin and evolution of topographic and bathymetric features that are generated by physical, chemical, or biological processes that operate at or near Earth's surface constitutes what is referred to as geomorphology. Why landscapes look the way they do, how to understand land forming and terrain history, as well as predicting the dynamics of the changes that occur near Earth's surface are questions that are addressed by geomorphology. Of particular interest is Earth's critical zone (CZ), which is (National Research Council 2001) the 'heterogeneous, near surface environment in which complex interactions involving rock, soil, water, air, and living organisms regulate the natural habitat and determine the availability of life-sustaining resources.' All terrestrial life is supported by the CZ. As Parsekian *et al* (2014) put it, the 'CZ is Earth's breathing skin: a life-supporting epidermis that reaches from the top of vegetation down through soil, weathered rock, and fractured bedrock.' It is through the CZ that the intact bedrock is transformed through physical, geochemical, hydrological, and biological processes into regolith, soil, sediments and, ultimately, solutes, processes that occur from the top of the subaerial biological zone to the base of the groundwater zone. The complex interactions among such processes, the rate at which they take place, and their response to natural and anthropogenic changes have been studied for decades, with the research spanning several scientific disciplines. In the United States a group of CZ Observatories (CZOs) has been formed, accompanied by a parallel network of global sites, which brings to the attention of scientistists and the public alike the importance attributed by the global Earth science community to understanding CZ processes.

doi:10.1088/978-0-7503-5698-5ch10

Human societies depend on nutrient cycles, erosion and landscape evolution, and water supply and quality, and the associated issues and problems are addressed by research on the CZ. Some of such issues are (Parsekian *et al* 2014): (1) how fast does bedrock weather into regolith and soil? (2) What controls the rate of rock weathering? (3) Where is water stored in the CZ? (4) How does CZ structure affect the exchange of water between surface and subsurface reservoirs? (5) Does the deep CZ play any role in supporting ecosystems and, if so, what? (6) How thick is the regolith? (7) How do climate, mechanical failure, chemical weathering, and biologically mediated processes regulate the thickness of regolith? (8) Given the ongoing climate change, how is CZ structure and function responding to the change?

Vegetation, soil, and surface water constitute the upper part of the CZ, which is readily accessible (see, e.g., Johnson *et al* 2012), whereas the deeper part in which many processes occur is more difficult to reach. The CZ does not have the same thickness everywhere. Its thickness can vary anywhere from millimeters or centimeters (in bedrock outcrops) to many tens of meters (in heavily weathered terrains). Lateral variability of lithologic, sedimentary, and hydrologic units of the CZ, which is a common and important characteristic of the CZ, is another important feature. Although drilling and augering are used for measuring deep CZ properties and obtain insights into the processes that occur there, they have limitations. The latter is not expensive, but is limited to the depth of refusal, invasive, and a few narrow, widely spaced holes, while the former is not only expensive, but it can only be used in limited locations where it is possible to use the required heavy drilling equipment for installing deep wells. At the same time, however, the same region of the CZ is critical to processes that support ecosystems, society, and the environment. Parsekian *et al* (2014) reviewed the application of geophysical methods to mapping the geometry of structural features of the CZ, such as regolith thickness, lithological boundaries, permafrost extent, snow thickness, and shallow root zones.

This chapter aims to address, from a perspective of percolation theory, several issues associated with the morphology of the near surface Earth in the CZ, including river basin architecture and tectonics, weathering depths over immense time scales, and soil depth and landslides, as well as explain why so many soils are only 1 m deep.

10.2 River networks

We began the discussion of river networks in chapter 2. Many aspects of geology, geomorphology, and hydrology, such as flood magnitude-frequency relations, the water cycle, tectonic response to erosion and its inverse, and several others are influenced by the organization of river drainage systems. Drainage organization is also key to the type of sediments that rivers deliver to sedimentary basins, since the spatial extent of the drainage basin also controls the sediment transport sources. Important factors that limit the volume of sediments transported include the rate of chemical weathering of bedrock and its conversion to soil (Dixon and Heimsath 2009, DiBiase and Heimsath 2012, Egli *et al* 2018), and the close relationship between soil erosion and soil production rates, which are mostly water flux-limited through chemical weathering (Egli *et al* 2018), or, equivalently (Maher 2010,

Stolze *et al* 2023), by residence times. Flow rates depend, of course, on specific drainage architecture, as well as climatic variables. Therefore, predicting the two-dimensional (2D) extent of a drainage system, and the rates of weathering of the bedrock and transport of sediments are the foundation for predicting the total volume and kind of sediments that are ultimately delivered to sedimentary basins. Indeed, dates for a reorganization of drainage systems are often extracted by dating changes in the provenance of deltaic sediments (Matthews *et al* 2001, Said *et al* 2015, Fielding *et al* 2018). In general, water fluxes, surface, and subsurface are viewed as primary agents of the evolution of both lateral and vertical dimensions of drainage basins, which are simply land areas that slope toward a single outlet, such as a river mouth, or points of higher infiltration, or evaporation rates.

Water in a river basin drains through many smaller watersheds to a large river. Factors that contribute to forming watersheds include tectonics, climate, vegetation cover, topography, shape, size, soil type, and land use in urban, agriculture, and natural areas (Rhoads 2020). In addition, characteristics of organization of a drainage system are quantified in many ways (Horton 1932, 1945, Schumm 1956, Strahler 1964, Scheidegger 1965, Shreve 1967). Kirchner (1993) first recognized the statistical regularities in drainage form and architecture, after which a range of principles that govern the structural features of river drainages were proposed (Maritan *et al* 1996, Rigon *et al* 1996, Pelletier 1999, Bejan and Errera 2011). The specific drainage reorganization processes that have been identified include head-ward erosion through spring sapping (Laity and Malin 1986, Baker *et al* 1990), capture (Young and Spamer 2001), basin fill and sill overtopping (Meek 1989, 1990, Hilgendorf *et al* 2020), or combinations of such mechanisms, the basis for which lies in combinations of erosion processes. Those that are driven by water can either be mostly chemical or physical in basis, surface or subsurface in location. In particular, many studies (Willgoose *et al* 1991, Tucker and Bras 2000, Gunnell and Harbor 2010, Willett *et al* 2014) focused on tectonic and climatic drivers of basin (re) organization.

While tectonic influence is assumed to be mainly disaggregating or breaking up, the convergence of groundwater flow fields and the flow rate is viewed as the primary factor in the organization. To understand the contrasting roles of the two, we only need to recall that development of surface structures, such as mountains and faults, disrupt surface and subsurface flow fields, whereas subsurface flow fields reorganize in a variety of ways in response to changing regional subsurface gradients, i.e., chemical dissolution and erosion that are linked to the surface hydrological processes and organization (Petroff *et al* 2013) through surface lowering, thereby influencing surface and subsurface convergence upstream, or through subsurface flow convergence and spring sapping (Laity and Malin 1986, Baker *et al* 1990), or by surface convergence further downgradient, the latter two of which both promote headward erosion. In the model that is described here (Hunt *et al* 2023), the influence of climate on drainage basin reorganization is expressed through groundwater flow rates, which are known to be proportional to the difference between precipitation and evapotranspiration (Maxwell *et al* 2016).

The interaction between the river and groundwater occurs primarily in two ways (Xiangjiang and Niemann 2006): streams can gain or lose water from inflow or outflow of groundwater through the streambed. When gaining water, the water table slopes down toward the stream, whereas when losing water, the water table slopes away from the stream. Therefore, in arid regions, erosional processes are favored for gaining reaches or streams, and depositional for losing reaches or streams (Grant 1948). Thus, losing reaches promote basin fill. In particular, depending on the groundwater flow rates, in arid zones where the surface run-off can be strongly elevation-dependent, both basin fill and sill erosion rates may be enhanced simultaneously. Even when relief is insufficient to produce a climatic signal in groundwater flow rates, in the opposite extreme of river impoundment by dams, failure, and, thus, reintegration, can occur at least as often by piping as by overtopping (Zhang *et al* 2016).

It is for these reasons that in the drainage (re)organization model that is descrbed here, it is assumed the process is controlled by the assembly of groundwater flow paths into optimal networks. The approach, which is scale-independent, is based on a hypothesis that quantifies the organization of optimal groundwater flow paths between two vertical planes—a divide and a river. This has similarity with the work of Bejan and Errera (2011) that optimized flow paths between a line and a point. The emphasis in the model presented in this section is, however, on the organization of subsurface flow paths through a 3D optimization. Thus, it may appear that the model considers headward erosion and capture, as opposed to basin fill and overtopping, as specific reorganization mechanisms, but one can only conclude that purely surface processes dominate integration rates when groundwater fluxes do not generate the appropriate time scale for the organization. Thus, a concrete analytical formulation based on groundwater flow speeds may help distinguish when groundwater processes are the limiting process on drainage development and/or reorganization rates, regardless of whether the drainage reorganization is from the bottom up (Young and Spamer 2001, Dickinson 2015) or from the top down (Spencer *et al* 2001, Repasch *et al* 2017).

Characterizing drainage basins has been studied intensively for decades, as it is important to many environmental issues, ranging from water management, to landslides, flood prevention, and aquatic dead zones. The evolution of river networks is a consequence of headward growth and branching away from escarpments in a substrate. The presence of variations in the lithology, structure, soil developments, and topography makes the local substrate strength a stochastic field. Scheidegger (1965, 1967) was probably one of the first investigators who proposed a quantitative model for the planar development of river networks, although as mentioned above, others, including Horton (1932, 1945), Schumm (1956), Strahler (1964), and Shreve (1967) had also attempted to model river networks. In Scheidegger's model that used a hexagonal lattice, channels were developed from an edge of the lattice that formed a line of basin outlets. At each 'time step,' an entire row of adjacent neighbors was simultaneously considered, with each of these nodes assigned at random a flow direction toward one of the two neighbors that were within the basin. The open boundary served as the location of the seed points, and the clusters of river basins were grown by a pixel layer at each time step. Although the model did produce patterns that were qualitatively similar to some of the patterns in actual river networks, it produced uniform headward growth, with the frontier of the cluster remaining smooth during development. In addition, each channel tended to follow a more or less straight path between its source and outlet.

10.2.1 Percolation models of connectivity of river metworks

Hack (1957) discovered that the principal stream length L_p scales with the drainage basin area S as

$$L_p \sim S^{D_H}, \qquad (10.1)$$

where D_H is a scaling exponent. Analyzing the available data, Hack estimated that, $D_H \approx 0.6$. Equation (10.1) is known as *Hack's law*, and has been explained in terms of fractal concepts (Tarboton *et al* 1988, Rinaldo *et al* 1993, 2014, Maritan *et al* 1996), and percolation (Stark 1991, Hunt 2016a). An attempt was also made (Reis 2006) to explain Hack's law in terms of the constructal law (Bejan and Lorente 2013) that suggests that the natural tendency of all flow systems is to form configurations that, over time, generate progressively greater flow. Stark (1991; see also Niemann *et al* 2001) assumed that at any time the next failure takes place at the weakest point. On a lattice this is tantamount to the invasion percolation (IP) model, already described earlier in this book. He also considered the physical fact that streams rarely bifurcate downstream and, therefore, introduced the additional constraint that closed loops cannot be formed, hence leading to the so-called self-avoiding invasion percolation model, and used the model to simulate the evolution of stream (river, drainage) networks.

Following the IP-based model of Stark (1991), Fehr *et al* (2009; see also Fehr *et al* 2011a, 2011b) proposed another IP-based model for extracting watersheds from landscapes. In their model, simulated on a $L \times L$ square lattice with fixed boundary conditions in the vertical (growth of the network) direction and periodic boundary conditions in the horizontal direction, the upper and lower sides of the lattice represented the sinks of two basins, one in the north (N) and the other in the south (S). The height of each site i was represented by h_i. For each site i the IP rule was applied such that the basin (N or S) to which site i belongs was one that the IP-invaded cluster reached first. Therefore, all sites of the lattice belonged to one of the two basins, and the interface between them defined the watershed. To speed up the computations of identifying the interface, an efficient sweeping strategy was used: (1) Initially, the sites were selected along a straight line that connects the sinks. Therefore, when the IP algorithm from two neighboring sites evolved to different sinks, a segment of the watershed was between them. (b) From then on, the sites were selected only in the neighborhood of the already known watershed segments in order to reveal more segments of the watershed, resulting at the end in the complete watershed.

Oliveira *et al* (2019) further refined the model by Fehr *et al*. Consider a rectangular lattice of size $L_x \times L_y$ in which, similar to Fehr *et al*'s model, the height of each site i is h_i. A height threshold height h^* is introduced such that, if $h_i > h^*$, then the ith site belongs to a cluster called the height cluster, which is composed of all connected sites with height larger than h^*. Otherwise, site i does not belong to any cluster. Oliveira *et al* (2019) used $h^*=0$, which for Earth corresponds to the sea level, and the height clusters define continents and islands on Earth. Thus, it is known *a priori* that they define drainage basins separated by several interface lines, whose specific sizes and shapes must be determined. The rules in Oliveira *et al*'s model were similar to those proposed by Fehr *et al*.

Oliveira *et al* (2019) studied two versions of the model. One was with the traditional periodic boundary conditions in the horizontal direction and unconventional periodic boundary conditions in the vertical direction, in order to simulate real landscapes. In the unconventional periodic boundary conditions, each site at the top (bottom) row is a neighbor of every other site in the top (bottom) row and represents a mapping of a sphere onto a lattice. In the second model fixed boundary conditions in both directions were utilized for synthetic landscapes. Simulations of Oliveira *et al* indicated that the perimeter and area distributions of the basins display long tails, extending over several orders of magnitude and following approximately power laws, with exponents that depend on the spatial correlations and are invariant under the landscape orientation, not only for terrestrial, but also lunar and Martian landscapes. The terrestrial and Martian results were statistically identical, hence suggesting that a hypothetical Martian river would be similar to the terrestrial rivers. The simulations suggested that Hack's law may have its origin purely in the maximum and minimum lines of the landscapes. Thus, IP-based models are able to explain the observed self-similarity in river networks and predicted structural quantities, such as D_H, as well as the network's fractal dimension, $D_f \simeq 1.82$ and other important quantities, such as the bifurcation, length and area ratios. All of Stark's results were consistent with the data for natural stream networks. His model produces tree-like structures. Although such structures are abundant in Nature, there are also many cases that contain closed loops.

Hunt, Ghanbarian, and Faybishenko (HGF) (2023) suggested that the correct value of the exponent D_H is D_{op}, the fractal dimension of optimal paths, already described in the previous chapters. The basis for their proposal was derived from the organization of optimal flow paths in the subsurface and their effects on drainage reorganization. They assumed that, above a threshold length scale, the surface expression of these optimal flow paths is the drainage network. It is already known that features of stream bifurcation and channel initiation in Florida are related to the convergence of groundwater flow paths (Petroff *et al* 2013), and that characteristics of amphitheaters on the Colorado Plateau can be traced to subsurface chemical weathering and erosion from groundwater flow (Laity and Malin 1986, Baker *et al* 1990). These results lend support to the proposal of HGF. It may seem surprising that the HGF proposal relates large-scale systems to pore-scale processes, but if one treats the subsurface as a complex network with heterogeneity on a wide range of scales, from the pore, through the core and facies, and on through structural and landscape to the tectonic scale, then the connection becomes clear. The second assumption of HGF is that the dominant surface flow paths are those for which the cumulative resistance is minimized—a quantification of the general concept that water chooses the path of least resistance. Such an optimization is assumed to be constrained to a thin, roughly horizontal layer. Therefore, one seeks an optimal path in 2D. For any Strahler stream order[1], it is also assumed that the path connects two lines, the next higher order stream and any divide. Under such assumptions, the actual length of a river L_r from, for example, a continental divide to its junction with streams of increasing order should follow the power law

$$L_r = C_1 \ell^{D_{op}},$$
(10.2)

where C_1 is a constant, and ℓ is the Euclidean connecting distance. In 2D one has, $D_{op} \simeq 1.21$. Since ℓ is known to relate to drainage basin area S according to the simple Euclidean geometry relationship, $\ell \sim S^{1/2}$, equation (10.2) also yields the Hack-type relationship given by equation (10.1).

Thus, the fundamental hypothesis is that the optimal 3D path model provides the tortuosity of the constructed interconnected groundwater flow paths with the lowest cumulative resistance. It is known that the length L_g of such a path that connects two perpendicular planes (the vertical extension of the two lines above) is proportional to the Euclidean separation of the planes ℓ, raised to the power D_{op}, which in 3D is equal to 1.43 (Sheppard *et al* 1999). Thus,

$$L_g = C_2 \ell^{1.43},$$
(10.3)

where C_2 is another constant whose magnitude is not important. Thus, combining equations (10.2) and (10.3), we obtain $L_g = C_3 \ell^{1.43/1.21} = C_3 \ell^{1.18}$, with C_3 being another constant. To make this relationship predictive involving space and time, one needs to include in a model fundamental length and time scales. To make such choices, we are guided by the analogy to the formulation of the scaling relationships for root growth in soil. As we describe in chapter 11, a spatio-temporal scaling relationship with similarity to equation (10.2) has been proposed to relate root length ℓ_R to root radial extent R_L [see equation (11.6)] with $\ell_R = C_4 L_R^{D_{\mid rmop}}$ where C_4 is a constant. In that case, the assumption that the root tip extension rate is constant in time leads to the spatio-temporal scaling equation given by

$$t = t_0 \left(\frac{x}{x_0} \right)^{D_{op}},$$
(10.4)

with t_0 and x_0 being fundamental time and length scales required by dimensional analysis. Thus, rearranging equation (10.4) yields,

$$x = x_0 \left(\frac{t}{t_0} \right)^{1/D_{op}},$$
(10.5)

---

[1] The Strahler number, or Horton–Strahler number of a mathematical tree is a numerical measure of its branching complexity. They were first developed in hydrology by Horton (1945) and Strahler (1952, 1957) as a way of measuring the complexity of rivers and streams, and are usually referred to as the Strahler stream order. They are used to define stream size based on a hierarchy of tributaries.

Making predictions based on such nonlinear relationships as equations (10.4) and (10.5) requires reference to particular scales, even though they are scale-free, at least over a wide range of spatial and temporal scales. Since the fractal nature of optimal paths in a heterogeneous medium extends from the smallest (pore) scale to the maximum extent of the critical zone CZ, HGF chose (Hunt *et al* 2021) the typical, or median, particle size to define the fundamental length scale x_0. A typical particle size was suggested to be the middle of the silt range, at about 10^{-5} m, or 30 μm. Then, in equation (10.4), the ratio x_0/t_0 represents an annual mean vadose zone flow rate. If vadose zone flow is not limited by the hydraulic conductivity of the shallow soil, it is related to the difference between climatic variables—precipitation and evapotranspiration, i.e., it is strongly climate dependent. Thus, to take into account the vadose zone flow rate, we rewrite equation (10.5) in the following form (Hunt 2017):

$$x = x_0 \left(\frac{t}{x_0/v_0} \right)^{1/D_{op}},$$

(10.6)

where v_0 is a characterisaitc flow rate. Estimates of v_0 range across Earth's climate systems from about 0.025 m yr^{-1} (near a minimum for appreciable growth of vascular plants) to about 25 m yr^{-1}. One can see that the predictions using equation (10.6) are sensitive to the flow rate v_0, but nearly insensitive to the fundamental length scale, $x_0/x_0^{0.83} = x_0^{0.17}$. Thus, an error in the choice of x_0 of a factor of 10^3 produces an error in length scale of a factor near 3.

Using equation (10.6) as an analogy for predicting the spatio-temporal scaling of river networks, retaining the fundamental spatial scale of about 30 microns, but with two distinctions, one can obtain a spatio-temporal scaling equation for the river length, $L_r = (1/C_3)L_g^{1/1.18} = (1/C_3)[t/(x_0/v_0)]^{1/1.18}$. One difference from equation (10.6) is in the exponent, which was proposed above to be derived from the quotient $1.43/1.21 = 1.18$. The second change is that the results for regional groundwater flow rates, v_G, is utilized, rather than that in the vadose zone, v_0. The range of typical values of v_G is 10 m yr^{-1}–20 m yr^{-1} for groundwater (Blöschl and Sivapalan 1995) with about an order of magnitude variability on either side, which would be something like 1.5 m yr^{-1} to 150 m yr^{-1}. According to Bloemendal and Theo (2018), high groundwater flow velocity implies $v_G > 25$ m yr^{-1}. As the United States Geological Survey (USGS) (2021) puts it, 'A velocity of 1 foot per day or greater is a high rate of movement for groundwater, and groundwater velocities can be as low as 1 foot per year.' The latter range corresponds to a range of 0.3 m yr^{-1} up to 110 m yr^{-1}. A somewhat narrower (and cleaner) range, 1 m yr^{-1}–100 m yr^{-1} is used to characterize the common spread in such flow rates, a range nearly the same as suggested by Blöschl and Sivapalan (1995). For example, desert regions in both southern California (Kulongoski *et al* 2003) and the Sudan (Gossel *et al* 2004) have known groundwater flow rates of about 1 m yr^{-1}. In Germany, however, typical reported values of groundwater flow velocities in the Rhein graben are closer to the upper limit cited by USGS, and can exceed that limit significantly[2]. Nevertheless,

---

[2] https://www.umwelt-online.de/regelwerk/cgi-bin/suchausgabe.cgi?pfad=/wasser/ltws/26b.htm&such=RdErl

when the flow rate for a given system is known, that rate should be utilized, in order to use equation (10.6) appropriately.

10.2.2 The data

To compare the predictions of the percolation model with the data, we first describe their source and characteristics.

General data at any scale: Headward erosion of streams in Costa Rica and Spain by distances 20 km and 80 km over 100 thousands years (kyr) and 320 kyr were documented, respectively, by Marshall *et al* (2003) and Mather *et al* (2002) (see also Dorsey and Roering 2006). Struth *et al* (2020) discussed the reorganization of the 170 km Suarez River basin in Colombia, including its piracy of additional smaller basins along its east side, over a 405 kyr period from its capture by the Magdalena. Specific distances for the smaller events were not, however, possible to extract. Fan *et al* (2018) demonstrated the reorganization of the Daotang basin within 80 kyr of the capture of Yihe River by the Chaiwen, adding 25 km^2 to the Yihe River drainage. Hack's law was used to predict a river length from the basin area. Goudie (2005) pointed out that Africa's river systems are some of the most ancient in the world, dating to the Mesozoic, but that Cenozoic reorganization of several, particularly the Nile, the Niger, and the Limpopo/Zambezi system, has been important. The Blue Nile and Niger are discussed by Hunt *et al* (2021). For the Limpopo/Zambezi system, Said *et al* (2015) pointed out that the southern Mozambique basin began filling with sediments much more rapidly about 25 million years ago (Ma) at both the Limpopo and Zambezi River deltas (see also Matthews *et al* 2001), but that superposed on that was a dramatic shift of sediment volume from the Limpopo to the Zambezi at 5 Ma. This was interpreted in terms of an onset of uplift at 25 Ma and a capture of the upper Limpopo by the Zambezi at the latter date. The time interval Is 20 Myr, the site of capture is upstream of Victoria Falls, and the length of the Zambezi above there is about 1100 km.

Wang *et al* (2020) summarized discussion of the drainage reorganization of North America. According to them, drainage of most of the United States was to the north in the early Cretaceous (mean age 122 Ma). By the late Paleocene (56 Ma) the drainage from roughly the present U.S.–Canada border was to the south, with a kind of paleo-Missouri–Mississippi system extending from Idaho through the middle of the continent, before turning south to the Gulf of Mexico. The present 5950 km long Missouri–Mississippi River is a model for the length of such a river. By the Eocene (mean age 45 Ma) a river drainage had formed parallel to the Rio Grande, which started in southern central present-day Arizona and flowed to the Gulf of Mexico, for which the present Rio Grande length (3016 km) is a reasonable model.

Rio Grande and Pecos: These are semi-arid areas in Southwestern United States. The development of the Rio Grande was discussed by Repasch *et al* (2017) in a top-down evolution, starting in the San Juan range of southern Colorado about 8 Ma. By about 5.7 Ma, it had arrived in the lower San Luis Basin of New Mexico, by 5.3 Ma, Albuquerque, by about 4.5 Ma, the Palomas Basin, and by 3.1 Ma, the Mesilla Basin near El Paso. The Rio Grande arrived at the Hueco Basin further southeast downstream by 2.06 Ma and reached the already existing Pecos River and, thus, the Gulf of Mexico by 0.8 Ma. Repasch *et al* (2017) also stated that at 5.3 Ma, the Pecos was divided into two separate segments, one flowing eastward to the Ogallala aquifer and one flowing northwards from the border of Texas. The two had integrated as far south as Texas by 4.5 Ma and the Pecos without the Rio Grande was fully integrated to the Gulf of Mexico by about 1.5 Ma. An important trigger for the integration of the Pecos may have been the onset of the southwest monsoon about 6 Ma. According to Sanford *et al* (2004), groundwater in the middle Rio Grande basin in Albuquerque contains a thin veneer of water with source from the Rio Grande itself overlying groundwater from mountains on both the east and west sides of the Rio Grande rift. The layer immediately below the young water from the river itself has a source in the Jemez Mountains, a little over 100 km to the north-northwest and a mean age of 19 000 years. This is consistent with a regional flow rate of $5 \, \mathrm{m} \, \mathrm{yr}^{-1}$. That it may be reasonable to use such a value over much of the Rio Grande course is suggested by the review of McMahon *et al* (2011) in which it was stated that late Pleistocene groundwater is common in the basins throughout the southwest and Great Plains, but exists further east only in confined aquifers.

Mojave River and other inland southern California drainages: The initiation of the Mojave River drainage system is believed to have occurred (Hillhouse and Cox 2000) about 3.8 Ma, and it became integrated to a length of 200 km by about 25 thousand years ago (ka). The Mojave River was dammed at Afton for 160 ky during pluvial climates before it finally breached the sill and advanced to the Soda Lake about 40 km downstream (Reheis *et al* 2012). Less than 10 thousands years (kyr) later, the Mojave River arrived in Dumont Lake, 50 km further, and likely reached Death Valley, nearly 150 km further downstream, but Holocene drying interrupted integration of this drainage system (Enzel *et al* 2003).

According to Hillhouse and Cox (2000), the known history of Mojave River in southern California began about 0.8 Ma, and is associated with the uplift of the transverse ranges through transpression along the San Andreas fault. From these mountain ranges, the Mojave flows first north and then east to the eastern Mojave Desert. In this region of mostly disconnected, internal drainage systems, the climate is currently arid, except in the mountains of the river's source. The Mojave River system is currently given about ~200 km in length, integrating several pluvial lake basins along its length. Times of the arrival of the Mojave River at various sites are mostly established by dating the bottom of lakebed sequences, or other deposits just below the lakebed sediments. It has been emphasized by several research groups (Enzel *et al* 2003, Reheis *et al* 2012, Garcia *et al* 2014) that that the expansion of the Mojave River system to the northeast likely required a much wetter climate than the current state, with rainfall much larger than at present. Overall, the evolutionary picture is of significant time periods with the Pacific storm track aimed either at northern California or even southern California (Enzel *et al* 2003, Reheis *et al* 2012). Kulongoski *et al* (2003) give groundwater flow rates along the merged alluvial fans north of the transverse ranges as ca. $1 \, \mathrm{m} \, \mathrm{yr}^{-1}$ over the last 20 000 years. Maxwell *et al* (2016) observed, however, that groundwater flow rates are proportional to the difference between precipitation P and evapotranspiration E. Given rapidly decreasing flow rates with depth (Koltzer *et al* 2019), higher flow rates are also associated with higher water tables. Pluvial precipitation in the Mojave desert has, however, been estimated to have been 1.6 to 4 times higher than at present and temperatures 3–8 degrees lower (Harvey *et al* 1999). The average of 1.6 and 4 is about 2.8, while the geometric mean is 2.53. The lower temperature likely increased the difference $P - E$ further. Thus, groundwater flow rates during such Pleistocene climates with their higher precipitation are likely to have been higher than $1 \ \text{m} \ \text{yr}^{-1}$ by a factor of 2–3 or larger. Furthermore, Enzel *et al* (2003) stated that, 'Reducing modern rates of evaporation by 50%, and doubling modern rainfall would result in a full lake almost at the elevation of the Lake Mojave shoreline.'

Amargosa River: This river originates in southwestern Nevada and flows southward through extreme eastern California before turning west, and then north into Death Valley. It is a mostly ephemeral stream, and has a history of feeding various lakes in the same (Tecopa) basin multiple times before finally advancing as far as the bottom of Death Valley. Reheis *et al* (2020) stated, 'The High lake reached the highest level achieved in the Tecopa basin, and it may have briefly discharged southward, but did not significantly erode its threshold. The High lake was followed by a long hiatus of as much as 300 k.y., during which there is evidence for alluvial, eolian, and groundwater-discharge deposition, but no lakes. We attribute this hiatus, as have others, to blockage of the Amargosa River by an alluvial fan (ca. 20 km) upstream near Eagle Mountain.' This discussion provides an estimate for a data point 300 k.y., 20 km. A more accurate estimate of the length scale might be 5 km, as the west side of Eagle Mountain is adjacent to the course of the Amargosa for about 5 km. Over much of its length, the Amargosa River flows through a climate similar to that of the Mojave River drainage. Since no additional information was found about groundwater flow rates, it was assumed that the values for the Mojave were probably reasonable first estimates for the Amargosa as well.

San Jacinto Mountains and River: Dorsey and Roering (2006) established the adjustment of a drainage on the east (desert) side of the San Jacinto Mountains (in Riverside County east of Los Angeles) to the uplift. The advantage is in the analogy of its position to that of the Mojave River, on the lee side of the mountains bounding the coastal plain in southern California, with strong similarity in climate, provenance, and relief. The measure involved is knick point migration, rather than drainage integration per se, which can be a disadvantage, as knick point migration incorporates a greater input from surface processes. Dorsey and Roering (2006) stated, 'The total distance of knick point migration is ~30 km, as measured along the Clark fault from the pre-SJFZ [San Jacinto fault zone] drainage divide at Borrego Mt. to the area of active stream capture points at the south edge of Burnt Valley. This distance appears to be a minimum because rocks on the NE side of the Clark fault are moving SE toward Borrego Mt. (on the SW side of the fault). An alternate measurement, from the pre-SJFZ divide on the NE side of the Clark fault to the area of modern stream captures at the edge of Burnt Valley, gives an along-fault distance of ~44 km, which are considered to bracket the total distance of knick point migration.' Thus, the mean of the two values, 37 km, was used in the comparison of the theory and data as the length scale. Assigning a time interval to the process adds uncertainty due to its connection with the uplift of the San Jacinto Mountains. The onset of the uplift of the San Jacinto Mountains in southern California is believed to have been triggered by the initiation of offset along the San Jacinto fault. This has been a challenge on account of the necessity to apportion offsets among many approximately parallel faults over the relatively recent geologic time scale of about 2.5 million years. While disagreement in the timing of the onset has persisted, with some inferring a 2.5 Ma onset (based on a smaller rate of relative motion), more recently the geophysical data have been more uniformly understood to imply ca. 1 Ma, or perhaps a few hundred thousand years earlier (Langenheim *et al* 2004, Janecke *et al* 2010).

A second inference, with greater uncertainty, can be made from the coastal side of the San Jacinto Mountains. The San Jacinto River flows through two lake basins to merge with the Santa Ana River, 68 km downstream from its source. However, its flow is only rarely sufficient to fill either basin, the San Jacinto (Wang *et al* 1995) or Lake Elsinore (three times since 1900 and 20 times since 1769; see Kirby *et al* 2007), and reach the Santa Ana River and, thereby, the ocean. This combination suggests a relatively recent integration of the San Jacinto River through its length (68 km). The climate on the coastal side of these mountain ranges is significantly wetter than on the desert side, though the San Jacinto Mountains overall are drier than the San Bernardino and San Gabriel ranges, where the headwaters of the Mojave River are located. Additionally, due to the rainshadow effect from the Santa Ana Mountains (reaching 1620 m), the San Jacinto basin is the driest inland valley in southern California.

Gila River and tributaries: This system represents the longest integrated drainage system within Arizona, excluding the through-flowing Colorado River. Its chief moisture source and permanent stream tributaries originate in the mountains of the Mogollon Rim, extending northwest to southeast from Arizona to New Mexico. The streams include the Verde River, Tonto Creek, and the Salt River. Larson *et al* (2020) stated that, 'A ca. 2.5 Ma age for the initiation of top-down integration of the Verde River from the upper Verde Valley into what are now downstream basins is consistent with the presence of a 3.3 Ma volcanic tephra.... The basins depicted here were formerly endorheic, but integrated within the last 0.2–2.8 Ma. The integration of these basins resulted in the modern through-flowing drainage networks of the (320 km) Salt (272 km), Verde, and Gila Rivers of central Arizona.' The same time frame was implicitly extended to the Salt River. However, the integration of the Gila River itself has been a much longer process, commencing between 15 Ma and 12 Ma (with a mean of 13.5 Ma) near the lower Colorado River and continuing to the present, where headward erosion is still occurring and, in discrete steps, lengthening the drainage into disconnected basins nearer the continental divide (Dickinson 2015). Its length is currently 1044 km. Dickinson (2015) described initiation of the incision of Quiburis Basin at about 5.75 Ma, the Safford Basin at about 3.5 Ma, and the Duncan Basin at about 2 Ma, in the process of headward erosion of the Gila River. He also stated that headward erosion of the Santa Cruz River was finished by about 2 Ma. These dates correspond fairly closely to dates given on his maps in his figures, which yield two data points on the Santa Cruz, two on the San Simon, and five on the San Pedro River. The date associated with the bifurcation at the confluence of the San Pedro and Gila Rivers, as well as the corresponding date for a bifurcation near the confluence of the Santa Cruz and Gila Rivers, are, however, missing. In neither case is it certain that the date is the same for the further evolution of both mainstem river and its tributary. It would, however, be unjustified to assume two different dates. The date for the former bifurcation was estimated by extrapolation down the San Pedro from the last known date to find 7.5 Ma, and the second by extrapolation down the Gila from the confluence with the San Pedro to find about 9.5 Ma. The remaining dates are read off Dickinson's figures 6 and 7. Upstream distances from the most recent dates given are also required. The distance from the site roughly midway between Duncan and Redrock to the Gila source was estimated at 300 km.

The oldest ages for spring water emerging near the Mogollon Rim in the Verde River catchment were reported by Beisner *et al* (2018) to be 4000–6000 years, and their paths were traced to the southern slopes of the San Francisco peaks and the Flagstaff area, about 50 km to the north, which yields a flow rate of about 10 m yr^{-1}, but the aquifer is a fractured limestone, and the precipitation at the source is nearly the largest in the state. The same recharge area, located about 80 km from Grand Canyon Village, and somewhat further from Cataract Canyon, was determined to be a source for ancient (<10 000 yr old) groundwater emerging from springs below the South Rim of the Grand Canyon (Solder *et al* 2020). This ancient water mixed in various proportions, up to 100%, with local groundwater, which suggests a regional groundwater flow rate of less than 8 m yr^{-1}.

The remaining groundwater in the vicinity of Tucson has been getting older as pumping has drawn down the water levels (Kalin 1994). Experiments from 1965 revealed ages of less than 2000 yr; by 1989, however, the maximum age had increased to over 6000 yr. In both cases, the greatest ages were found along the axis of the valley. The middle of the measured aquifer lies about 10 km from the recharge sources in the Tucson Mountains to the southwest and about 15 km from the Catalina and Rincon Mountains to the Northeast. Therefore, pertinent flow rates range from about 2 m yr^{-1} to about 10 m yr^{-1}. Kalin (1994) reported flow rates near the center of the valley to be about 8 m yr^{-1}. The relatively large flow rate under Tucson is consistent with its proximity to the Catalina Mountains, with maximum annual precipitation near the summit of nearly 90 cm (Whittaker and Niering 1975).

Data are also available for groundwater flow rates in the middle San Pedro basin. Hopkins *et al* (2014) stated, 'Groundwater in the lower basin fill aquifer (semi-confined) was recharged at high elevations in the fractured bedrock and has been extensively modified by water-rock reactions (increasing F and Sr, decreasing ^{14}C) over long time scales (up to 35 000 years B.P.).' Since these mountains are 15 km to 20 km from the San Pedro River (Cordova *et al* 2015), one obtains an estimate of the flow rate of about 0.5 m yr^{-1}. Further north, in the lower San Pedro river basin groundwater ages (Robertson 1992) are closer to 8kyr B.P. to 15 kyr B.P., but the mountain ridges are closer too, at typical distances of 10 km to 15 km from the valley bottom or ca. 1 m yr^{-1}. Thus, in this area, flow rates of about 1 m yr^{-1} are common, an estimate identical to what was measured in the upper Mojave watershed.

Eastoe and Towne (2018) provided a literature survey of groundwater ages in 34 alluvial basins in the Basin-and-Range Province and the Transition Zone of Arizona. Late Pleistocene water is known in six basins below the Mogollon Rim, from Golden Valley in the northwest through Wikieup, Wickenburg, and Phoenix, to Tucson, as well as in three basins along the next lineament northeast, Tonto, Safford, and Duncan. Safford and Duncan indeed lie along the Gila River. A Pleistocene age longer than 10 ka suggests broadly similar regional flow rates all along the dropped down margins of Arizona's topographic high at the Mogollon Rim.

10.2.3 Comparison of the predictions with the data

Figure 10.1 compares the predictions of the percolation model with the data discussed above, which reveals clearly a relationship between the scaling of river drainages and root systems. The flow rates governing river lengths appear, however, to be larger than those for vegetation, in accord with the aforementioned comparisons of regional groundwater flow rates with vadose zone flow rates.

To improve clarity and distinguish better between potential influences climate and tectonics, figure 10.2 focuses on the upper right corner of figure 10.1, hence applying specifically the range of regional subsurface flow rates given by the USGS (1 m yr^{-1} to 100 m yr^{-1}) to obtain minimum and maximum predictions for river length together with a linear spatio-temporal scaling relationship from tectonics. A qualitative designation of climatic conditions is also included. Thus, drainage basins were assessed as humid, neutral, or wet. Specific classifications are given in table 10.1. Unlisted river systems were considered neutral. Areas classified as neither 'humid' nor 'dry' include four rivers on the west side of India, the Blue Nile, rivers that drain the north side of the Tibetan Plateau (the Yellow River), a river in Italy and one of the river systems from Spain. The Almanzora River, designated as 'arid,' is located in what is currently the most arid region of Spain, its southeast. In some of these 'neutral' river drainages the headwaters are in humid climates, while the lower reaches in arid climates.

According to figure 10.2, arid drainages are found above, but mostly near, the tectonic rate, almost never below it. Such arid drainages include the Gila River, its tributaries, and the southeastern California river drainages. Consider, for example, the Santa Cruz River that does not really reach the Gila River (Dickinson 2015). Moreover, a critical scarcity of water exists in the San Pedro drainage somewhat further east, while the Mojave River only reaches the end of its current drainage when rainfall in southern California is especially heavy, and even the Gila River rarely flows all the way to the Colorado River. These are not purely due to the arid climate, as water impoundment for human use also plays an important role in current surface water depletion. At the time scales investigated, the rate of drainage integration for these rivers, measured as a spatial velocity, barely exceeds typical extensional tectonic velocities. Thus, the rate of integration is close to the rate at

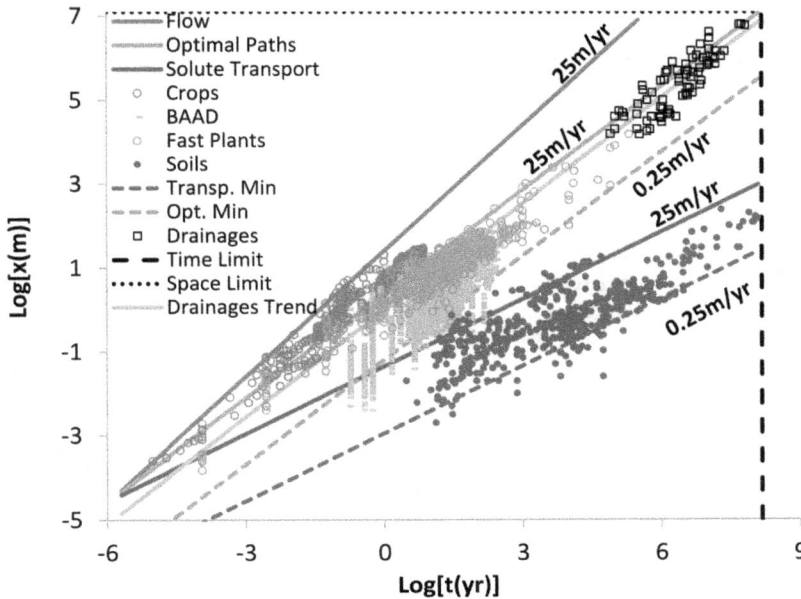

Figure 10.1. Logarithmic plot of the data for plant height or root radial extent as a function of time for 'fast plants,' (relatively rapidly growing tree species, such as Sequoias and Eucalypts), Biometric and Allometric Database (referred to as BAAD), crop heights (without water or nutrient limitations), and soil depths, as well as predictions of equations analogous to equation (10.6), but with the exponent 1/1.87, appropriate to solute transport (soil predictions) where 1.87 represents the fractal dimension of the backbone of percolation cluster; 1 (appropriate for crop height predictions), or 1/1.21, where 1.21 is the fractal dimension of the optimal paths or 'biological transport,' appropriate for vegetation. The upper and lower bounds of 25 m yr^{-1} and 0.25 m yr^{-1} are reasonable for unsaturated zone flow, relevant for transpiration or infiltration. Note that the scaling exponent for vegetation growth (1/1.21) is very similar to the power law extracted from the data for drainage system development (1/1.18). Drainage basin data are consistent with a range of flow rates between 1 m yr^{-1} and 100 m yr^{-1}, as designated typical by the USGS. The designations 'time limit' and 'space limit' correspond to the period of a Wilson tectonic cycle and the linear extent of a supercontinent. Thus, drainages have at most a time of about one Wilson cycle to develop, and can achieve a length that is, at most, about 1 supercontinent extent. Reproduced from Hunt *et al* (2023). CC BY 4.0.

which basins can be pulled apart. In an even drier climate, subsurface flow rates would presumably be too small to accomplish the integration and the river drainages would become discontinuous. The argument does not incorporate the tendency for surface processes, such as wind transport of sand, or gravity-based elevation diffusion, to fill in channels. It is, therefore, not surprising that the Gila River, particularly its southern tributaries, as well as the Mojave and Amargosa Rivers from California, represent the slowest rates of basin integration that could be found. At yet longer time scales in, for example, the Sahara Desert, even river drainages that were once organized, disaggregate (McCauley *et al* 1986, Ghoneim *et al* 2005), particularly with the existing low flow rates (Gossel *et al* 2004).

We point out that, as another important conclusion drawn from figure 10.2, several drainage basins located largely within regions of arid climate—the Pecos

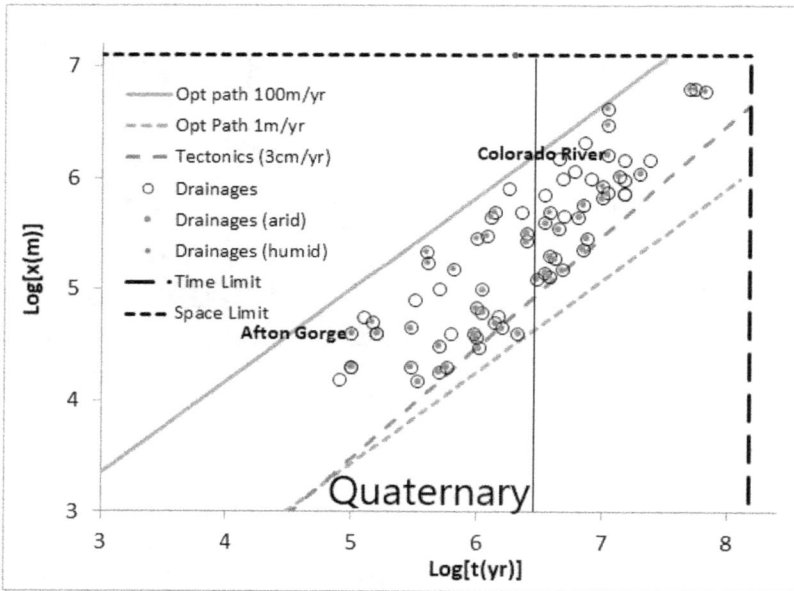

Figure 10.2. The upper right corner of figure 10.1, with crops, vegetation and soil depth removed, but with limiting regional groundwater velocities taken from the U.S. Geological Survey (2021). More arid and more humid climatic conditions in the corresponding river drainages are distinguished from each other according to the designations from table 10.1. Also included is a typical continental-scale tectonic rate of about 3 cm yr^{-1} and the designations 'Time Limit' and 'Space Limit.' The time limit of about 150 Ma represents the interval since the break-up of the supercontinent Pangaea, while the space limit of about 12 000 km represents the linear dimension of the land mass of Pangaea. An approximate median groundwater flow rate of about 20 m yr^{-1}, a rapid tectonic rate of about 6 cm yr^{-1}, and the regression line from the observed draines would all meet very nearly at the intersection of 'Time Limit' and 'Space Limit.' Reproduced from Hunt *et al* (2023). CC BY 4.0.

River, the Colorado River, and the Rio Grande, as well as, in particular, the Afton Gorge segment of the Mojave River—became integrated more rapidly than other drainages with similar aridity, and more rapidly than some in much more humid climates. This anomaly may be partly due to the locations of their sources in areas of significantly larger values of the difference $P-E$, such as the Rocky Mountains and the southern California transverse ranges. Such factors may also contribute to a greater role of lake spillover in the process of drainage reorganization, which has been invoked in each case (Meek 1989, 1990, Spencer *et al* 2001, Crow *et al* 2021). In these drainages, the oldest dates marking the position of the rivers are, indeed, found furthest upstream. However, this observation does not necessarily exclude alternate mechanisms of drainage integration in the same drainage basin predominantly due to subsurface flow. One expects that uncertainty may sometimes be resolved by considering the actual rates of drainage integration.

As discussed above, more detailed data for both flow rates and drainage integration in the cases of the Mojave and Gila River drainages are available. Therefore, the two of them are analyzed in greater detail. Figure 10.3 presents the

Table 10.1. General climatic conditions within river drainages.

Humid	Dry
Aare	Colorado
Rhine	Morocco
Suarez	Almanzora
Amazon	Verde
Yangtze	Salt
Dadu	Amargosa
Katonga	Mojave
Meuse	Gila
Cahabon	San Pedro
Mississippi	Santa Cruz
Niger	San Simon
Orinoco	San Jacinto
Costa Rica	Zambezi
Rio Grande[1]	Rio Grande[2]
Gunnison	

[1]Paleo Rio Grande analog (before 40 Ma). [2]Modern Rio Grande (after 8 Ma).

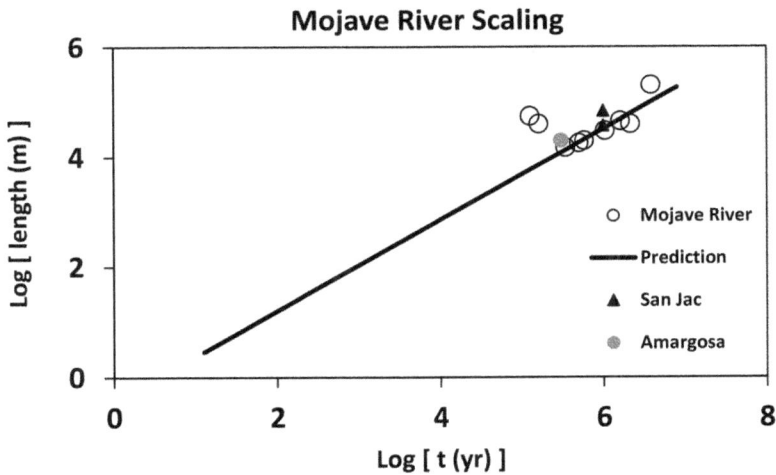

Figure 10.3. Scaling of length versus time for Mojave Desert Rivers. The California rivers shown are as follows. The Mojave and the Amargosa; San Jac (with two data points) that stands for two separate rivers flowing off the San Jacinto Mountains, a river flowing down the east side of the southern San Jacinto mountains toward Anza-Borrego Desert State Park and the San Jacinto River, on the west side of the San Jacinto Mountains. The first three examples are all in similar climates on the desert side of the southern California mountains. The fourth is in a somewhat wetter climate on the west side of the San Jacinto Mountains, but the desert side of the Santa Ana Mountains. In the case of the three Mojave River points that are well above the predicted power law, the drainage integration below Barstow is interpreted as having been driven primarily by surface processes with higher fundamental rates. Reproduced from Hunt *et al* (2023). CC BY 4.0.

lengths of the Mojave River sections that were integrated over particular time scales against those times. Equation (10.6) is used for prediction, and a pore-scale flow rate of $2.5\,\text{m yr}^{-1}$ is utilized, which is generally compatible with $P-E$ greater than current observations by at least a factor 2.5, since the known groundwater speeds are close to $1\,\text{m yr}^{-1}$. The prediction, except for three data points, is in excellent agreement with observation. The two large discrepancies represent the time for the Mojave River to reach Lake Manix from the vicinity of Barstow, and the time required for the Mojave River to overtop the sill at Lake Afton and arrive downstream at Soda and Silver Lakes. As pointed out by Meek (1989, 1990) and Reheis *et al* (2012), the latter event is particularly considered to have a completely different mechanism than what can be explained through groundwater flow. The overall development of the Mojave over the 3.8 Myr time span is also somewhat faster than predicted, largely because of the two counterexamples identified, which suggests the potential application of equation (10.6) for diagnosis of chief modes of drainage basin development. However, the potential confounding role of uncertainties in dating should also be considered, because a much earlier arrival of the Mojave River in the Barstow area than the mean of the range of the dates given is also possible, shifting the discrepancy between our prediction and observation upstream to the section between Victorville and Barstow.

Note that the time scale for the upstream knick point migration on the desert side of the San Jacinto Mountains is also predicted nearly exactly by equation (10.6), although both the Amargosa evolution and the San Jacinto River development are both somewhat underestimated. The underestimation of the Amargosa River length may be due to applying equation (10.6) to predict the entire distance to Tecopa Lake from the alluvial fans at Eagle Mountain, instead of merely across the fan. The San Jacinto River drainage basin, on the coastal side of the California peninsular ranges, receives at least double the rainfall that is measured on the desert side and for the Mojave River drainage, which means that using the groundwater flow rates for the Mojave basin would be expected to lead to an underestimation by a factor on the order of 2.

As for the Gila River and associated drainages, figure 10.4 presents the data for the scaling of drainage basin evolution in the desert southwest of the United States, which appear in agreement with the scaling law predicted by equation (10.6). Moreover, the fundamental flow rates that appear in the equation ($2.5\,\text{m yr}^{-1}$ for southern California and $6\,\text{m yr}^{-1}$ for the Gila and its northern tributaries), which define time scales in terms of the basic network size and the flow rate, are in reasonable agreement with a gradient in such flow rates from larger than $0.5\,\text{m yr}^{-1}$ on the southern margins of the Gila Basin to $10\,\text{m yr}^{-1}$ on the northern margins. As wetter climates in the Mojave during the Pleistocene were cited above to justify employing a subsurface flow rate of $2.5\,\text{m yr}^{-1}$, instead of the more recently observed $1\,\text{m yr}^{-1}$, it is important that there is also evidence for the existence of lakes in the enclosed basins of south-eastern Arizona over similar time frames (30 000 ky B.P to 10 000 ky B.P.) (Waters 1989), in particular Lake Cochise in the Wilcox basin. Thus, use of a larger flow rate than is currently observed is appropriate. The Rio Grande, however, for which evidence cited above exists,

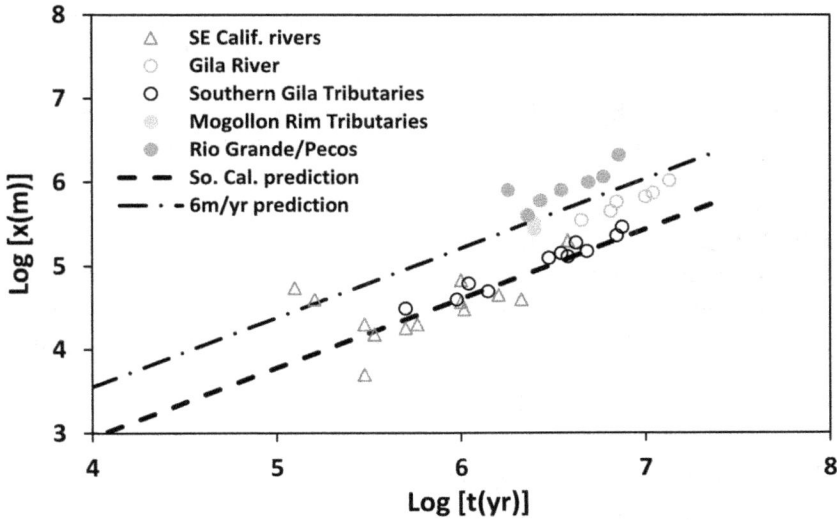

Figure 10.4. Gila and left and right (south and north) side drainage organization according to time scales. Note that the south side tributaries follow very nearly the same trend as the results of the Mojave River system. The north side tributaries and the Gila River itself are consistent with a scaling function that has the same exponent, but a somewhat larger flow rate, nearly 6 m yr^{-1}, instead of 2.5 m yr^{-1}. The Rio Grande and Pecos Rivers are, however, significantly above the 6 m yr^{-1} prediction. Reproduced from Hunt *et al* (2023). CC BY 4.0.

that a 5 m yr^{-1} regional groundwater flow rate may be appropriate, lies well above the prediction. Any one or combination of the following three factors may play a role: (1) the orientation of the rift is more favorable to drainage basin expansion than in the Basin and Range province, where the principle relief lies athwart the drainages; (2) surface hydrologic processes had a greater relative importance; and (3) groundwater flow rates were higher at relevant times in the past.

Note that two of the data points for the Mojave River, which account for the advance of the Mojave from Barstow to Lake Manix, and from Afton Lake to Soda Lake, are well above the remaining points. These can, as already suggested, be associated with a different process of drainage reorganization, sill overtopping, promoted by Meek (1989, 1990) and supported by Reheis *et al* (2012) and Hilgendorf *et al* (2020). When the two northern tributaries of the Gila, the Salt and the Verde are addressed, however, their positions on the graph do not indicate a significant departure from the prediction of equation (10.6). Although they are slightly higher than the points for the mainstem Gila River, the present data do not allow for a definite distinction between overtopping and headward erosion, particularly since groundwater flow rates tend to increase toward the north. Dickinson (2015) and Skotnicki *et al* (2021) raised questions regarding mechanisms of these processes, which may be more readily and certainly resolved, if: (1) the present percolation model is accurate enough; (2) paleodata to infer groundwater flow rates that are sufficiently precise can be obtained; or (3) more simply, though not necessarily conclusively, if geologic dating methods reveal progressions in age.

Thus, summarizing this section, application of a percolation model to a fundamental problem in geology and geomorphology, namely, the spatio-temporal scaling of river basin organization, was proposed based on principles tested within the disciplines of soil physics and hydrology. The model is based on a physical definition of a principle of scale-independence of hydrological processes and, in particular, the tendency of water to 'seek' paths of least resistance, in order to utilized concepts and ideas from the modern theory of percolation across scales (Sahimi 2011, Hunt *et al* 2014). In order to predict river length as a function of time subsequent to tectonic or other events triggering drainage basin reorganization, a spatio-temporal scaling relationship for transport time scales of mass transported along fractal paths of least resistance was used, namely, the so-called optimal paths defined by percolation theory. The predictions are in very good agreement wih a great deal of data at global scale.

10.3 Steady-state versus unsteady-state soil production and the implications for long time scales

Due to the vast time scales involved, soil production is a complex process to model (Huggett 1998). The modeling is also hampered by the fact that little is directly known or even knowable, on account of the huge time scales involved. Although attempts have been made to define soil production as a continuous process that starts at the Earth's surface (Jenny 1941), they have had mixed success since for periods less than a few decades it becomes increasingly difficult to develop criteria for what is and what is not soil. Soil depths tend to increase over time non-linearly and non-uniformly (Huggett 1998, Heimsath *et al* 1997, 2009). In the main, the shallower the soil, the more rapidly soil depths increase, although it has been known since the 19th century (Gilbert 1877) that soil production associated with shallow soils covering bedrock can be so slow that it is insufficient to sustain such soils, which was in fact confirmed much more recently (Heimsath *et al* 2009). Such considerations have led to the concept of *humped* soil production function, one that first increases with increasing depth but, then, beyond a certain depth, decreases. Even the fundamental processes that are involved in soil production vary, depending on whether the soil is developing within recently deposited or exposed unconsolidated media, such as alluvium, glacial moraines or landslide deposits, or, alternatively, from bedrock, whether fractured or not. In the latter case an entire array of physical weathering processes that work in tandem with chemical processes (Dietrich *et al* 2003, Anderson and Anderson 2010, Roering 2008, Braun *et al* 2016) is essential to help generate soil-size particles, and to generate the flow paths through, for example, fracturing of the bedrock that facilitates the entry of surface waters to the medium. In addition, soil production rates vary for other reasons as well, such as variations with surface curvature (Heimsath *et al* 2001a, 2001b) and slope (Montgomery and Brandon 2002), water content, upflow area (Liu *et al* 2013), and precipitation. Moreover, a wide range of biological processes also affect soils.

According to Dokuchaev (1948), five factors that contribute to soil formation are climate, topography, time scale, organisms and parent material. Regardless, using the argument that biogeochemical weathering is critical to the formation of soil provides the basis for the functional form of soil production function, and involves the respiration of plants and other organisms that provide CO_2 for, in simplified form, the Urey reaction (Urey 1952), i.e., the reaction between calcium silicates, exposed by erosion, with atmospheric CO_2 that produces carbonates. Another complication is the fact that soil is transported along the surface by incremental, such as soil creep (see, e.g., Heimsath *et al* 2001a, 2001b) and catastrophic, such as landslides or solifluction, surface processes, as well as by overland flow, by plants through treethrow or wind scraping, and animals through burrowing (Anderson and Anderson 2010). Soil erosion or deposition can be caused by divergence in the surface soil flux. Surface transport that depends spatially on distance to a divide, slope or substrate heterogeneity (Roering 2008) can also lead to spatially variable erosion rates, which also vary with surface convergence and flow routing. Depending on the functional relationship of such processes with the slope angle, one can generate convexo-concave or planar slopes (Roering 2008).

Since soil production is critical to understanding landscape evolution, and biogeochemical weathering is also an important aspect of soil production, attempts have been made for developing coupled geomorphological and pedogenic models that can, in principle, take into account all the processes relevant to hillslope evolution and can be used over any time scale, without assuming steady-state conditions, which is, nevertheless, a popular concept in the literature. As pointed out by Phillips (2010), however, steady-state soil depths are a 'convenient fiction,' because steady state is unlikely to be achieved, particularly where soil erosion and production rates are small. Although the question of how to distinguish between steady- and unsteady-state (non-equilibrium) landscapes is of interest in geomorphology, the reliability of the steady-state assumption remains a particularly relevant issue, since it allows one to address the issue of soil removed by erosional processes, or added by new production.

Soil formation is limited by chemical weathering—disintegration of soil by chemical reactions—since plants and vegetables cannot use bare rock to grow, but need the cations in the minerals that are transported by flowing water. In turn, chemical weathering is limited by the rate of transport of the cations—the solute in flowing water. Thus, solute transport plays a fundamental role in the processes that considered here. This was exploited by Yu and Hent (2017a) who developed a theoretical approach based on the concepts of percolation theory that relates soil formation rates to chemical weathering rates, and solute transport rates in the same soil. Their theoretical approach is valid if the solute transport rates is not equal to that of the flowing water, which is the case in Gaussian solute transport. Solute transport in highly heterogeneous porous media is characterized by heavy-tailed distribution of solute arrival times and, therefore, is not Gaussian (Sahimi 1987). The implication is that the traditional advection-dispersion equation is inadequate for describing solute transport in the typically heterogeneous geological media that weather to form soils. Following the earlier work of Sahimi *et al* (1983, 1986), Sahimi and Imdakm (1988), Hunt *et al* (2011), Ghanbarian-Alavijeh *et al* (2012), and Sahimi (2012) develop an entire solute arrival time distribution.

Briefly, a pore-scale distribution of conductances connecting neighboring pores is assumed, and the probability that an optimal flow path between planes separated by a distance x can be connected, which passes through no conductance smaller than some arbitrary value, is calculated. Such a procedure is based on the cluster statistics of percolation theory, described in chapter 4, and transforming variables from cluster volume to cluster length, and from percolation probability to conductance value. Then, the topology of the path so defined in terms of percolation variables, together with the distribution of conductance values on that path, is used to calculate the expected arrival time, $t(g)$, for solutes traveling over such a cluster characterized by a minimum conductance g. Finally, $t(g)$ is used to transform the result for the probability that a finite system is characterized by a particular controlling conductance to a probability that it gives rise to a particular arrival time. While the procedure is a bit easier to understand than the corresponding one for calculation of a spatial solute distribution at any particular instant in time, the two procedures are analogous. It is the latter spatial distribution on which reaction rate predictions are based. In particular, the temporal derivative of the mean solute position gives a solute flux, which is used interchangeably with a solute velocity.

We first note that Braun *et al* (2016) distinguished between three weathering regimes: one controlled by reaction kinetics at large flow rates; a second one controlled by advective transport at intermediate flow rates, and the third one that is controlled by diffusion at very slow flow rates. The assessment is generally correct, but we should keep in mind that the diminishing solute velocity with time tends to raise the importance of advection relative to reaction kinetics with increasing time scales even when flow rates are large (Yu and Hunt 2017b). The question of when weathering rates are limited by solute transport is usually addressed using the Damköhler number, Da, which is the ratio of a solute advection time to a reaction time, with transport limitations being relevant when Da >1. Yu and Hunt (2017b) calculated Da explicitly for non-Gaussian transport and showed that, over time scales exceeding days, chemical weathering is practically always limited by transport, rather than reaction kinetics, and that the chemical weathering rates measured by Salehikhoo *et al* (2013) can be predicted accurately when the Damköhler number was much larger than one. Yu and Hunt (2017b) did not, however, address the possible complications at very low Péclet number—the ratio of convective and diffusive transport rate—considered by Braun *et al* (2016) as leading to a dominance of the effects of diffusion on chemical weathering and soil formation. This perspective is in agreement with the fact that the present theoretical approach can predict soil formation rates at time scales up to 50 Myr (Hunt and Ghanbarian-Alavijeh 2016).

In the past, solute transport was used to model reaction processes in porous media in the long-time limit, and was utilized to predict silicate weathering rates and laboratory experiments on reactive solute transport (Hunt *et al* 2015), principally from the Hanford site, Washington State. The comparison showed that a single-medium model provides simultaneous agreement with all the experimental data considered. Furthermore, the same percolation conditions, namely, 3D flow connectivity and saturated (or saturating) conditions, were appropriate for all experiments, except one (Liu *et al* 2008) that required 2D flow connectivity and unsaturated conditions (Hunt *et al* 2015), which was interpreted as having been performed under conditions of predominant wall flow, a known complication for coarser Hanford site soils in which large particles near the walls generate highly permeable flow paths with such large pores that it is difficult to maintain local conditions of full saturation. Thus, in principle and with only a single exception, all the experimental data would fit on the same universal curve, given by dx/dt (see below), which is provided by percolation theory for saturated (or wetting) conditions, and 3D network connectivity. The same conditions were assumed by Yu and Hunt (2017a), with the theoretical equations used being from the simple scaling theory of percolation (Sahimi 1987, Lee *et al* 1999, Sheppard *et al* 1999).

10.3.1 The percolation model

Sahimi (1987) argued that, in porous media that are sufficiently disordered that water flow is dominated by the critical paths (see chapter 4), properties of solute transport follow the percolation power laws. Following his work, Lee *et al* (1999) studied solute transport in porous media and showed that

$$t \sim x^{D_{bb}}, \tag{10.7}$$

where D_{bb} is the fractal dimension of the percolation backbone, the flow-carrying part of the percolation clusters. Here, t is the time, and x is the transport distance. Equation (10.7) is completely similar to equations (10.4)–(10.6), except that in the context of soil formation, D_{bb} is associated with downward water fluxes, and for 3D connectivity it should be equal to either 1.87 for saturated conditions, or 1.86, for wetting conditions (Sheppard *et al* 1999), a negligible distinction. Sahimi and Mukhopadhyay (1996) discussed limitations on this result arising from certain classes of long-range correlations in the medium itself, but we ignore it here. The fact that $D_{bb} > 1$ implies retardation in the solute transport, a result of the topological complexity of the preferred flow paths near the percolation threshold that define the water flow rate, and also have orders of magnitude lower cumulative resistance than non-preferred paths. For dimensional consistency, we rewrite equation (10.7) as

$$x = x_0 \left(\frac{t}{t_0}\right)^{1/D_{bb}}, \tag{10.8}$$

with x_0 being the fundamental spatial scaling factor, as a network node separation, or the median particle size, d_{50}, and t_0 the ratio of d_{50} and the mean annual vertical flux at the pore scale. If the flow is 2D, such as along a fracture plane, or along the walls of a cylindrical core, under saturated conditions $D_{bb} \simeq 1.64$, as mentioned earlier in this book. If, however, flow is constrained to 2D surfaces, both in the field (Sahimi *et al* 1998, Glass *et al* 1998) and in experiments, unsaturated conditions are common, in which case, $D_{bb} \simeq 1.21$. For unsaturated 3D flow, $D_{bb} \simeq 1.46$ for drying, and $D_{bb} \simeq 1.87$ for wetting conditions. Hence, summarizing all of these, we have

$$t = \begin{cases} t_0 \left(\dfrac{x}{x_0} \right)^{1.46} & \text{3D drainage} \\[2em] t_0 \left(\dfrac{x}{x_0} \right)^{1.87} & \text{3D imbibition of saturated} \end{cases} \qquad (10.9)$$

and

$$t = \begin{cases} t_0 \left(\dfrac{x}{x_0} \right)^{1.64} & \text{2D saturated} \\[2em] t_0 \left(\dfrac{x}{x_0} \right)^{1.21} & \text{2D unsaturated} \end{cases} \qquad (10.10)$$

In equations (10.7)–(10.10), t is the time required for solutes to be transported a distance x, x_0, as before, is a fundamental length scale, which is assumed to be a pore separation or particle size, and t_0 is a fundamental time scale, defined by the fluid flow rate, $v_0 = x_0/t_0$, through a characteristic pore, which was discussed above. Note that if the fundamental length scale x_0 is as large as meters, then, in accord with human-introduced solutes (Hunt and Ewing 2016), the velocity of solute transport is the same as that of water at scales up to a few meters, implying that Gaussian transport models are presumably valid.

Since $x(t)$ represents a total solute transport distance, $dx(t)/dt$ would be a solute velocity. The rate of chemical weathering, as well as the soil production function, as functions of increasing time have been argued to be $dx(t)/dt$ (Hunt $et\ al$ 2015, Hunt and Ghanbarian-Alavijeh 2016). Thus, in the absence of erosion, the soil depth x as a function of time t is simply the integral of the soil production function over the time that the medium has been in place, which yields the solute transport distance as a function of time; that is, for 3D saturated or imbibition condition one has

$$x = x_0 \left(\frac{t}{t_0} \right)^{0.53}. \qquad (10.11)$$

(where, $0.53 = 1/1.87$). It should, however, be pointed out that equation (10.11) is not the fundamental equation for soil modeling, because it contains implicitly a specific history of the medium, but does not address the soil production as a function of its current state. As discussed above, the characteristic particle size x_0 in equation (10.11) is best estimated if we take it to be the median particle size, d_{50}. Defining t_0 under field conditions is more difficult, since one would need to determine precipitation, evapotranspiration, porosity and the difference of run-on and run-off at the particular site where the soil is produced, which will ultimately bring in the necessity of routing surface water, but is beyond the present scope. Moreover, in order to treat soil erosion, one must first derive an equation for the rate of soil production, $\mathcal{R}_s$, given by

$$\mathcal{R}_s = \frac{dx}{dt} = 0.53\left(\frac{x_0}{t_0}\right)^{-0.47} = 0.53\mathcal{I}\left(\frac{x}{x_0}\right)^{-0.87}, \tag{10.12}$$

where $\mathcal{I} = x_0/t_0$ is the deep infiltration rate, implying that it is the actual downward flux of the CO_2^- carrying water that is critical to the weathering reaction, which can be affected by climate (precipitation), plants (transpiration), surface water routing (run-on less run-off), and the medium hydraulic conductivity. In principle, t in equation (10.12) is the age of soil, but under steady-state condition such identification loses its meaning, because in that context the history of the soil has been lost. Under steady-state condition, one may interpret t as a particle residence time, which does not necessarily correspond to the soil age, even when steady state has not been reached. To include the erosion rate $\mathcal{E}(t)$, equation (10.12) is modified to

$$\mathcal{R}_s - \mathcal{E}(t) = \frac{dx}{dt} = 0.53\mathcal{I}\left(\frac{x}{x_0}\right)^{-0.87} - \mathcal{E}(t), \tag{10.13}$$

which can be solved numerically, if the function $\mathcal{E}(t)$ is known.

The solution of equation (10.13) under steady-state condition is simple:

$$x = x_0\left(0.53\frac{\mathcal{I}}{\mathcal{E}}\right)^{1.15}, \tag{10.14}$$

Note that the important roles of vegetation and temperature are subsumed in $\mathcal{I}$, which in the absence of overland flow is proportional to $(P-E)$, where P is the precipitation rate, and E is the actual evapotranspiration. In the present analysis, $(P-E)$ is used to estimate $\mathcal{I}$ and to explain why, at least to first approximation, this is a reasonable approach. The mean terrestrial P is reported to be between 850 mm and 1100 mm, with a mean of 975 mm (Willmott et al 1994), although Lvovitch (1973) reported its value to be 834 mm. A mean of the two estimates is 905 mm. Lvovitch (1973) also estimated the global mean value of E to be 65% of P [the factor 0.65 will actually be derived in chapter 11 based on a percolation model], so that $(P-E)$ is approximately 35.5% of P, which is 321 mm yr^{-1}. To justify this estimate, one should note that transformation of an incident atmospheric water flux to a pore-scale flow velocity requires division by the porosity ϕ, which is typically about 0.4. Run-off is estimated to average about 23% of precipitation worldwide. Thus, $(P-E$ run-off) is only about $0.13P$ and, therefore, the expected net (deep) infiltration rate $\mathcal{I}$ is approximate $0.13P/\phi = 0.13P/0.4 = 294$ mm yr^{-1}, which compares well with the earlier estimate, 321 mm yr^{-1}. We should, however, keep in mind that areas with a higher fraction of P lost to E, such as Australia, may have a lower run-off fraction, implying that the approximation will underestimate $\mathcal{I}$, and will introduce scatter in the predictions. Higher temperatures increase E and reduce soil development if the environment is water-limited to begin with. In addition, the soil depth is proportional to the particle size and nearly equal to the ratio of two rates, namely, infiltration and erosion. Reasonable values are $x_0 = 30$ μm and a net (deep) infiltration 1 m yr^{-1}, which lead to a steady-state soil depth of slightly over 2 m. But a decrease of the net infiltration rate to 0.1 m yr^{-1} yield (keeping the other parameters constant) a steady-state soil depth of about 15 cm. Therefore, the typical 10–100 cm depths estimated by the Heimsath and co-workers (1997, 2001a, 2001b, 2009) for climates with precipitations ranging from 200 mm to 2 m (and evapotranspiration accounting for something more than half of the precipitation) are in general agreement with the present predictions of the values of steady-state soil depths, although, as shown below, the overall agreement is not verified in many specific cases.

It remains to estimate the particle size. Soil ternary diagrams distinguish particles, in order of increasing size, clay, silt and sand. A middle silt particle size would, in the absence of any information regarding soil texture at a given site, be the best estimate for a median particle size. The size of silt particle varies from 2 to 63 μm (U.S. Geological Survey), with a mean value of 32 μm, or a geometric mean of 11 μm. In the following analysis and comparison with the data, if information on soil particle size was not available, a value of 30 μm was used for the median particle size, particularly since semi-arid regions tend to have somewhat larger particles. According to Sanderman and Amundson (2009), however, most of the soils at Tennessee Valley—the most humid region of those considered—fit into the clay loam category, with an average 36% sand, 24% silt and 40% clay. Thus, the median particle size should be slightly below the median silt particle of 11 μm. Accordingly, for this site only, 10 μm was used as the median particle size.

To calculate soil depth $x(t)$ under unsteady-state condition, equation (10.13) is integrated,

$$x = \int_0^t \mathcal{R}_s(\zeta)d\zeta = \int_0^t \left\{ 0.53\mathcal{I}\left[\frac{x(\zeta)}{x_0}\right]^{-0.87} - \mathcal{E}(\zeta)\right\}d\zeta, \qquad (10.15)$$

which, for a general $E(t)$, does not have an analytical solution, and is integrated numerically. Soil depths assuming steady-station condition are predicted using equation (10.14). In addition, one needs 'soil ages' to make prediction under unsteady-state conditions. In the context of the derivation of the model, it is more conducive to understanding the issues if we replace soil age with a typical soil residence time, since available information for 'soil ages' is limited. Thus, a method (Ivy-Ochs and Kober 2008) in cosmogenic nuclides dating was used to determine the 'ages' that are missing based on the concentration of beryllium isotope ^{10}Be and the aluminum isotope ^{26}Al, published along with the soil depths at the rest of the five sites. The 'age' of soil was calculated through the following equation,

$$C_I(t) = \left(\frac{\mathcal{P}}{\lambda}\right)[1 - \exp(-\lambda t)]. \qquad (10.16)$$

Here, $C_I(t)$ is the isotope concentration at time t, $\mathcal{P}$ is the nuclide production rate at the sampling site, with $\mathcal{P} = 6$ and 36.8 for, respectively, ^{10}Be and ^{26}Al, at sea level and high latitude ($>60°$) (Heimsath $et\ al$ 2000), $\lambda = \ln 2/t_{1/2}$ is the radioactive decay constant, with $t_{1/2}$ being the half-life of the isotopes and equal to 500 000 years and 701 000 years for ^{10}Be and ^{26}Al, respectively, and t is the exposure time of the bedrock where the concentration of nuclides was determined, which is assumed to be the same as the time period over which the soil was forming. The initial concentrations of the isotopes at time $t = 0$ was assumed to be zero. At $t = 0$, soil begins to form at the exposed rock surface due to weathering, and at the same time cosmogenic nuclides within the rock at depth h from the rock surface (depth $= 0$) starts to accumulate following equation (10.16). Over a certain time period t, as long as the material at depth h from the surface of the soil is still in rock form and has not been weathered, accumulation of the nuclides should follow the same equation, regardless of how much rock has been altered into soil on top of it. Therefore, the exposure age calculated based on equation (10.16) based on the concentration of nuclide $C_l(t)$ determined within the bedrock at depth h corresponds to the time period over which the soil was forming. Table 10.2 compares the calculated times t based on equation (10.16) with published values of 'age' at Nunnock River and Frogs Hollow (Heimsath *et al* 2000, 2001a). The discrepancies for 16 out of 19 data points of the two datasets are less than 1%. The highest discrepancy is only 2.85%.

Table 10.2. Verification of calculated soil residence time based on concentrations of cosmogenic nuclides with published data. The data are from Heimsath *et al* (2000, 2001a). Published age is published soil residence time by Heimsath *et al* (2000, 2001a).

Be (atom g^{-1})	Al (atom g^{-1})	Predicted average age (kyr)	Data (kyr)
Frogs Hollow			
103 217	592 709	16.75	16.60
99 159	608 001	16.62	16.5
202 519	1 059 721	31.62	31.2
392 381	2 505 639	68.45	66.6
422 049	2 637 396	72.92	70.9
Nunnock River			
513 300	3 059 000	87.01	86.81
113 700	624 100	18.07	18.02
107 100	601 500	17.20	17.16
249 100	1 353 000	39.69	39.58
168 300	1 080 000	29.01	28.94
148 000	829 300	23.80	23.74
195 700	1 087 000	31.42	31.34
234 600	1 402 000	39.15	39.22
262 900	1 503 000	42.98	42.86
87 590	522 300	14.47	14.53
319 000	1 870 000	52.98	53.23
166 500	1 012 000	27.91	27.80
229 000	1 308 000	37.42	37.32
152 800	931 600	25.63	25.55

10.3.2 Comparison with data

The predicted soil depths, under both steady- and unsteady-state conditions have been compared with field data at five sites, namely, Snug, Brown Mountain, Tin Camp Creek, Frogs Hollow, and Nunnock River in southeastern Australia, and two sites in California (San Gabriel Mountain) and in Tennessee (Tennessee Valley), reported by Heimsath *et al* (1997, 2000, 2001a, 2006, 2009, 2012). Individual infiltration rates $\mathcal{I}$ at each study site are presented in table 10.3 (taken from Hunt and Ghanbarian-Alavijeh 2016). Erosion rates along with the soil depths (Heimsath *et al* 1997, 2000, 2001a, 2006, 2009, 2012) for individual soil sample were used as values for $\mathcal{E}$ in equation (10.14).

Figure 10.5 compares the predicted soil depths under steady-state condition with the field data at five Australian sites, where the bounds on prediction are taken from

Table 10.3. Precipitation P and actual evapotranspiration E rates at the study sites.

Site	P (m yr^{-1})	E (m yr^{-1})	$(P{-}E)$
Nunnock River	0.71	0.6	0.11
Frogs Hollow	0.7	0.6	0.1
Tin Camp Creek	1.4	1.1	0.3
Snug	0.91	0.65	0.26
Brown Mountain	0.71	0.65	0.06
Tennessee Valley	1.2	0.6	0.6
San Gabriel Mountains	0.81	0.35	0.46

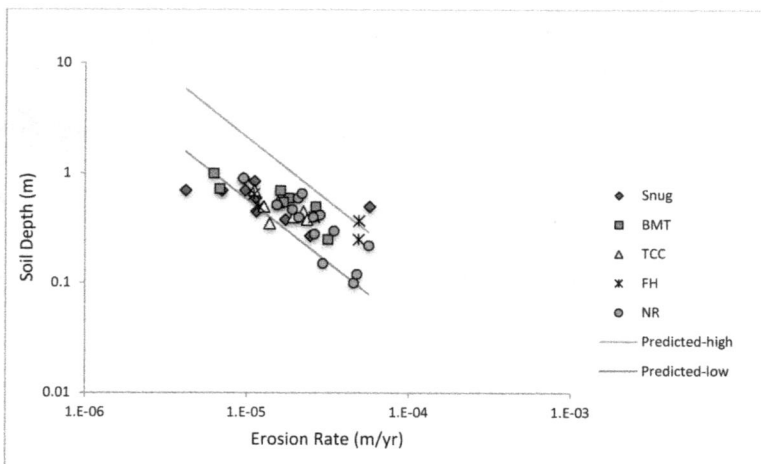

Figure 10.5. Predicted soil depth assuming steady-state conditions at five study fields in southeastern Australia, with the upper and lower bounds being, $\mathcal{I} = 0.31$ m yr^{-1} and 0.1 m yr^{-1}, and particle size of 30 μm. Data are from Heimsath *et al* (2000, 2001a, 2006, 2009). Yu and Hunt (2017a) John Wiley & Sons. Copyright 2017 John Wiley & Sons, Ltd.

relevant variability in infiltration among the five study sites. The highest infiltration rate, $\mathcal{I} = 0.31$ m yr^{-1} at Snug, and the lowest, $\mathcal{I} = 0.1$ m yr^{-1} at Frogs Hollow, were used to calculate the upper and lower bounds of the theoretical soil depths. The majority of the data points are within the predicted limits and follow the scaling predicted by equation (10.14). There are, however, a few data points (some soils at Snug, BMT and TCC) that not only do not agree with predicted bounds, but also with the scaling relation, especially at small erosion rates, indicating possible unsteady-state condition at lower erosion rates.

To examine any possible unsteady-state condition, the relevant parameters, including published data for soil depths and erosion rates, the corresponding soil residence time calculated using equation (10.16), and the limits of infiltration rates among sites were used in equation (10.15) to predict soil depths assuming unsteady-state condition. The results are shown in figure 10.6, which indicates that better agreement with the data is obtained at slower erosion conditions without assuming steady-state condition. It should be emphasized that although unsteady state was assumed, steady-state condition can be reached at high erosion rate when 'soil age' t is larger than the time t_s it takes for production rate $\mathcal{R}_s$ to diminish to equal the erosion rate $\mathcal{E}$. Therefore, the predicted soil depths shown in figure 10.6 do not exclude possible steady-state conditions, which is also indicated by the predictions lining up on a straight line, having overlap with the predictions assuming steady state, but not shown in the figure at large erosion rates. The better agreement between the theoretical predictions and field data for soils at small erosion rates without assuming steady-state condition indicates that these soils are probably not

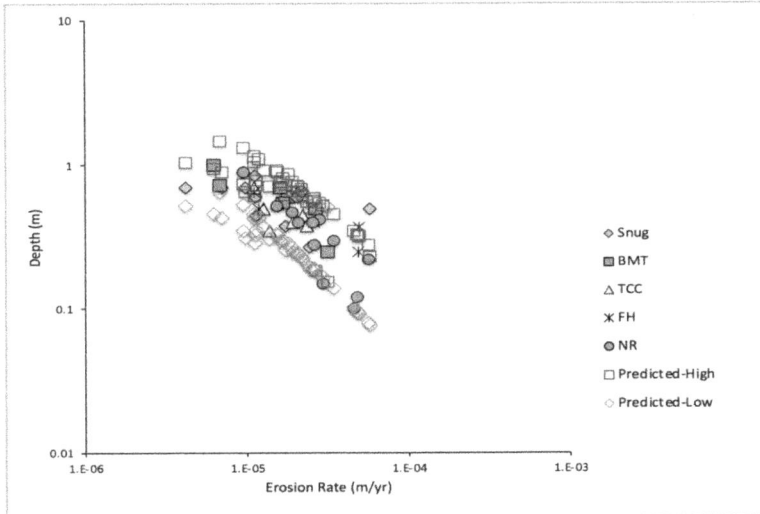

Figure 10.6. Predicted soil depth assuming unsteady-state conditions at five study fields in southeastern Australia, with the upper and lower bounds being, $\mathcal{I} = 0.31$ m yr^{-1} and 0.1 m yr^{-1}, and particle size of 30 μm. Data are from Heimsath *et al* (2000, 2001a, 2006, 2009). Yu and Hunt (2017a) John Wiley & Sons. Copyright 2017 John Wiley & Sons, Ltd.

at equilibrium, while the consistency between predictions assuming both steady- and unsteady-state conditions and observed soil depths at large erosion rates, suggest that most of these soils have reached equilibrium condition.

Comparisons of the predictions and data for San Gabriel Mountain (SGM) and Tennessee Valley (TV) are shown in figures 10.7 and 10.8, respectively. For SGM, the infiltration rate was determined from the evapotranspiration E and precipitation map across the United States (Sanford and Selnick 2013). Due to the wides ranges of

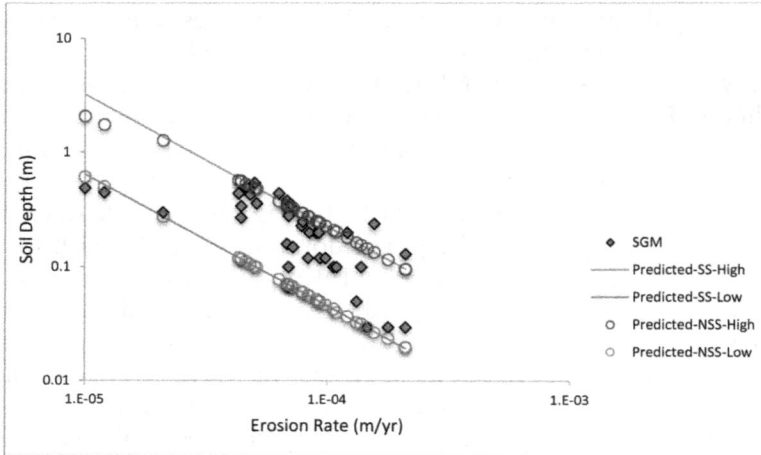

Figure 10.7. Predicted soil depth assuming steady-state (SS) and non-steady-state (NSS) conditions at San Gabriel Mountain, with upper bound and lower bound being $0.44 \, \mathrm{m \, yr^{-1}}$ and $0.11 \, \mathrm{m \, yr^{-1}}$, and particle size of $30 \, \mu\mathrm{m}$. Data from Heimsath *et al* (2012). Yu and Hunt (2017a) John Wiley & Sons. Copyright 2017 John Wiley & Sons, Ltd.

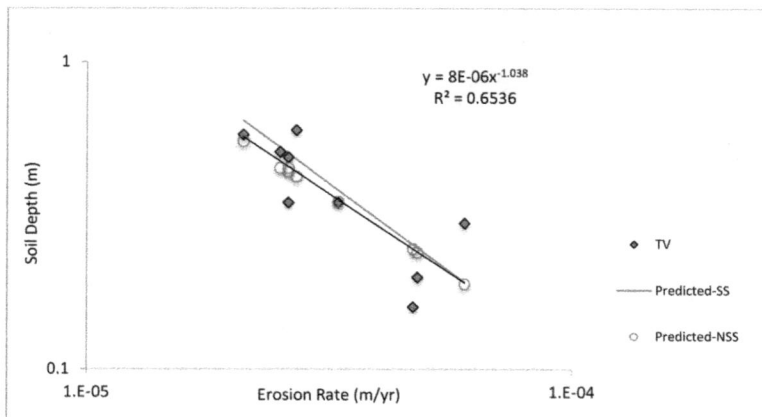

Figure 10.8. Predicted soil depth assuming steady-state (SS) and non-steady-state (NSS) conditions at Tennessee Valley, with infiltration being $0.6 \, \mathrm{m \, yr^{-1}}$ and particle size of $10 \, \mu\mathrm{m}$. Data from Heimsath *et al* (1997). Yu and Hunt (2017a) John Wiley & Sons. Copyright 2017 John Wiley & Sons, Ltd.

of evapotranspiration, 0.31–0.4 m yr^{-1}, and precipitation, 0.51–0.75 m yr^{-1}, showing on the map, and the various microclimates across Los Angeles County, where SGM is located, the infiltration rate at SGM was determined as a range with an upper bound of $0.75 - 0.31 = 0.44$ m yr^{-1} and a lower of $0.51 - 0.4 = 0.11$ m yr^{-1}. For the TV site, typical particle size was 10 μm, and the infiltration rate is what is shown in table 10.2. As the comparisons indicated, soils at both SGM and TV sites are more likely to have reached steady-state condition, as there is good agreement between the predictions and the data assuming both conditions. Since the calculations for soil depth at unsteady state include the scenario of steady-state condition, the discrepancy between the two cases would get smaller as soil approaches steady state, and diminish to zero when steady-state conditions are reached. In addition, the majority of the data for the SGM site are confined by theoretical boundaries, and for the TV site the data are distributed around the theoretical soil depth with a power of -1.038, compared with -1.149 of equation (10.14), indicating that it is likely that soils at TV have also reached steady-state condition.

The discrepancy between the predictions and field data for the two site may also be for reasons other than the failure of steady-state assumption, as indicated by the consistency between predicted soil depths assuming both steady- and unsteady-state conditions, and the data. Possible reasons causing the spread could be as folows. (1) Specific characteristics of the field samples, such as the variations in particle size and infiltration rate, as uniform parameters were used at each study site for prediction. (2) Other surface processes that cause fluctuation in the erosion rate. There are factors that could cause variations in soil transport along the surface, and by taking a constant erosion rate, the model only deals with ideal circumstances. (3) The uncertainty in erosion rate $\mathcal{E}$ and the time period t over which the soil is forming. Since equation (10.16) deals with exposed rock surface, nuclide production rate at any given depth is probably lower than that at the surface, leading to a possible underestimation of t that is used in the model, and the published erosion rates that are used in the model were calculated based on the assumption of steady-state condition, which might not be the actual values. (4) Other physical and biological processes that affect soil formation that are not taken into account in the model. These reasons pose major difficulties in predicting soil depth accurately, as the deviations from the $y = x$ line indicates in figure 10.9. In this figure an exponent of 0.76 was estimated based on the plotting of the prediction against data at all seven study sites. Since shallow soil depth is more vulnerable to outside forcing, such as trampling from animals, the removal of the three shallowest soil samples at depth 0.03 m resulted in closer conformance with linearity (an exponent of 0.84 not shown in figure 10.9). Thus the accuracy is not sufficient to certify correctness of the model. But, in addition to such difficulties, one main reason that the accuracy of the model cannot be comprehensively assessed is the lack of data.

Summarizing this section, a model of soil production was developed that is tied to solute transport-limited chemical weathering, and was used to address the question of whether the common assumptions regarding steady-state conditions of soil columns are justified. The study indicates that steady-state conditions are more nearly attained at higher erosion rates, consistent with an earlier suggestion by

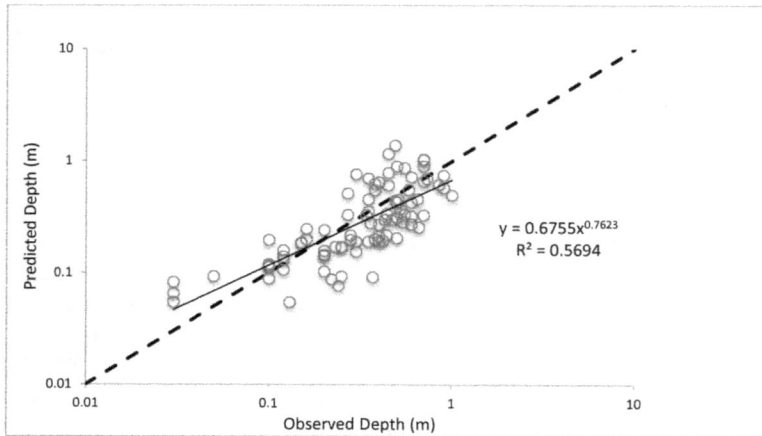

Figure 10.9. Observed soil depth versus predicted soil depth at all seven study sites. Infiltration rate varies between sites. The SGM site has infiltration rate ranging from 0.11 to 0.44 m ye^{-1}, and an average of 0.27 m yr^{-1} was used. The root mean square deviation is 22.6%. Yu and Hunt (2017a) John Wiley & Sons. Copyright 2017 John Wiley & Sons, Ltd.

Phillips (2010), but that this does not necessarily translate to other studies or other regions. In less arid regions, for example, soil production tends to be larger, and steady-state conditions can be attained at shorter time intervals. The predictions of the model are also limited by the incomplete information about variations of the relevant parameters, including the particle size and net infiltration rates. Moreover, there are uncertainties contributed by the time period for soil formation and erosion rates, either as theoretically estimated or in the published data based on the concentration of cosmogenic nuclides. Neglecting the effect of other physical and biological processes may also affect the accuracy of the prediction. Given the fact that solute transport in the soil is so slow (and keeps slowing down with time), it may still be one of the major limiting factors of soil formation when all other relevant processes are considered.

10.4 The 'mystery' of one meter-deep soils

What is a typical soil depth? One naturally thinks that the typical depth should be, at least in a narrow sense, a mean soil depth. While the actual values of soil depths vary from zero to tens of meters, the typical soil depth that many authors have reported is about 1 m. Batjes (1996) studied the 4353 soils in the World Inventory of Soil Emission (WISE) and used it to build up the database of United Nations Educational, Scientific, and Cultural Organization (UNESCO) with 106 soil types presented on its soil map, assuming a characteristic soil depth of 1 m. Montgomery (2007) reported a mean soil depth of 1.09 m, with a mean soil depth of 2.74 m for native vegetation, and 2.01 m for soil production areas. Hillel (2005) suggested that 1 m is a typical soil depth. Let us consider the issue using the percolation model described above (Yu *et al* 2017).

Consider equation (10.14) again, modified slightly to account for the effect of porosity ϕ,

$$x = x_0\left(0.53\frac{\mathcal{I}}{\phi\mathcal{E}}\right)^{1.15}, \tag{10.17}$$

Assume that $x_0 = 30\ \mu$ m, the size of a typical silt particle. Silt is the middle (geometric mean) particle size class in all soil classification schemes, and 30 μm is the middle (arithmetic mean) of the silt range. The same value is also the geometric mean of the individual arithmetic means of the three principal soil particle classes, namely, clay, silt, and sand. This particular length scale relates most closely to parent material, whether the soil is weathering from a bedrock with a specific mineral size, or is forming on, for example an alluvial deposition. As discussed above, to calculate a mean infiltration rate, we must consider not only the precipitation, but also the water lost to evaporation and transpiration, as well as what runs off along the surface. These variables relate to climate, the hydraulic conductivity of the substrate, and to the role of plants in the water cycle. It was already mentioned that Schlesinger and Jasechko (2014) estimated that, globally, transpiration constitutes 61% of actual evapotranspiration E, and returns approximately 39% of precipitation P to the atmosphere and, therefore, $E \approx 0.39/0.61 = 0.64P$, while Lvovich (1973) estimated that the global mean precipitation is 834 mm, and that a global mean of 24% of P travels to streams by overland flow, leaving only 11% of P for deep infiltration. The mean terrestrial P is, as mentioned in the last section, between 850 mm and 1100 mm, with a mean of 975 mm. Sixty four percent of 975 mm is 624 mm, leaving 351 mm for $(P–E)$. However, 11% of 975 mm is only 102 mm. On any local site, however, the difference between the run-on and the run-off can be either positive or negative. Thus, these estimates suggest that the amount of water reaching the base of the soil should be a column of water somewhere between 102 mm and 351 mm. Alternatively, one can consider the mean global E over cold, temperate, and tropical, forested and non-forested, regions. Using the six different values given by Peel et al (2010) for these biomes yields $E = 654$ mm, fairly close to the value inferred from Schlesinger and Jasechko (2014), and implying $(P–E) = 321$ mm. The actual infiltration rate is $\mathcal{I}/\phi$. Assuming a typical porosity of 0.4 leads to $\mathcal{I}/\phi$ being between 255 mm yr^{-1} and 878 mm yr^{-1}, or between 225 mm yr^{-1} and 735 mm yr^{-1} (using data reported by Lvovich), which average to 566 mm yr^{-1}, or 480 mm yr^{-1}, depending on the particular estimates applied. A typical erosion rate of about $\mathcal{E} = 30$ m/Myear $\approx$ [(1 m Myr^{-1}) (1000 m Myr^{-1})]$^{0.5}$, is obtained from the geometric mean of the range of erosion rates discussed earlier. Thus, using $x_0 = 3 \times 10^{-5}$ m, $\mathcal{I}/\phi = 806$ mm yr^{-1}, and $\mathcal{E} = 30$ m Myr^{-1}, and the first range of $\mathcal{I}$ values given, the result for x is 0.48 m $< x <$ 1.81 m, while for the second range of the $\mathcal{I}$ values, one obtains 0.42 m $< x <$ 1.53 m. Thus, both the arithmetic and geometric means of both ranges cluster around 1 m.

The ratio $\mathcal{I}/E$, raised to the power 1.15, has the potential to produce the greatest variations in soil depths, but it is quite insensitive to precipitation P, since both $\mathcal{I}$ and $\mathcal{E}$ tend to increase with increasing P. Dunne et al (1991) reported a linear relationship between $\mathcal{I}$ and P, which is in general agreement with the aforementioned fact that E is $0.5P$. Reiners *et al* (2003) reported a linear relationship between P and $\mathcal{E}$. They utilized a rainfall gradient at similar temperatures across the Cascade Mountains in Washington State, United States. Thus, one should expect roughly constant soil depths, and the ratio $\mathcal{I}/E$ should, in the absence of steep topography, remain relatively invariant. Data reported by Sanford and Selnick (2013) reveal a tendency for the fraction of precipitation lost to E to increase with increasing temperature, particularly in conjunction with aridity. Thus, the conclusion of Heimsath *et al* (2010) that, '[t]he suite of results from different field sites [in Australia] indicates that erosion rates generally increase with increasing precipitation and decreasing temperature,' indicates that the processes of soil formation may be dependent on evapotranspiration. Consequently, the water potentially available for either infiltration or overland flow, $(P–E)$, serves as a predictor of E and soil formation, rather than simply P. Thus, Yu *et al* (2017) hypothesized that both the numerator and denominator in equation (10.17) would be proportional to $(P–E)$, implying that weather conditions (within a specific climatic zone) would have far less influence on soil depth than commonly assumed. $\mathcal{I}$ and $\mathcal{E}$ can, however, be expected to have a complementary dependence on the partitioning of water to overland flow, which brings in the effect of topography. The relationship between the potential evaporation, evapotranspiration, precipitation, and run-off was considered in great detail by Budyko (1974) and others.

Topography may affect erosion rates. Regions with steeper topography tend to have higher overland flow and, thus, higher erosion rates, which results in lower infiltration and soil formation rates. Burbank *et al* (2003), for example, reported that erosion rates and precipitation in the Himalayan mountains were not correlated, and attributed their anomalous data to the strong tendency for the precipitation to decline where the slope increases. Notably, however, compared with erosion, the declining precipitation with increasing slope should lead to a diminution in soil production, and a higher probability of bedrock exposure, as is indeed the case in the Himalayan region. Divergent topography, with concomitant divergence in surface water flux and, therefore, soil transport, produce thinner soils than convergent topography, as noted by Heimsath *et al* (1999), a tendency intensified by steeper topography generally. Thus, the reasoning presented here, though it may be accentuated in reality by lateral soil transport, does not depend on such transport, and is merely a consequence of the greater infiltration rates in topography that is convergent and not so steep that soil covering is missing entirely.

Consider the dependence of soil depth on slope angle, as exhibited by the data from the San Gabriel Mountains (Heimsath *et al* 2012) and data reported by Norton and Smith in 1930, as reported in Jenny (1941). In this case, one has, $\mathcal{I}/\phi = 0.55$ m yr^{-1}, $E = 35$ m Myr^{-1}, and $P = 0.81$ m yr^{-1}. If equation (10.17), together with the known value of $(P–E)$ run-off and a 30 μm median particle diameter, is used, in order to address the slope dependence of soil depth, it requires a slope-dependent erosion rate. Mongomery and Brandon (2002) reported in their figure 1 an empirical function for the slope angle-dependence of erosion rates in the Olympic Mountains in Washington, which is given by

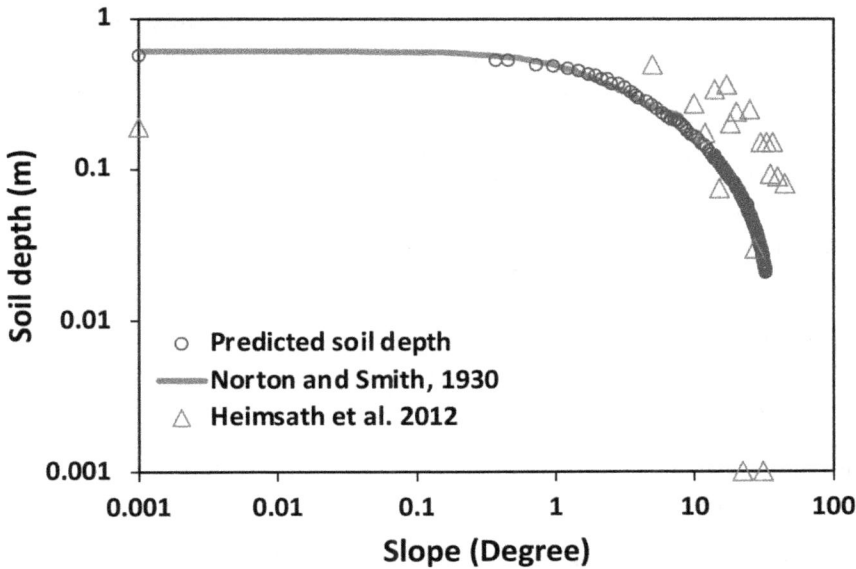

Figure 10.10. Predicted soil depth versus data as a function of the slope. The Norton and Smith data were reported by Jenny (1941), but extend only to a slope of 8°. The Heimsath *et al* (2012) data extend over a range of 6°–32°. Reproduced from Yu *et al* (2017). CC BY 4.0.

$$\mathcal{E} = \mathcal{E}_0 + \frac{kS}{1 - (S/S_0)^2}, \qquad (10.18)$$

where S is the slope in degrees, $S_0 = 40°$ is a limiting hillslope, $\mathcal{E}_0$ is the background erosion rate due to chemical weathering whose reported values vary from 0.016 to 0.059 mm yr^{-1} for six major drainage basins in the Olympics Mountains, with its fitted value in equation (10.18) being 0.05 mm yr^{-1}, and $k = 0.6$ mm yr^{-1} is a rate constant. Incorporating equation (10.18) in equation (10.17) makes it possible to use the latter to predict the slope angle-dependence of soil depth. The comparison is shown in figure 10.10. To make equation (10.17) predictive in this regard, the input of the erosion rate function, equation (10.18), is critical, which tends to produce a rapid reduction in soil depths to nearly zero, if slopes of about 30 degrees are exceeded (since zero values cannot be plotted on a logarithmic graph, such values were converted to 0.001 for both axes). Despite the considerable scatter in the field values, it appears that the predictions of equation (10.17) capture the essential trends accurately. Note that the use of either 20 μm or 30 μm for the fundamental particle size will result in an overestimation of the soil depth at zero slope, but an underestimation at larger slopes, since the latter values are deeper than the zero slope depths.

10.5 Soil depth and landslides

Steep topography is an important aspect of geomorphology. The morphologic characteristics of landscapes reflect the complex feedback between tectonics and climate-driven processes in sculpting the topography. Tectonically active landscapes are often exposed to natural hazards, such as landslides, debris flows, floods, and earthquakes, phenomena that typify threshold processes that have significant influence on topography. There are still many unresolved questions about threshold landscapes. They range from the conceptual (DiBiase *et al* 2012) issues, such as whether such landscapes respond meaningfully to changes in erosion rates, to the practical (see, e.g., Montgomery and Brandon 2002, herefater referred to as MB) problems, such as how to predict such erosion rates. The key question is, what constitutes steady state in a threshold landscape? Several worldwide studies demonstrate an equivalence between soil production and soil erosion. In a threshold landscape, such an equivalence can only reflect a spatial or long-term average. As discussed above, the relevance of steady-state landscapes has been called into question, even though an equivalence of soil production and soil erosion rates is often assumed. Nevertheless, as the discussion above indicated that, while slowly evolving landscapes in arid continental interiors are unlikely to be in steady state, tectonically active regions are much more likely to conform to steady-state conditions, at least if erosion processes were largely gradual. The same conclusion was reached in a study of Braun *et al* (2016).

Beginning with Gilbert (1877), many studies have addressed quantitative understanding of the relationship between erosion rate and topographic elements, including hillslope gradient, topographic relief, hilltop curvature, and drainage density (see e.g., Ahnert 1970, Binnie *et al* 2007, DiBiase *et al* 2010, 2012, Hurst *et al* 2012, Montgomery and Brandon 2002, Roering 2008). General conclusions have, however, been slow to emerge. Ahnert (1970) reported a linear relation between erosion rate and mean local relief at midlatitude drainage basins. Several other studies demonstrated, however, that the linear relationship breaks down as the mean slope increases and approaches a threshold angle of stability S_c (see, e.g., Carson and Petley 1970, Binnie *et al* 2007, DiBiase *et al* 2010, Montgomery and Brandon 2002, Ouimet *et al* 2009) at which downslope sediment fluxes can become infinite (Roering *et al* 2007). In this case, sediment flux switches from creep-related process to mass wasting (DiBiase *et al* 2012), and landslides can occur such that hillslope lowering prevents the hillslope from becoming steeper than S_c, and erosion rate and topographic relief become decoupled (Burbank *et al* 1996, Montgomery 2001, Montgomery and Brandon 2002). The fundamental characteristics of the empirical erosion formula given by the last authors incorporate effects from both gradual erosion and landsliding and is utilized in the percolation model here.

Accurate modeling of landslides and the processes that lead to them is important not only to a general understanding of geomorphology of soil, but also to prediction and diagnosis of the resulting hazards. According to Claessens *et al* (2007), 'High annual rainfall, steep slopes, deforestation, high weathering rates, and slope material with a low shear strength or high clay content are considered the preparatory causal factors for mass movements.' To predict when landslides may be imminent, one must also take into account the moisture history of slopes, modeling the spatial distribution of moisture, pore pressure, and soil strength (Dietrich *et al* 1995), address the effect of Coulomb failure and friction forces (Dietrich *et al* 2007), and consider many combinations thereof (Claessens *et al* 2007). Nevertheless, soil depth is often considered (Iida 1999, Okimura 1987) to be the most important parameter for predicting the risk of shallow landsliding. Thus, understanding soil depth development on steep topography may be useful for understanding and predicting the frequency of landsliding occuurences in threshold landscapes. But modeling threshold processes, let alone landscapes, entails addressing additional theoretical and practical challenges.

In the last section, the percolation model addressed the evolution of soil depth with gradual erosion processes. In this section the investigation is broadened to include mean soil depths between landslides in threshold landscapes. In particular, it is shown that, for the power-law dependence on soil depth of the soil production (SPR), the mean soil depth is goverened by an analytical result that is almost identical to that for the steady state soil depth for gradual erosion. Thus, for the same erosion rate, the predicted mean soil depth is scarcely dependent on whether the erosion is gradual or by a threshold process. Yu *et al* (2019) proposed to predict the mean soil depths in both landslide-dominated landscapes and those dominated by gradual erosion processes, such as soil creep, as long as one has specific evidence for the magnitude of the relevant erosion processes. It is clear that soil depth varies with hillslope gradient and, therefore, one requires either suitable measurements of erosion rates or an accurate model for it with hillslope gradient. To this end, Yu *et al* (2019) adopted the empirical model proposed by Montgomery and Brandon (2002), equation (10.18), together with solution of equation (10.13) as the SPR function.

10.5.1 Analysis of gradual erosion

As discussed earlier in this chapter, if a time-dependent denudation rate $D(t) = \mathcal{E}(t)$ is included in the percolation model, one obtains the governing equation for net formation rate of soil, which is equal to the time derivative of its depth dx/dt, given by equation (10.13) or, in integrated form, equation (10.15). In the simplest case, $D(t)$ is assumed to be constant, although clearly such climate change as occurred at the beginning of the Holocene produced detectable changes in landscape character-istics in many locations worldwide. $D(t)$ is often written as the sum of a slope-independent $\mathcal{E}_0$ and slope-dependent term kS; see equation (10.18). $\mathcal{E}_0$ is then usually conceptualized as a loss of soil volume due to the removal of soil products by chemical weathering, which may also be time dependent, since, as long as steady state is not reached, chemical weathering and soil production are time dependent. The term with linear dependence on slope is consistent with a steady-state solution of topography where soil production is proportional to the negative of the landscape curvature (see, e.g., Heimsath *et al* 1999). This result is then coupled with the inference that soil production on the top of hills is greater than that at the bottom, with an associated lateral transport of soil from the top to the bottom. For steeper slopes and threshold topographies, more complex forms of erosion rates from physical processes, such as dependence on depth, may be considered (Roering 2008). Regardless of the particular form of the slope dependence, whenever the long-term erosion rate can be considered time independent, $D(t) = D$, equation (10.13) becomes

$$\frac{dx}{dx} = 0.53\left(\frac{\mathcal{I}}{\phi}\right)\left(\frac{x}{x_0}\right)^{1-D_{bb}}, \tag{10.19}$$

where $D_{bb} \simeq 1.87$, and the infiltration rate $\mathcal{I}$ has been replaced by its actual value in soil, namely, $\mathcal{I}/\phi$. The solution x_{ss} of equation (10.19) at steady state is simple:

$$x_{ss} = x_0\left(0.53\frac{\mathcal{I}}{\phi D}\right)^{1/(D_{bb}-1)} = \left(0.53\frac{\mathcal{I}}{\phi D}\right)^{1.15}, \tag{10.20}$$

which is the same as equation (10.14), corrected for the effect of the porosity ϕ. If the assumption of steady state is relaxed, one has equation (10.15), but corrected for the effect of porosity, namely,

$$x = \int_0^t \left\{0.53\left(\frac{\mathcal{I}}{\phi}\right)\left[\frac{x(\zeta)}{x_0}\right]^{-0.87} - D(\zeta)\right\}d\zeta. \tag{10.21}$$

10.5.2 Analysis of threshold erosion

If landslides dominate the erosion, the approach based on purely gradual erosion is not valid. Consider, first, how to address a problem in which gradual processes are small enough to be negligible, in order to understand the overall effects of threshold erosion processes. An initial assessment is based on the magnitude of the contribution of landslides to the total denudation rate, compared with the contribution due to gradual processes. If the former is much larger than the latter, then, to first approximation, the gradual processes can be neglected, which gives rise to the question of how the effect of an unsteady erosion rate would contribute to the mean soil development. This can be determined by using equation (10.8) relate t_l, a typical time between landslides, and x_l, a typical soil depth at which landslides occur, in which case, $x_l/t_l = D$. Neglecting gradual erosion processes, and using, $t_0 = \phi d/\mathcal{I}$, one obtains

$$x_l = x_0\left(\frac{\mathcal{I}}{\phi D}\right)^{1/(D_{bb}-1)} = x_0\left(\frac{\mathcal{I}}{\phi D}\right)^{1.15}. \tag{10.22}$$

The only difference between equations (10.20) and (10.22) is the factor $0.53^{-1.15} \approx 2.05$, which, if it is approximated as a factor of 2, equation (10.22) would imply that when the majority of erosion occurs by landsliding, the maximum soil depth that is typically reached before a landslide occurs is twice the depth that would be attained if the same erosion rates were developed in a steady process. But the apparent contrast with gradual erosion processes is reduced, if one considers the temporal mean soil depth. An average depth over the time interval $0 < t < t_l$ is,

$t_l^{-1}(t_l^{1.53}/1.53) = t + l^{0.53}/1.53$, implying that the use of equation (10.21) to predict an average depth, instead of the final depth, would require substitution of $0.65t_l^{0.53}$ for $t_l^{0.53}$. Note that a numerical factor of 0.65 is not too different from the factor $1/D_{bb} = 0.53$. We assume that the mean soil depth over similar slopes in a specific geographical region characterized by an unsteady, but mean, denudation rate d will be the same as the temporal mean calculated above.

Finally, we note that if we substitute equation (10.18) for $D = \mathcal{E}$ in equation (10.20), we obtain for x,

$$x = x_0 \left\{ \frac{0.53\mathcal{I}}{\phi \left[\mathcal{E}_0 + \dfrac{kS}{1 - (S/S_c)^2} \right]} \right\}^{1.15}, \qquad (10.23)$$

which is obtained based on the assumption that D is independent of time. The predictions of equation (10.23) are highly sensitive to the parameters $\mathcal{I}$, x_0, and D, with the uncertainty in the predicted depth equal or greater than the uncertainty in the parameters. The latter uncertainties will turn out to be as large as a factor of 2. Equation (10.23) has six physical parameters: x_0, which is assumed (see, e.g., Yu *et al* 2017) to be a median particle size, d_{50}; $\mathcal{I}$, which is given as the deep infiltration rate; ϕ, the soil porosity; $\mathcal{E}_0$, a background erosion rate due to chemical weathering that takes into account mass lost in solution; k, the constant of proportionality, and S_c, a critical slope angle. $\mathcal{I}$ depends not only on climatic variables, but also on surface water routing. The only useful indication of how to take the run-on and run-off into account on any specific sites is found in global and continental estimates. In particular, Lvovitch (1973) cites the global proportionality, $\mathcal{I} = 0.23P$, with about a 50% variability in the numerical prefactor.

In addition, there is no general information to help one determine either $\mathcal{E}_0$ or k. Regarding S_c, the soil in the Olympic Mountains is only sustained from shallow landsliding at slope angles below 25° (Montgomery and Brandon 2002), even though the threshold angle is stated to be 40°. Landslide scars are also observed at the Apennine Mountains site. Salciarini *et al* (2006) studied slopes that ranged in overall steepness from 17.3° to 43.7°, which suggests a similar critical slope angle of 40°–45°, particularly since the data suggest a decrease in soil thickness by a factor 10 for slope angle of 25°–30°. These are are not greatly different from one another, implying that the original value of 40° for the Olympic Mountains could perhaps be reasonably used more generally, except in the the San Gabriel Mountains where a more suitable estimate can be made. In fact, whenever no other viable options were available, all the original erosion parameter values from the Olympic Mountains will be used in the comparison between the predictions and the data.

10.5.3 Data for landslide recovery

Let us now describe the sites whose data are used in the comparison with the predictions of the percolation model. Two sets of comparisons will be made. The first one is comparison of the predictions of equation (10.20) or (10.21) with the data, while the second comparison will be between the predictions of equation (10.23) with the data. The data for the first set of comparisons are as follows.

Taranaki Peninsula, North Island, New Zealand: This site extends southwestward from the city of Stratford on the southwestern coast of New Zealand. The Taranaki hill country 30 km to the east of Stratford is the site of the study by Trustrum and de Rose (1988) of landslide recovery. The yearly mean precipitation of the site is 1873 mm, with the substrate being a silty sandstone. The Taranaki Basin, which contains the only proven hydrocarbon reserves of New Zealand, extends from the hill country along the Peninsula to offshore and, due to its economic significance, it has been carefully characterized. Armstrong *et al* (1998) studied the evolution of porosity in the Taranaki province in order to deduce the exhumation magnitudes and erosion rates of the region. The study was carried out on and near the Taranaki Peninsula that is located between 39°S and 39.5°S and at about 174°E. Armstrong *et al* (1998) too state that the overlying rocks are silty sandstones. The porosity of these sandstones vary from 50% offshore (173°E) to between 15% and 20% on the eastern margin near 175°E (Armstrong *et al* 1998). The information was used by Armstrong *et al* (1998) to infer exhumation magnitudes and corresponding denudation rates of 400 m Myr^{-1} at the eastern margin and even 900 m Myr^{-1} in the northeast. The southernmost onshore regions have a denudation rate of 100–200 m Myr^{-1}. On the other hand, McBeath (1977) stated (in reference to the Taranaki Basin sandstones) that, 'The average porosity of gas-bearing sands penetrated by the wells is 18.8%.' He also noted that early nineteenth century European settlers discovered oil seeping from the surface in this region, and also presented a table that quotes 15% as the average porosity of the petroleum-bearing rocks. In the comparisons below, value 18.8% was used for the porosity. The median particle diameter for a silt is 15 μm, while for a sand it is approximately 115 μm (Skaggs *et al* 2011). An arithmetic mean of 65 μm of 15 and 115 μm is used in the comparison, since a silty sand should be closer to a sand than a silt. A similar result of 62 μm is obtained using a geometric mean of 70% sand and 30% silt, for example. The fraction of precipitation infiltrating is again assumed to be 23% (Lvovitch 1973).

Rossberg, Switzerland: This site is in the northern part of the European Alps, whose geology is characterized by subalpine molasse. The alluvial deposits are part of the Lower Freshwater Molasse (Lower Freshwater Molasse) of the Oligocene. Due to its regular, geological structure with oblique, southward dipping layers, the southern side of the Rossberg Mountain has been susceptible to landslides. Historic and prehistoric landslides are known and were dated (Egli and Fitze 2001, Keller 2017). The erosion data for the Rossberg site were accessed from Bundesamt für Umwelt (2015), while soil characteristics were obtained from Meili (1982) and Bundesamt für Landwirtschaft (2012).

San Gabrial Mountains, California: Instead of using purely statistical means to estimate the parameters for the SGM that Montgomery and Brandon (MB) (2002) present, all the observations of MB are taken into account. The contribution from a background erosion to soil production is suggested to be limited to 170 m Myr^{-1} or less for slopes of less than about 30°, for which MB state that, 'Samples collected across this more slowly eroding, convex-up part of the landscape define a robust soil-production function, with its maximum equaling 170 ± 10 m Myr^{-1},' while landsliding contributes up to an additional 200 m Myr^{-1} at steeper slopes, for a total erosion of about 370 m Myr^{-1}. Moreover, 'As morphology shifts from convex-up to planar, slope gradients increase and we observed the soil mantle transition from being ubiquitous to becoming increasingly patchy. We focused sampling of steep (average slope $>30°$) hillslopes on smooth, locally divergent ridges away from any landslide scars, thus ensuring that our ^{10}Be concentrations represent SPRs. Importantly, we observed that soil patches on threshold slopes, although typically thin (< 20 cm) and coarse grained, are clearly produced locally and are not colluvial accumulations. SPRs from saprolite under these thin to nonexistent soils are among the highest such rates ever reported, and exceed the maximum soil production predicted from the low-relief soil pits by up to a factor of four, with a predicted maximum rate of 370 ± 40 m Myr^{-1}.' Heimsath et al (2012) also reported a steepest observed slope of 45° that contained soil, meaning that the threshold slope angle is likely larger than 40°.

Sanford and Selnick (2013) reported a precipitation range for the SGM of 51–75 cm, with the middle being 63 cm. The fraction corresponding to infiltration is, according to Lvovitch (1973), 23%, yielding 14.5 cm yr^{-1}. Division by a typical porosity of 0.4 gives 36 cm yr^{-1}. Soils in SGM, at least on hillslopes, are mainly loams (Rulli and Rosso 2005), with median particle size ranging from 20 to 40 μm (Skaggs et al 2011). Thus, 30 μm was taken as a typical particle size. As noted, the range of reasonable values for each of the parameters is roughly a factor of 2, though, for example, adding variability in soil texture would increase this uncertainty.

10.5.4 Data for soil depth as a function of slope

The sites whose data are used for comparison with equation (10.23) as as follows.

Appenine Mountains, Italy: Soil in this region are (Salciarini et al 2006) mainly talus and are much coarser than the SGM. The particle size is estimated based on the hydraulic conductivity, 10^{-5}–10^{-2} m s^{-1}. at the site. Based on the information provided by Aqtesolv (nd), United States Geological Survey, and United States Department of Agriculture, comparable hydraulic conductivities are found within the category of coarse sand, which has particle size range from 600 to 2000 μm. The geometric mean of 1095 μm is taken as the typical particle size.

Sterling, Colorado: Soils at this site are mainly fine loamy and fine silty sand with percentage of sand ranging from 42% to 54% (Moore et al 1993), similar to the loamy texture in the SGM. Thus, same particle sizes as those for the SGM were also used. When depth values are unavailable at slope close to zero, the deepest soil depths obtained (mostly in the valleys) in the field for each site are used as estimations of soil depths at zero slope. These values may be affected by landslide deposits, but they are the only possible values to choose for normalization to zero slope.

Plastic Lake, Canada: Buttle *et al* (2004) studied this site, but did not mention its soil texture. The particle size was simply taken to be 30 μm, the typical particle size of soil, which is calculated from the geometric mean of individual arithmetic means of the three principal soil particle classes, clay, silt, and sand. It is also known that silt is the middle particle size (geometric mean) class in soil classification schemes that has a mean value of 32 μm. Thus, a 30 μm of typical particle size would be a reasonable choice in the absence of any information regarding soil texture at Plastic Lake.

Data Reported by Norton and Smith (1930): Their data were reproduced by Jenny (1941), and are used in the comparison. There is a slight upturn in the curve at the largest slope angles, which is attributed to a graphing error of the times.

10.5.5 Comparison of predictions of the percolation model with data

As the first step, the value of t_l is compared with t_{ss}, the time to reach steady state. Since a power-law decay is used for soil production rate, the usual rigorous definition of a time scale consistent with an exponential decay is not possible, but one can determine a minimum time to reach steady state by setting the erosion-free soil depth at t_{ss} equal to the depth at steady state. Thus, using $x = x_0(t_{ss}/t_0)^{1/D_{bb}} = x_0[(x_0/t_0)/(DD_{bb})]^{1/(D_{bb}-1)}$, one obtains

$$t_{ss} = \frac{x_{ss}}{DD_{bb}}, \qquad (10.24)$$

where x_{ss} is the steady-state soil depth given above. The two limiting cases of landslide-dominated and gradual erosion-dominated landscapes can now be compared by comparing t_{ss} and t_l. Using equation (10.24) and the definition of t_l given above, it is clear that t_{ss}/t_l is a ratio of rates of erosion by landsliding and gradual processes. Thus, when $t_{ss} \gg t_l$, landslide erosion rates are much larger than the gradual contributions, and landslides occur long before soil development approaches the steady state caused by the gradual erosion processes. In this case, equation (10.8) for soil depth in between landslides, which leads to a well-defined invertible relationship for $x_l(t_l)$ that relates typical landslide occurrence intervals to a unique soil depth, implying that one can predict the soil depth at which a given slope becomes unstable.

On the other hand, when $t_l \gg t_{ss}$, $x_l \approx x_{ss}$, because steady state is virtually always reached before a landslide develops. Moreover, since x_{ss} is independent of time, it is impossible to generate a unique time–depth relationship. The comparatively rare process of landsliding always occurs at the same depth in a given location—unless, for example, climate changes—but at times that are completely unpredictable from depth measurements, making landsliding effectively a purely random process to which the present analysis has no predictive capability. Moreover, due to the spatially variable infiltration rate, soil depths will also be spatially variable, even at the same slope angle, but the variability still does not permit the present approach to develop an associated distinction in how close the respective slopes are to failure. Since the importance of landsliding increases more rapidly with increasing slope angle than do gradual processes, both limits, $t_l \gg t_{ss}$ and $t_{ss} \gg t_l$, may develop in the same geographic region, possibly even on the same hillslope. But, if a total denudation rate D can be identified that is valid across the range of slope angles found in a given region, this relationship can be used to predict soil depth x, whether this represents a steady state value or an average value. Note that, in the third limit when $t_l = t_{ss}$, it is possible to estimate slope stability, but its accuracy is significantly decreased by the reduced sensitivity of depth to time.

Figure 10.11 presents a schematic diagram predicted by the percolation model, given the three scenarios. The parameters used $\mathcal{I} = 0.5$ m yr^{-1}, $x_0 = 100$ μm, and $\phi = 0.4$. In the first case, the figure shows the result for a gradual erosion rate of 100 m Myr^{-1} and no landsliding. For the second, the model indicates the appearance of the soil development, including an erosion rate by landsliding of 100 m Myr^{-1} and zero gradual erosion. If a rate of 100 m Myr^{-1} for both gradual and landsliding processes is used for the third case, it leads to $t_{ss} \simeq t_l$. Furthermore, in the case of pure landsliding, soil depth increases according to the power law with an exponent of 0.53 in time between landslides and, then, an abrupt loss of the entire soil column at each landslide. This choice is, of course, an idealization. It is not necessarily true that every landslide removes the entire soil column, particularly not at every point. In the final case, equation (10.21) must be used to predict soil deepening following landslides. For this choice of the parameters, the soil depth is

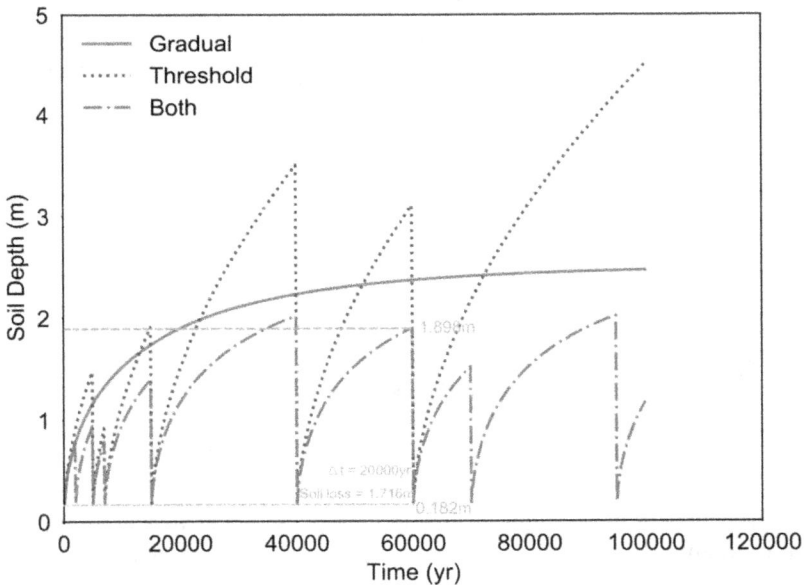

Figure 10.11. Schematic illustration of soil depths with and without disturbance of shallow landsliding, as well as background erosion over time. Red curve includes 100 m Myr^{-1} soil loss from gradual processes, but no landsliding; blue curve includes 100 m Myr^{-1} soil loss from landsliding, but no soil loss from gradual processes; while the green curve includes 100 m Myr^{-1} soil loss from each of the gradual and catastrophic processes. Time of blue curve is generated aperiodically at ages of 5000, 7000, 15 000, 40 000, and 60 000 years. Time of green curve is generated at ages of 2000, 5000, 7000, 15 000, 40 000, 60 000, 70 000, and 95 000 years. Yu *et al* (2019) John Wiley & Sons. Copyright 2019. American Geophysical Union. All Rights Reserved.

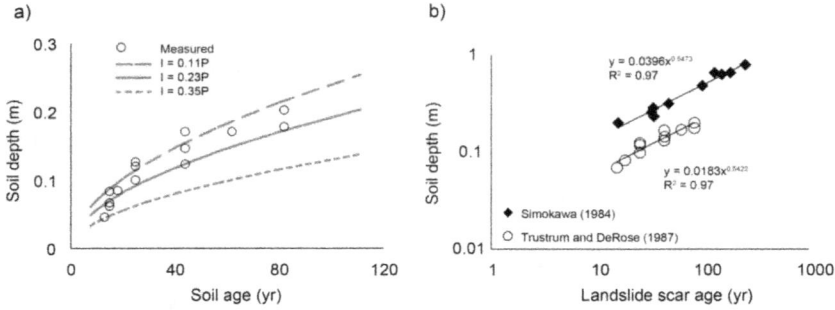

Figure 10.12. (a) Estimated soil depth development in time at the Taranaki site in New Zealand, compared with observed soil depths. Short dashed curve corresponds to $\mathcal{I} = 0.11P$, long dashes to $\mathcal{I} = 0.35P$, and the solid line to $\mathcal{I} = 0.23P$. (Note inversion of the correspondence in the original figure legend.) (b) Scaling of soil development with age of landslide scars. First data point at 4 years, which is well above the trend line, was omitted. For data from Trustrum and De Rose (1988), first data point at 13 year, well below the trend line, is neglected. Multiple soil depths at 15 years are averaged to reduce scatter. Yu *et al* (2019) John Wiley & Sons. Copyright 2019. American Geophysical Union. All Rights Reserved.

typically approaching a steady state when the landslide occurs, but the variability in landslide recurrence intervals means that the steady state is sometimes reached and sometimes not at all.

Next, consider the Taranaki hill country site in New Zealand. Figure 10.12(a) shows that by using the parameters given in the literature for porosity, 18.8%, precipitation, 1.873 m yr^{-1}, and particle diameter, 65 μm in equation (10.8), good agreement with the field data is obtained. The only parameter used that was not site-specific but a global mean, was the fraction of precipitation that infiltrates, which is between 11% and 35% (Lvovitch 1973), which bound the data reasonably well, although, $\mathcal{I} = 0.11P$ may be too low.

Equation (10.20) can be used to calculate a steady-state depth for each of the denudation rates given by Armstrong *et al* (1998). For 400 m Myr^{-1}, the result is 0.66 m, which compares well with the stated average hillslope depth of 0.7 m. The smallest onshore denudation rates reported by Armstrong *et al* (1998) are 100 to 200 m Myr^{-1}, which predict depths of 1.47 and 3.27 m, respectively, and compare well with the data reported by Trustrum and de Rose (1988) for accumulation regions, 1.5 m, and for foot slopes and swales, 3 m. The predicted soil depth for the 900 m Myr^{-1} denudation rate is 0.26 m, which is somewhat smaller than the 30–40 cm of soil overlying presettlement landslide scars, but, nevertheless, suggests that the variation in denudation rates is accommodated by the prevalence or rarity of landsliding. Note that regression of the four predicted and observed steady-state soil depths, if the interpretations are valid, yields a slope of 1.06 and an R^2 value of 0.992. Extrapolating equation (10.8) with the given parameters to a depth of 35 cm yields a time of 312 years, while using

$$x = x_0 \ln\left(\frac{\mathcal{R}_m t}{x_0}\right), \tag{10.25}$$

suggested by Heimsath *et al* (1997, 2012) (in which $\mathcal{R}_m$ is the maximum soil production rate, obtained from extrapolation of experimental soil production data to zero soil thickness) to predict a steady-state time yields $t_{ss} = 389$ years for the denudation rate of 900 m Myr^{-1}. Equation (10.25) was derived based on assuming, $\mathcal{R}_s = dx/dt = \mathcal{R}_m \exp(-x/x_s)$, where x_s (similar to x_0) is a parameter to be estimated. In usual applications, steady-state conditions are invoked to equate $\mathcal{R} = D$, denudation rate, from which one obtains

$$x = x_s \ln\left(\frac{\mathcal{R}_m}{D}\right).$$ (10.26)

Thus, estimates of the percolation model for landslide recurrence intervals are between 2 and 2.5 times the period since European settlement, which are in general agreement with the conclusions of Trustrum and de Rose (1988) that this time exceeds 80 years, and probably the time since settlement.

For the site in Hamada city in Shinlane Prefecture, Japan (Shimokawa 1984), no information is available to suggest plausible values for $\mathcal{I}$ or d_{50}. If the first data point in each of the first two datasets is omitted, each conforms closely to equation (10.8), as shown in figure 10.12(b). The first data point reported by Shimokawa (1984) lies well above the trend, with the first data point of Trustrum and de Rose (1988), equally below. The resulting powers, 0.54 and 0.55 (versus the predicted value of 0.53), and relatively high R^2 values of 0.89 and 0.97, suggest that, at both the Japan site and the New Zealand site, the time dependence of equation (10.8) is confirmed.

Figure 10.13 compares the predictions for the Rossberg site in Swiss Alps. Using equation (10.21), the predicted soil depths compare well with observed soil depths. The percolation model slightly overestimates soil depth for surface ages less than 1000 years and probably slightly underestimates the soil depth for older soils, but the trend displays a logarithmic function, rather than a power law, although the R^2 value (based on the regression curve of plotted values and the trend) for the power

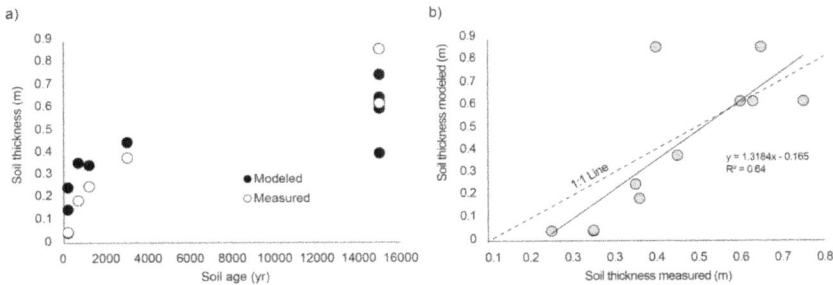

Figure 10.13. (a) Postlandslide soil development at Rossberg, Switzerland, as compared with theoretical predictions, which are made without the use of adjustable parameters and include a significant background soil loss from gradual erosion processes, noted particularly at time scales exceeding about 5000 years. (b) Comparison between predicted and measured values. The 1:1 relationship is given by the dashed line. Yu *et al* (2019) John Wiley & Sons. Copyright 2019. American Geophysical Union. All Rights Reserved.

law is higher. Equation (10.21) does, however, predict power-law behavior only at times short enough that it can be approximated by equation (10.8).

Next, the properties of the measured soil production (SPR) function for the erosion rates are used in order to be able to use equation (10.18), suggested by Montgomery and Brandon (2002), to analyze the SGM. The estimates are, $E_0 = 40$ m Myr^{-1}, $S_c = 50°$ (close to 45°, suggested by the Heimsath $et~al$ 2012), and $k = 2.3$ m Myr^{-1}, keeping in mind that the slopes are expressed in degrees. Figure 10.14 shows the comparison between predicted and observed SPRs. The predicted SPR reaches 170 m Myr^{-1} at about 32.5° (close to the suggested cutoff of 30) and 370 m Myr^{-1} at 42°, in agreement with the assumed value of $S_c = 45°$. The value of about 600 m Myr^{-1} at 45° is in accord with the only data point (594 m Myr^{-1} at the same slope) for slopes greater than 40°.

The predictions based on the exponential formulation of the SPR, equation (10.26), are presented next, after which those provided by the percolation model are described. To evaluate the accuracy of equation (10.26) for the data for soil depth reported by Heimsath $et~al$ (2012), estimates of x_s and $\mathcal{R}_m$ must be obtained. Heimsath $et~al$ (2012) reported two different values for both $\mathcal{R}_m$, 170 and 370 m Myr^{-1}, and x_s for the two different slopes, though the two values of x_s are similar at about 30 cm. Restricting to the single value, $\mathcal{R}_m = 506$ m Myr^{-1}, and choosing $x_s = 11.3$ cm, lead to the best fit with a slope of essentially one and an intercept of zero. The value of R^2 is unaffected by the choice of parameters. The estimate, $\mathcal{R}_m = 506$ m Myr^{-1} is larger than all the experimentally determined values, except one, 594 m Myr^{-1}, which leads to predicting negative soil depth in that case. However, even 506 m Myr^{-1} is larger than the largest value characteristic of the

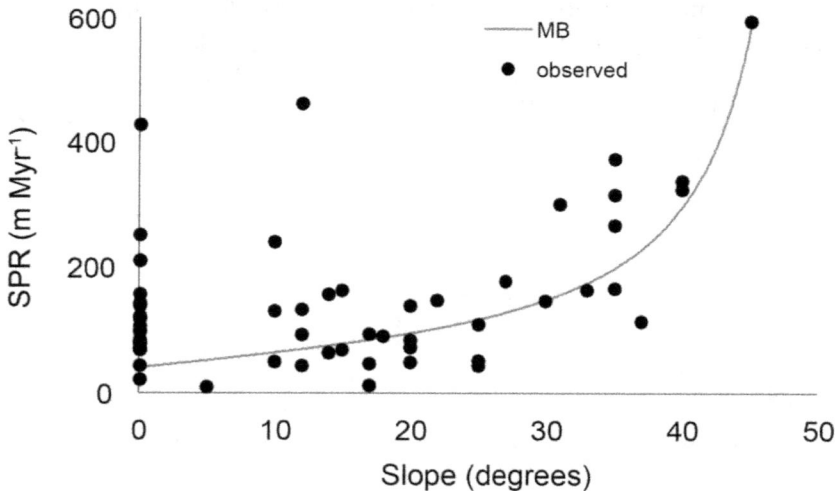

Figure 10.14. Fit of Montgomery–Brandon (MB) empirical relationship for erosion as a function of slope angle to data (Heimsath $et~al$ 2012) for soil production rates as a function of slope angle for San Gabriel Mountains. Yu $et~al$ (2019) John Wiley & Sons. Copyright 2019. American Geophysical Union. All Rights Reserved.

highest slopes (370 m Myr^{-1}) reported by Heimsath *et al* (2012). The comparison is shown in figure 10.15(a). Note that the use of a single SPR function requires a larger value of both the maximum SPR and a more rapid decrease in SPR with depth than is reported by Heimsath *et al* (2012), but the discrepancies are not large.

On the other hand, the predictions of the percolation model are compared with data in figure 10.15(b). Even though no adjustable parameters are used, on average the predictions are off by only about 14%, while the R^2 value is noticeably greater

a) Exponential fit

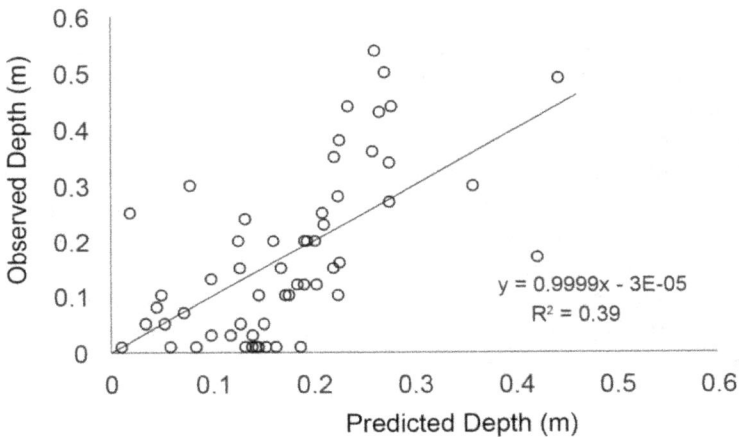

$y = 0.9999x - 3E\text{-}05$
$R^2 = 0.39$

b) Percolation prediction

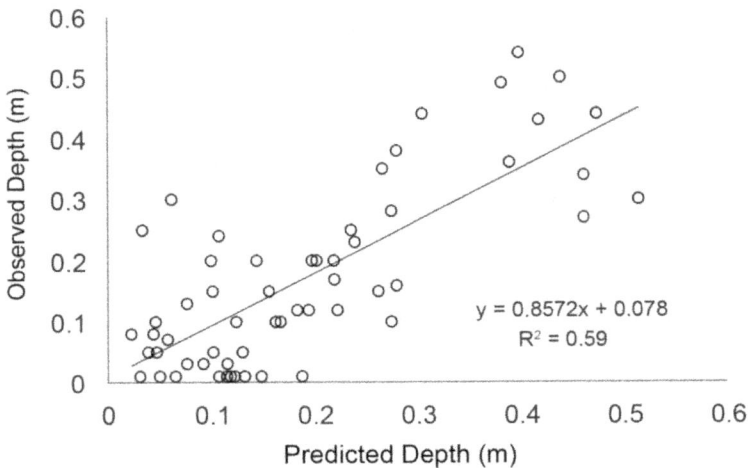

$y = 0.8572x + 0.078$
$R^2 = 0.59$

Figure 10.15. (a) Optimized fit for soil depths as a function of soil production rates using exponential model of Heimsath *et al* (2012) and a single maximum soil production rate. (b) For the same site and data, comparison of percolation predictions for soil depths using observed soil production rates for San Gabriel Mountains. Yu *et al* (2019) John Wiley & Sons. Copyright 2019. American Geophysical Union. All Rights Reserved.

than obtained for the exponential function, 0.59, instead of 0.39. For three sites in the above the extremely low measured SPR values have been replaced by values expected from equation (10.18). This was done for three reasons: first, the measured SPR values lead to soil depths that are much too large. Second, the measured SPR values are much smaller than any others with similar slopes and, third, the regression for the SPR as a function of soil depth given in Heimsath *et al* (2012) is not consistent with the three sites either. In the first argument, the three SPR values correspond to erosion rates of 12, 21, and 10 m Myr^{-1} for slopes with angles 17°, 0°, and 5°, respectively. Equation (10.18), with best fit parameters for the SGM site, yields erosion rates of 84, 40, and 52 m Myr^{-1}, however, for the three slope angles. If the values obtained from the fit of equation (10.18) are used for the erosion rates at these the sites, then adding them back to the analyzed data is completely consistent with the remaining data points. Note that adding the three points produces only a small change in the R^2 value, from 0.583 to 0.586. Moreover, the Heimsath regression for slopes less than 30°, $\mathcal{R}_m = 170\exp(-0.031x)$ yields 32 m Myr^{-1} for the deepest soils measured at $x = 54$ cm, which is only 20% smaller than the background denudation rate of 40 m Myr^{-1}, which was used to obtain the best fit parameters for equation (10.18), but triple the smallest SPR values measured.

Comparison of the predictions of the two theories with data is shown in figure 10.16. The most obvious distinction is the negative curvature of the predictions of the exponential model of SPR. An alternative representation is to plot both predicted and observed depths for each site as functions of slope angle.

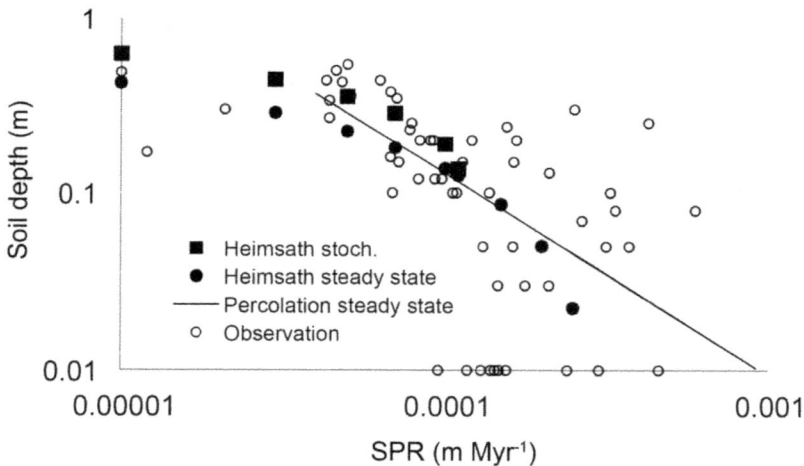

Figure 10.16. Comparison of the exponential model with percolation prediction and data for the San Gabriel Mountains. The parameters of the exponential model were chosen in order to generate similar predicted soil depths as obtained from percolation concepts, in order to emphasize the difference in curvature from the two models. The maximum SPR = 300 m Myr^{-1} and $x_0 = 25$ cm were applied, in each case within 20% of the values Heimsath *et al* (2012) used for steeper slopes. Description of terms: Heimsath stoch.: stochastic means in this context that all erosion is due to landsliding. Heimsath steady state: according to equation (10.26). Yu *et al* (2019) John Wiley & Sons. Copyright 2019. American Geophysical Union. All Rights Reserved.

This is shown in figure 10.17. Although the soil depth is represented as a function of slope angle, it was calculated point-by-point from the observed, rather than modeled, SPR. The reason that the R^2 value of 0.59 for the percolation model is higher than 0.39 for the exponential model is visible. The advantage is associated with the wider range of predicted soil depths at any given slope using the percolation

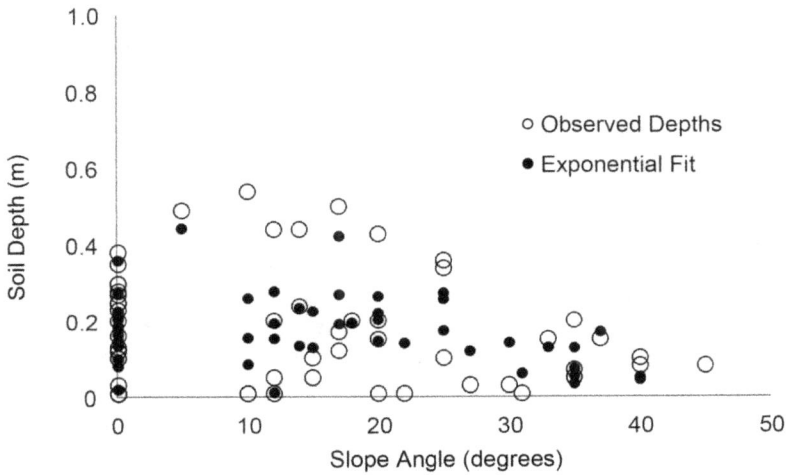

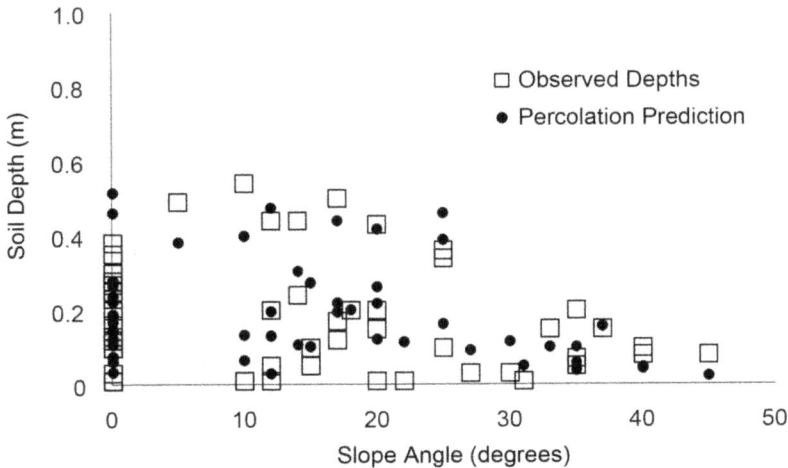

Figure 10.17. (a) Comparison of predicted and observed soil depths as a function of slope angle using the exponential model and the best fit parameters given in the text. (b) Comparison of predicted and observed soil depths using percolation theory for chemical weathering and best estimates for input parameters as discussed in text. Yu *et al* (2019) John Wiley & Sons. Copyright 2019. American Geophysical Union. All Rights Reserved.

model on account of its greater sensitivity to the SPR. Thus, the exponential model predictions tend to lie more nearly in the center of the envelope of measured soil depths at a given slope angle.

It may be useful to use equation (10.18), proposed by Montgomery and Brandon (2002) for the erosion rate, which was tested on the actual SPR data for the SGM, to predict a steady-state soil depth across the SGM sites by both percolation and exponential representations, as well as the percolation model for maximum soil depth attained between landslides. The results are shown in figure 10.18. For most of the range of slope angles, the distinction between the steady-state percolation model and the exponential model is small, but it should be noted that the percolation parameters were determined from other measurements, rather than from an optimal agreement with experiment. If the maximum percolation prediction using the Montgomery–Brandon slope erosion rate function and the full time between landslides is indeed reliable, one could infer that the eight (of a total 55) sites above this curve are susceptible to landslides, except that Heimsath *et al* (2012) noted specifically that they tried to avoid this eventuality by picking sites with rounded topography near ridge crests. Nevertheless, images of the two sites shows their close proximity to landslide scars. It should also be noted that the maximum soil depth from the stochastic solution seems to provide an upper limit for soil depths only at slopes larger than approximately 10°–15°, roughly in accord with the observation that such landsliding is dominant only on slopes greater than 30°. At slope angles smaller than 10°, the upper bound for soil depths is more nearly consistent with the steady state percolation calculation, a factor two smaller.

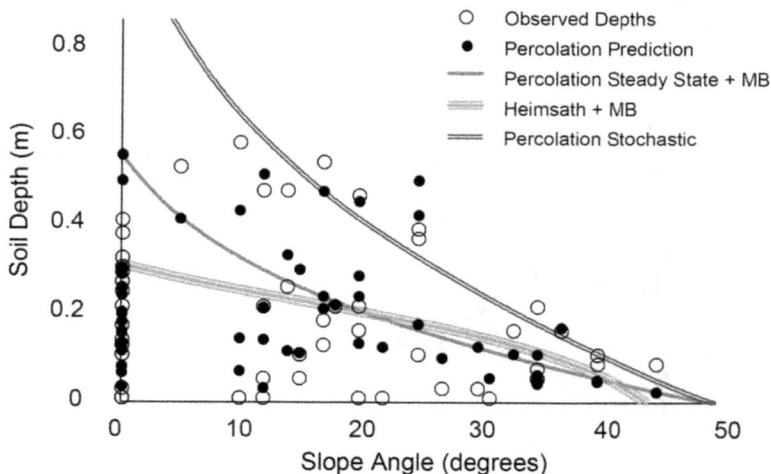

Figure 10.18. Comparison of exponential fit and percolation prediction for soil depths as a function of slope angle. MB means Montgomery and Brandon (2002). Percolation predictions were made with measured values of soil production for erosion rates. Percolation steady state are according to equation (10.24). Heimsath implies predictions according to equation (10.26) with MB equation substituted for the denudation rate, D. Percolation stochastic: equation (10.22) using MB empirical equation. Yu *et al* (2019) John Wiley & Sons. Copyright 2019. American Geophysical Union. All Rights Reserved

Figure 10.19 presents a comparison of the slope-dependent development of soils in three studies of gradual soil erosion by Moore *et al* (1993), Buttle *et al* (2004), and Norton and Smith (1930) as collected by Jenny (1941), as well as the Appenine Mountains where soil erosion is predominantly by landsliding (Salciarini *et al* 2006). Equation (10.23) with the same erosion parameters as for the Olympic Mountains was used to predict the soil depths. However, in each of the four cases the data were multiplied by a constant factor, representing distinctions in particle size and infiltration rates, in order to obtain closest agreement with equation (10.24). A study on the sensitivity of the comparison to the values of the constants indicated that they are accurate within about 5%.

Summarizing this section, percolation theory provides a model of soil formation that can be used to predict soil recovery after landslides. In its steady state form, it can also be used to derive an expression for mean soil depth as a function of slope by adopting an appropriate relationship between erosion rate and slope angle. Addressing the slope-dependence of soil depths introduces additional possibilities to test the predictions, as well as additional uncertainties in the input parameters. At steep slopes, where erosion of soil is mass-wasting dominated or susceptible to shallow landsliding, the steady-state predicted soil depth may be interpreted as a temporal and/or spatial mean soil depth. The predicted maximum soil depth neglecting gradual erosion, but evaluated at landslide recurrence intervals, is a

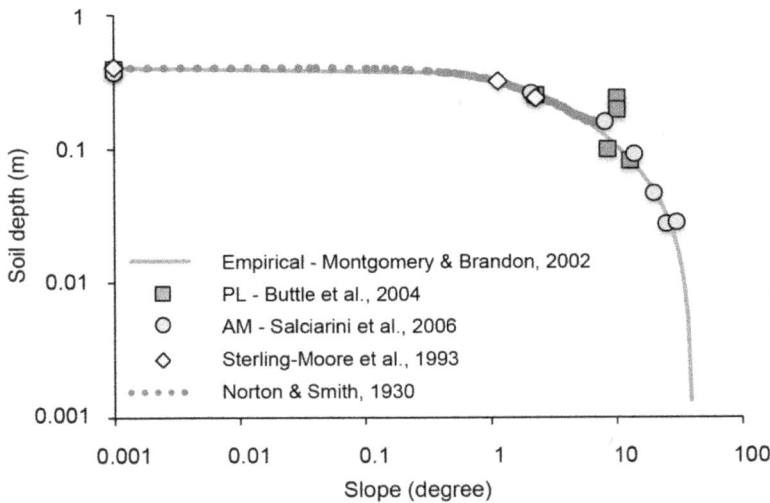

Figure 10.19. Comparison of predicted soil depth and observed soil depth at three study sites exhibiting gradual erosion, namely, Plastic Lake (PL), Sterling, and Norton and Smith (1930). At one of the four sites, Apennine Mountains (AM) (Salciarini *et al* 2006), erosion is chiefly a result of landsliding. Data from AM is adjusted by a factor of 54 due mainly to coarser particle size and, but to a lesser extent, also a greater infiltration rate. Constant numerical factor applied to data from remaining sites in order to isolate magnitude and shape of soil depth function. Erosion function from Olympic Mountains with original parameters from Montgomery and Brandon (2002) is used for all sites. Yu *et al* (2019) John Wiley & Sons. Copyright 2019. American Geophysical Union. All Rights Reserved

factor of two larger than the steady-state soil depth, predicted for the same total erosion rate. The mean soil depth in threshold landscapes is not greatly different from the case in which steady-state models are appropriate, as long as the total erosion rate is the same. Using a single published relationship for the erosion rate as a function of slope angle, it is possible to make quasi-universal prediction of mean soil depth as a function of slope angle.

References

Ahnert F 1970 Functional relationship between denudation relief and uplift in large mid-latitude drainage basins *Am. J. Sci.* **268** 243

Anderson R S and Anderson S P 2010 *Geomorphology: The Mechanics and Chemistry of Landscapes* (Cambridge: Cambridge University Press)

Aqtesolv nd http://www.aqtesolv.com/aquifer-tests/aquifer_properties.html (accessed 21 September 2017)

Armstrong P A, Allis R G, Funnell R H and Chapman D S 1998 Late Neogene exhumation patterns in the Taranaki basin (New Zealand): Evidence from offset porosity-depth trends *J. Geophys. Res.* **103(B12)** 30269

Baker V R, Kochel R C, Laity J E and Howard A E 1990 Spring sapping and valley network development *Geo. Soc. Am. Special Papers* **252** 235

Batjes N H 1996 Total carbon and nitrogen in the soils of the world *Eur. J. Soil Sci.* **47** 151

Beisner K R, Gardner W P and Hunt A G 2018 Geochemical characterization and modeling of regional groundwater contributing to the Verde River, Arizona between Mormon Pocket and the U.S.G.S. Clarkdale gage *J. Hydrol.* **564** 99

Bejan A and Errera M 2011 Deterministic tree networks for fluid flow: geometry for minimal flow resistance between a volume and one point *Fractals* **5** 685

Bejan A and Lorente S 2013 Constructal law of design and evolution: Physics, biology, technology, and society *J. Appl. Phys.* **113** 151301

Binnie S A, Phillips W M and Summerfield M A 2007 Tectonic uplift, threshold hillslopes, and denudation rates in a developing mountain range *Geology* **35** 743

Bloemendal M and Theo O 2018 ATES systems in aquifers with high ambient groundwater flow velocity *Geothermics* **75** 81

Blöschl G and Sivapalan M 1995 Scale issues in hydrological modelling: a review *Hydrol. Proc.* **9** 251

Braun J, Mercier J, Guillocheau F and Robin C 2016 A simple model for regolith formation by chemical weathering *J. Geophys. Res.-Earth Surf.* **121** 2140

Budyko M I 1974 *Climate and Life* (New York: Academic)

Bundesamt für Landwirtschaft 2012 *Digitale Bodeneignungskarte der Schweiz* (Bern: Eidgenössisches Volkswirtschaftsdepartement EVD)

Bundesamt für Umwelt 2015 *Hydrologischer Atlas der Schweiz* (Bern: Geographisches Institut der Universität Bern)

Burbank D W, Blythe A E, Putkonen J, Pratt-Sitaula B, Gabet E, Oskin M, Barros A and Ojha T P 2003 Decoupling of erosion and precipitation in the Himalayas *Nature* **426** 652

Burbank D W, Leland J, Fielding E, Anderson R S, Brozovic N, Reid M R and Ducan C 1996 Bedrock incision, rock uplift and threshold hillslopes in the northwestern Himalayas *Nature* **379** 505

Buttle J M, Dillon P J and Eerkes G R 2004 Hydrologic coupling of slopes, riparian zones and streams: an example from the Canadian Shield *J. Hydrol.* **287** 161

Carson M A and Petley D J 1970 Existence of threshold hillslopes in denudation of landscape *Ins. Br. Geograph. Trans.* **49** 71

Church M and Mark D M 1980 On size and scale in geomorphology *Prog. Phys. Geograph.* **4** 302

Claessens L, Knapen A, Kitutu M G, Poesen J and Deckers J A 2007 Modelling landslide hazard, soil redistribution and sediment yield on the Ugandan footslopes of Mt. Elgon *Geomorphology* **90** 23

Cordova J T, Dickinson J E, Beisner K R, Hopkins C B, Kennedy J R and Pool D R 2015 Hydrology of the Middle San Pedro Watershed, Southeastern Arizona, U.S Geological Survey, Scientific Investigations Report 2013–5040. Prepared in Cooperation With the Arizona Department of Water Resources, Menlo Park (U.S. Department of the Interior, U.S. Geological Survey, U.S. Department of the Interior)

Crossey L C, Karlstrom K E, Dorsey R, Pearce J, Wan E and Beard L S 2015 Importance of groundwater in propagating downward integration of the 6-5 Ma Colorado River system: geochemistry of springs, travertines, and lacustrine carbonates of the Grand Canyon region over the past 12 Ma *Geosphere* **11** 660

Crow R S, Schwing J, Karlsrom K E, Heizler M, Pearthree P A and House P K 2021 Redefining the age of the lower Colorado River, southwestern United States *Geology* **49** 635

DiBiase R A and Heimsath A M 2012 Hillslope response to tectonic forcing in threshold landscapes *Earth Surf. Proc. Landf.* **289** 134

DiBiase R A, Heimsath A M and Whipple K X 2012 Hillslope response to tectonic forcing in threshold landscapes *Earth Surf. Process. Landforms* **289** 134

DiBiase R A, Whipple K X, Heimsath A M and Ouimet W B 2010 Landscape form and millennial erosion rate in the San Gabriel Mountains, CA *Earth Planet. Sci. Lett.* **289** 134

Dickinson W R 2015 Integration of the Gila River drainage system through the Basin and Range province of southern Arizona and southwestern New Mexico *Geomorphology* **236** 1

Dietrich W E, Bellugi D, Sklar L S, Stock J D, Heimsath A M and Roering J J 2003 Geomorphic transport laws for predicting landscape form and dynamics *Prediction in Geomorphology* ed R M Iverson and P Wilcock (Washington, DC: American Geophysical Union) p 103

Dietrich W E, McKean J, Bellugi D and Perron T 2007 The prediction of shallow landslide location and size using multidimensional landslide analysis in a digital terrain model *Proc. of the 4th Int. on debris-flow hazards mitigation: Mechanics, prediction and assessment (DFHM-4)* (Amsterdam: IOS Press)

Dietrich W E, Reiss R, Hsu M-L and Montgomery D 1995 A process-based model for colluvial soil depth and shallow landsliding using digital elevation data *Hydrol. Process.* **9** 383

Dixon J L and Heimsath A M 2009 The critical role of climate and saprolite weathering in landscape evolution *Earth Surf. Proc. Landf.* **34** 1507

Dobbs S C, McHargue T, Malkowski M A, Gooley J T, Jaikla C, White C J *et al* 2009 Are submarine and subaerial drainages morphologically distinct? *Geology* **47** 1093

Dokuchaev V V 1948 *Russian Chernozem—Selected works of V V Dokuchaev, Volume I. Moskva, 1948* (translated from the Russian by Israel Program for Scientific Translations, Jerusalem, 1967)

Dorsey R J and Roering J J 2006 Quaternary landscape evolution in the San Jacinto fault zone, peninsular ranges of Southern California: transient response to strike-slip fault initiation *Geomorphology* **73** 0169–555X

Dunne T, Zhang W and Aubrey B F 1991 Effects of rainfall, vegetation, and microtopography on infiltration and run-off *Water Resour. Res.* **27** 2271

Eastoe C and Towne D 2018 Regional zonation of groundwater recharge mechanisms in alluvial basins in Arizona: interpretation of isotope mapping *J. Geochem. Expl.* **194** 134

Egli M and Fitze P 2001 Quantitative aspects of carbonate leaching of soils with differing ages and climates *Catena* **46** 35

Egli M, Hunt A G, Dahms D, Raab G, Derungs C, Raimondi S *et al* 2018 Prediction of soil formation as a function of age using the percolation theory approach *Front. Environ. Sci.* **28** 108

Enzel Y, Wells S G and Lancaster N 2003 Late Pleistocene lakes along the Mojave River, southeast California *Geol. Soc. Am. Special Papers* **368** 61

Fan N, Chu Z, Jiang L, Hassan M A, Lamb M P, Liu X *et al* 2018 Abrupt drainage reorganization following a Pleistocene river capture *Nat. Commun.* **9** 3756

Fehr E, Andrade J S Jr, daCunha S D, da Silva L R, Herrmann H J, Kadau D, Moukarzel C F and Oliveira E A 2009 New efficient methods for calculating watersheds *J. Stat. Mech.* **P09007**

Fehr E, Kadau D, Andrade J S Jr and Herrmann H J 2011a Impact of perturbations on watersheds *Phys. Rev. Lett.* **106** 048501

Fehr E, Kadau D, Araújo N A M, Andrade J S Jr and Herrmann H J 2011b Scaling relations for watersheds *Phys. Rev.* E **84** 036116

Fielding L, Najman Y, Millar I, Butterworth P, Garzanti E, Vezzoli G *et al* 2018 The initiation and evolution of the river Nile *Earth Planet. Sci. Lett.* **489** 166

Garcia A L, Knott J R, Mahan S A and Bright J 2014 Geochronology and paleoenvironment of pluvial Harper Lake, Mojave Desert, California, USA *Quat. Res.* **81** 305

Ghanbarian B, Hunt A G, Ewing R P and Sahimi M 2013 Tortuosity in porous media: a critical review *Soil Sci. Soc. Am.* **77** 1461

Ghanbarian-Alavijeh B, Skinner T E and Hunt A G 2012 Saturation dependence of dispersion in porous media *Phys. Rev.* E **86** 066316

Ghoneim E, Benedotti M and El-Baz F 2005 An integrated remote sensing and GIS analysis of the Kufrah paleoriver, eastern Sahara *Geomorphology* **140** 242

Gilbert G K 1877 Report on the geology of Henry Mountains, U.S. Geological and geographical survey of the Rocky Mountain region (Government Printing Office)

Glass R J, Nicholl M and Yarrington L 1998 A modified invasion percolation model for low-capillary number immiscible displacements in horizontal rough-walled fractures: influence of local in-plane curvature *Water Resour. Res.* **34** 3215

Gossel W, Ebraheem A M and Wycisk O 2004 A very large scale GIS-based groundwater flow model for the Nubian sandstone aquifer in Eastern Sahara (Egypt, northern Sudan and eastern Libya) *Hydrogeol. J.* **12** 698

Goudie A S 2005 The drainage of Africa since the Cretaceous *Geomorphology* **67** 437

Grant U S 1948 Influence of the water table on beach aggradation and egradation *J. Marine Res.* **7** 655

Gray D M 1961 Interrelationships of watershed characteristics *J. Geophys. Res.* **66** 1215

Gunnell Y and Harbor D J 2010 Butte detachment: how pre-rift geological structure and drainage integration drive escarpment evolution at rifted continental margins *Earth Surf. Proc. Landforms* **35** 1373

Hack J T 1957 *Studies of Longitudinal Profiles in Virginia and Maryland* (Washington DC: Geological Survey Professional Paper 294-B)

Harvey A M, Wigand P E and Wells S G 1999 Response of alluvial fan systems to the late Pleistocene to Holocene climatic transition: contrasts between the margins of pluvial Lakes Lahontan and Mojave, Nevada and California, USA *Catena* **36** 255

Heimsath A M, Chappell J, Finkel R C, Fifield K and Alimanovic A 2006 Escarpment erosion and landscape evolution in southeastern Australia *Geological Society of America, Special Paper* **398** 173

Heimsath A M, Chappell J, Dietrich W E, Nishiizumi K and Finkel R C 2000 Soil production on retreating escarpment in southeaster Australia *Geology* **28** 787

Heimsath A M, Chappell J, Dietrich W E, Nishiizumi K and Finkel R C 2001a Late quaternary erosion in southeastern Australia: a field example using cosmogenic nuclides *Quat. Int.* **83–5**

Heimsath A M, Chappell J and Fifield K 2010 Eroding Australia: rates and processes from Bega Valley to Arnhem Land *Australian Landscapes* **vol 346** ed P Bishop and B Pillans (Geological Society) p 225

Heimsath A M, DiBiase R A and Whipple K X 2012 Soil production limits and the transition to bedrock-dominated landscapes *Nat. Geosci.* **5** 210

Heimsath A M, Dietrich W E, Nishiizumi K and Finkel R C 1997 The soil production function and landscape equilibrium *Nature* **388** 358

Heimsath A M, Dietrich W E, Nishiizumi K and Finkel R C 1999 Cosmogenic nuclides, topography, and the spatial variation of soil depth *Geomorphology* **27** 151

Heimsath A M, Dietrich W E, Nishiizumi K and Finkel R C 2001b Stochastic processes of soil production and transport: erosion rates, topographic variation and cosmogenic nuclides in the Oregon coast range *Earth Surf. Process. Landforms* **26** 531

Heimsath A M, Fink D and Hancock G R 2009 The 'humped' soil production function: eroding Arnhem Land, Australia *Earth Surf. Process. Landforms* **34** 1674

Heimsath A M, DiBiase R A and Whipple K X 2012 Soil production limits and the transition to bedrock-dominated landscapes *Nat. Geosci.* **5** 210

Hilgendorf Z, Wells G, Larson P H, Millett J and Kohout M 2020 From basins to rivers: Understanding the revitalization and significance of top-down drainage integration mechanisms in drainage basin evolution *Geomorphology* **352** 107020

Hillel D 2005 Soil: Crucible of life *J. Nat. Resour. Life Sci. Educ.* **34** 60

Hillhouse J W and Cox B F 2000 *Pliocene and Pleistocene Evolution of the Mojave River, and Associated Tectonic Development of the Transverse Ranges and Mojave Desert, Based on Borehole Stratigraphy Studies* (Victorville, CA: US Department of the Interior, US Geological Survey)

Hopkins C B, McIntosh J C, Eastoe C, Dickinson J E and Meixner T 2014 Evaluation of the importance of clay confining units on groundwater flow in alluvial basins using solute and isotope tracers: the case of Middle San Pedro Basin in southeastern Arizona (USA) *Hydrogeol. J.* **22** 829

Horton R E 1932 Drainage basin characteristics *Trans. Am. Geophys. Union* **13** 350

Horton R E 1945 Erosional development of streams and their drainage basins; hydrophysical approach to quantitative morphology *Geol. Soc. Am. Bull.* **56** 275

Huggett R J 1998 Soil chronosequences, soil development, and soil evolution: a critical review *Catena* **32** 155

Hunt A G 2016a Possible explanation of the values of Hack's drainage basin, river length scaling exponent *Non-lin. Proc. Geophys.* **23** 91

Hunt A G 2016b Spatio-temporal scaling of vegetation growth and soil formation from percolation theory *Vadose Zone J.* **15** 1

Hunt A G 2017 Spatiotemporal scaling of vegetation growth and soil formation: explicit predictions *Vadose Zone J.* **16** 1

Hunt A G and Ewing R P 2016 *Handbook of Groundwater Engineering* ed J H Cushman and D TartakovskyD (London: Taylor and Francis)

Hunt A G, Ewing R P and Ghanbarian B 2014 *Percolation Theory for Flow in Porous Media* (Berlin: Springer)

Hunt A G, Faybishenko B and Ghanbarian B 2021 Non-linear hydrologic organization *Non-lin. Proc. Geophys.* **28** 599

Hunt A G, Ghanbarian B and Faybishenko B 2023 A model of temporal and spatial river network evolution with climatic inputs *Water* **5** 1174570

Hunt A G and Ghanbarian-Alavijeh B 2016 Percolation theory for solute transport in porous media: geochemistry, geomorphology, and carbon cycling *Water Resour. Res.* **52** 7444

Hunt A G, Ghanbarian-Alavijeh B, Skinner T E and Ewing R P 2015 Scaling of geochemical reaction rates via advective solute transport *Chaos* **25** 075403

Hunt A G, Skinner T E, Ewing R P and Ghanbarian-Alavijeh B 2011 Dispersion of solutes in porous media *Eur. Phys. J.* B **80** 411–32

Hunt A G, Skinner T E, Ewing R P and Ghanbarian-Alavijeh B 2012 Dispersion of solutes in porous media *Eur. Phys. J.* B **80** 411

Hurst M D, Mudd S M, Walcott R, Mikael A and Yoo K 2012 Using hilltop curvature to derive the spatial distribution of erosion rates *J. Geophys. Res.* **117** F02017

Iida T 1999 A stochastic hydro-geomorphological model for shallow landsliding due to rainstorm *Catena* **34** 293

Ivy-Ochs S and Kober F 2008 Surface exposure dating with cosmogenic nuclides: Eiszeitalter und Gegenwart *Quat. Sci. J.* **57** 179

Janecke S U, Dorsey R J, Forand D, Steely A N, Kirby S M, Lutz A T *et al* 2010 High geologic slip rates since early Pleistocene initition of the San Jacinto and San Felipe fault zones in the San Andreas fault system: Southern California, USA *Geol. Soc. Am. Spec.* **25** 4752010

Jenny H 1941 *Factors of Soil Formation: A System of Quantitative Pedology* (New York: McGraw-Hill)

Johnson T C, Slater L D, Ntarlagiannis D, Day-Lewis F D and Elwaseif M 2012 Monitoring groundwater-surface water interaction using time-series and time-frequency analysis of transient three-dimensional electrical resistivity changes *Water Resour. Res.* **48** W07506

Kalin R M 1994 *The Hydrogeochemical Evolution of the Groundwater of the Tucson Basin with Application to 3-Dimensional Groundwater Flow Modelling* (Tucson, AZ: University of Arizona)

Keller B 2017 Massive rock slope failure in Central Switzerland: history, geologic-geomorphological predisposition, types and triggers, and resulting risks *Landslides* **14** 1633

Kirby M, Lund S, Anderson M and Bird B 2007 Insolation forcing of Holocene climate change in Southern California: a sediment study from Lake Elsinore *J. Paleolimnol.* **38** 395

Kirchner J W 1993 Statistical inevitability in Horton's laws and the apparent randomness of stream channel networks *Geology* **21** 591

Networks on Networks (Second Edition)

Koltzer N, Scheck-Wenderoth M, Cacace M, Frick M and Bott J 2019 Regional hydraulic model of the Upper Rhine Graben *Adv. Geosci.* **49** 197

Kulongoski J T, Hilton D R and Izbicki J A 2003 Helium isotope studies in the Mojave Desert, California: implications for groundwater chronology and regional seismicity *Chem. Geol.* **202** 95

Laity J E and Malin M 1986 Sapping processes and the development of theater headed valley networks on the Colorado Plateau *Geol. Soc. Am. Bull.* **96** 203

Langenheim V E, Jachens R C, Morton D M, Kistler R W and Matti J C 2004 Geophysical and isotopic mapping of preexisting crustal structures that influenced the location and development of the San Jacinto fault zone *Geol. Soc. Am. Bull.* **116** 1143

Larson P H, Dorn R I, Skotnicki J, Seong Y B, Jeong A, Deponty J *et al* 2020 Impact of drainage integration on basin geomorphology and landform evolution: a case study along the Salt and Verde rivers, Sonoran Desert, USA *Geomorphology* **371** 107439

Lee Y, Andrade J S, Buldyrev S V, Dokholoyan N V, Havlin S, King P R, Paul G and Stanley H E 1999 Traveling time and traveling length in critical percolation clusters *Phys. Rev. E* **60** 3425

Liu C, Zachara J M, Qafoku N P and Wang Z 2008 Scale-dependent desorption of uranium from contaminated subsurface sediments *Water Resour. Res.* **44** W08413

Liu J X, Chen H, Lin H, Liu H and Song H 2013 A simple geomorphicbased analytical model for predicting the spatial distribution of soil thickness in headwater hillslopes and catchments *Water Resour. Res.* **49** 7733

Lvovitch M I 1973 The global water balance *U.S. National Committee for the International Hydrological Decade* (National Academy of Science Bulletin) p 28

Maher K 2010 The dependence of chemical weathering rates on fluid residence time *Earth Planet. Sci. Lett.* **294** 101

Maritan A, Rinaldo A, Rigon R, Giacometti A and Rodriguez-Iturbe I 1996 Scaling laws for river networks *Phys. Rev. E* **53** 1510

Marshall J S, Idleman B D, Gardner T W and Fisher D M 2003 Landscape evolution within a retreating volcanic arc Costa Rica *Central Am. Geol.* **31** 419

Mather A E, Stokes M and Griffiths J S 2002 Quaternary landscape evolution: a framework for understanding contemporary erosion Spain *Land Deg. Dev.* **13** 89

Matthews A, Lawrence S R, Mamad A V and Fortes C 2001 Mozambique basin may have bright future under new geological interpretation *Oil Gas. J.* **2** 70

Maxwell R, Condon L E, Kollet S J, Maher K, Haggerty R and Forrester M M 2016 The imprint of climate and geology on the residence times of groundwater *Geophys. Res. Lett.* **43** 701

McBeath D M 1977 Gas-condensate fields of the Taranaki basin, New Zealand *J. Geol. Geophys.* **20** 99

McCauley J F, Breed C S, Schaber G G, McHugh W P, Bahayissawi C V, Haynes V *et al* 1986 Paleodrainages of the Eastern Sahara, the radar rivers revisited (SIR-A/B implicatins for a Mid-Tertiary Trans-African drainage system) *IEEE Trans. Geosci. Remote Sensing* **24** 678

McMahon P B, Plummer L N, Böhlke J K, Shapiro S D and Hinkle S R 2011 A comparison of recharge rates in aquifers of the United States based on groundwater-age data *Hydrogeol. J.* **19** 779

Meek N 1989 Geomorphologic and hydrologic implications of the rapid incision of afton canyon, Mojave desert *Geology* **12** 7

Meek N 1990 Late quaternary geochronology and geomorphology of the Manix Basin, San Bernardino County, California *PhD Thesis* Los AngelesUniversity of California

Meili R B 1982 Untersuchungen zur Bodenentwicklung im Bergsturzgebiet des Rossbergs *Unpublished Diploma Thesis* Geograph. Institut Univ. Zürich

Montgomery D R 2001 Slope distributions, threshold hillslopes, and steady-state topography *Am. J. Sci.* **301** 432

Montgomery D R 2007 Soil erosion and agricultural sustainability *Proc. Natl Acad. Sci. USA* **104** 13268

Montgomery D R and Brandon M T 2002 Topographic controls on erosion rates in tectonically active mountain ranges *Earth Planet. Sci. Lett.* **201** 481

Montgomery D R and Dietrich W E 1992 Channel initiation and the problem of landscape scale *Science* **255** 826

Moore D, Gessler P E, Nielsen G A and Peterson G A 1993 Soil attribute prediction using terrain analysis *Soil Sci. Soc. Am. J.* **57** 443

National Research Council 2001 *Basic Research Opportunities in Earth Science* (Washington, DC: National Academies Press)

Niemann J D, Bras R L, Veneziano D and Rinaldo A 2001 Impacts of surface elevation on the growth and scaling properties of simulated river networks *Geomorphology* **40** 37

Norton E A and Smith R S 1930 The influence of topography on soil profile character *J. Am. Soc. Agron.* **22** 251

Okimura T 1987 Investigation and countermeasure of surface failures *Chisitsu to Tyosa* **33** 22

Oliveira E A, Pires R S, Oliveira R S, Furtado V, Herrmann H J and Andrade J S Jr 2019 Auniversal approach for drainage basins *Sci. Rep.* **9** 9845

Ouimet W B, Whipple K X and Granger D E 2009 Beyond threshold hillslope: channel adjustment to base-level fall in tectonically active mountain ranges *Geology* **37** 579–82

Parsekian A D, Singha K, Minsley B J, Holbrook W S and Slater L 2014 Multiscale geophysical imaging of the critical zone *Rev. Geophys.* **53** 1

Peel M C, McMahon T A and Finlayson B L 2010 Vegetation impact on mean annual evapotranspiration at a global catchment scale *Water Resour. Res.* **46** W09508

Pelletier J D 1999 Self-organization and scaling relationships of evolving river networks *J. Geophys. Res.* **104** 7359

Petroff A P, Devauchelle O, Seybold M and Rothman D H 2013 Bifurcation dynamics of natural drainage networks *Philos. Trans. Royal Soc.* **371** 365

Phillips J 2010 The convenient fiction of steady-state soil thickness *Geoderma* **156** 369

Porto M, Havlin S, Schwarzer S and Bunde A 1997 Optimal path in strong disorder and shortest path in invasion percolation with trapping *Phys. Rev. Lett.* **79** 4060

Reheis M C, Bright J, Lund S P and Miller D M 2012 A half-million year record of paleoclimate from the Lake Manix core, Mojave Desert, California *Palaeogeograph. Palaeoclim. Palaeoecol.* **366** 11

Reheis M C, Caskey J, Bright J, Paces J B, Mahan S, Wan E *et al* 2020 Pleistocene lakes and paleohydrologic environments of the Tecopa basin, California: constraints on the drainage integration of the Amargosa River *Geol. Soc. Am. Bull.* **132** 1537

Reiners P W, Ehlers T A, Mitchell S G and Montgomery D R 2003 Coupled spatial variations in precipitation and long-term erosion rates across the Washington Cascades *Nature* **426** 645

Reis A H 2006 Constructal view of scaling laws of river basins *Geomorphology* **78** 201

Repasch M, Karlstrom K, Heizler M and Pecha M 2017 Birth and evolution of the Rio Grande fluvial system in the past 8Ma: progressive downward integration and the influence of tectonics, volcanism, and climate *Earth Sci. Rev.* **168** 113

Rhoads B 2020 *The Dynamics of Drainage Basins and Stream Networks* (Cambridge: Cambridge University Press)

Rigon R, Rinaldo A, Rodriguez-Iturbe I, Bras R L and Ijjasz-Vasquez E 1996 Optimal channel networks: a framework for the study of river basin morphology *Water Resour. Res.* **29** 1635

Rinaldo A, Rigon R, Banavar J R, Maritan A and Rodriguez-Iturbe I 2014 Evolution and selection of river networks: Statics, dynamics, and complexity *Proc. Natl Acad. Sci. USA* **111** 2417

Rinaldo A, Rodríguez-Iturbe I, Rigon R, Ijjasz-Vasquez E and Bras R L 1993 Self-organized fractal river networks *Phys. Rev. Lett.* **70** 822

Roberts G 2019 Scales of similarity and disparity between drainage networks *Geophys. Rese. Lett.* **46** 3781

Robertson F N 1992 Radiocarbon dating of a confined aquifer in southeast Arizona *Radiocarbon* **34** 664

Roering J J 2008 How well can hillslope evolution models 'explain' topography? Simulating soil transport and production with highresolution topographic data *Geol. Soc. Am. Bull.* **120** 1248

Roering J J, Perron J T and Kirchner J W 2007 Functional relationships between denudation and hillslope form and relief *Earth Planet. Sci. Lett.* **264** 245

Rulli M C and Rosso R 2005 Modeling catchment erosion after wildfires in the San Gabriel Mountains in the southern California *Geophys. Res. Lett.* **32** L19401

Sahimi M 1987 Hydrodynamic dispersion near the percolation threshold: scaling and probability densities *J. Phys.* A **20** L1293

Sahimi M 2011 *Flow and Transport in Porous Media and Fractured Rock* 2nd edn (New York: Wiley-VCH)

Sahimi M 2012 Dispersion in porous media, continuous-time random walks, and percolation *Phys. Rev.* E **85** 016316

Sahimi M, Davis H T and Scriven L E 1983 Dispersion in disordered porous media *Chem. Eng. Commun.* **23** 329

Sahimi M, Hashemi M and Ghassemzadeh J 1998 Site-bond invasion percolation with fluid trapping *Physica* A **260** 231

Sahimi M, Hughes B D, Scriven L E and Davis H T 1986 Dispersion in flow through porous media: I. One-phase flow *Chem. Eng. Sci.* **41** 2103

Sahimi M and Imdakm A O 1988 The effect of morphological disorder on hydrodynamic dispersion in flow through porous media *J. Phys.* A **21** 3833

Sahimi M and Mukhopadhyay S 1996 Scaling properties of a percolation model with long-range correlations *Phys. Rev.* E **54** 3870

Said A, Moder C, Clark S and Ghorba B 2015 Creatceous-Cenozoic sedimentary budgets of the southern Mozambique Basin: implications for uplift history of the South African Plateau *J. Afr. Earth Sci.* **109** 1

Salciarini D, Godt J W, Savage W Z, Conversini P, Baum R L and Michael J A 2006 Modeling regional initiation of rainfall induced shallow landslides in the eastern Umbria Region of central Italy *Landslides* **3** 181

Salehikhoo F, Li L and Brantley S L 2013 Magnesite dissolution rates at different spatial scales: the role of mineral spatial distribution and flow velocity *Geochim. Cosmochim. Acta* **108** 91

Sanderman J and Amundson R 2009 A comparative study of dissolved organic carbon transport and stabilization in California forest and grassland soils *Biogeochemistry* **92** 41

Sanford W E, Plummer L N, McAda D P, Bexfield L M and Anderholm S K 2004 Hydrochemical tracers in the middle Rio Grande Basin, USA: 2. Calibration of a groundwater-flow model *Hydrogeol. J.* **12** 389

Sanford W E and Selnick D L 2013 Estimation of evapotranspiration across the conterminous United States using a regression with climate and land-cover data *J. Am. Water Resour. Assoc.* **49** 217

Scheidegger A E 1965 The algebra of stream order numbers *U.S. Geol. Survey Prof. Paper* **52** 187–9

Scheidegger A E 1967 On the topology of river nets *Water Resour. Res.* **3** 103–6

Schlesinger W H and Jasechko S 2014 Transpiration in the global water cycle *Agric. For. Meteorol.* **189** 115

Schumm S A 1956 Evaluation of drainage system and slopes in badlands at Perth Amboy, New Jersey *Geol. Soc. Am. Bull.* **67** 597

Sheppard A P, Knackstedt M A, Pinczewski W V and Sahimi M 1999 Invasion percolation: new algorithms and universality classes *J. Phys.* A **32** L521

Shimokawa E 1984 A natural recovery process of vegetation on landslide scars and landslide periodicity in forested drainage basins *Symposium on Effects of Forest Land Use on Erosion and Slope Stability* ed C L O'Loughlin and A J Pierce (Honolulu, HI: East-West Center, University of Hawaii) pp 99–107

Shreve R L 1967 Infinite topologically random channel networks *J. Geol.* **75** 178

Skaggs T H, Arya L M, Shouse P J and Mohanty B 2011 Estimating particle-size distribution from limited soil texture data *Sci. Soc. Am. J.* **65** 1038

Skotnicki S J, Seong Y B, Dorn R I, Larson P H, DePonty J, Jeong A *et al* 2021 Drainage integration of the Salt and Verde rivers in a Basin and Range extensioal landscape, central Arizona, USA *Geomorphology* **374** 107512

Solder J E, Beisner K R, Anderson J and Bills D J 2020 Rethinking groundwater flow on the South Rim of the Grand Canyon, USA: characterizing recharge sources and flow paths with environmental tracers *Hydrogeol. J.* **28** 1593

Spencer J E, Pearthree P A, Young R A and Spamer E E 2001 Headward erosion versus closed-basin spillover as alternative causes of Neogene capture of the ancestral Colorado River by the Gulf of California *Proc. of a Symp. Held at Grand Canyon National Park in June, 2000* (Grand Canyon Association) p 215

Stamos C L, Cox B F, Izbicki J A and Mendez G O 2003 Geologic Setting, Geohydrology and Ground-Water Quality near the Helendale Fault in the Mojave River Basin, San Bernardino County, California *Water-Resources Investigations Report 03-40697208-24 Washington, DC U.S. Department of the interior, U.S. Geological Survey*

Stark C P 1991 An invasion percolation model of drainage network evolution *Nature* **352** 423

Stolze L, Arora B, Dwivedi D, Steefel C, Li Z, Carrero S *et al* 2023 Aerobic respiration controls on shale weathering *Geochim. Cosmochim. Acta* **340** 172

Strahler A N 1952 Hypsometric (area-altitude) analysis of erosional topography *Geolog. Sec. Am. Bull.* **63** 1117–42

Strahler A N 1957 Quantitative analysis of watershed geomorphology *Eos. Trans. Am. Geophys. Union* **38** 913–20

Strahler A N 1964 Geology. Part II. Quantitative geomorphology of drainage basins and channel networks *Handbook of Applied Hydrology* **vol 4** ed V T Chow (New York: Wiley)

Struth L, Giachetta E, Willett S D, Owen L A and Teson E 2020 Quaternary drainage network reorganization in the Colombian eastern cordillera plateau *Easrth Surf. Process. Landforms* **45** 1789

Tarboton D G, Bras R L and Rodriguez-Iturbe I 1988 The fractal nature of river networks *Water Resour. Res.* **24** 1317

Trustrum N A and De Rose R C 1988 Soil depth-age relationship of landslides on deforested hillslopes, Taranaki, New Zealand *Geomorphology* **1** 143

Tucker G E and Bras R L 2000 A stochastic approach to modeling the role of rainfall variability in drainage basin evolution *Water Resour. Res.* **36** 1953

Urey H C 1952 On the early chemical history of the Earth and the origin of life *Proc. Natl Acad. Sci. USA* **38** 351

U.S. Geological Survey, *General Facts and Concepts About Ground Water*; https://pubs.usgs.gov/circ/circ1186/html/gen_facts.html (accessed 3 April 2021)

Wang C, Mortazavi B, Liang W K, Sun N Z and Yeh W W-G 1995 Model development for conjunctive use study of the San Jacinto Basin California *J. Am. Water Res. Assoc.* **31** 227

Wang H, Gurnis M and Skogseid J 2020 Continent-wide drainage reorganization in North America driven by mantle flow *Earth Planet. Sci. Lett.* **530** 115910

Waters M R 1989 Late Quaternary lacustrine history and palaeoclimatic significance of pluvial Lake Cochise, southeastern Arizona *Quat. Res.* **32** 1

Whittaker R H and Niering W W 1975 Vegetation of the Santa Catalina Mountains Arizona, V. Biomass, production, and diversity along the elevation gradient *Ecology* **56** 71

Willett S D, McCoy S W, Perron J T, Goren L and Chen C-Y 2014 Dynamic reorganization of river basins *Science* **343** 1248765

Willgoose G, Bras R L and Rodriguez-Iturbe I 1991 Results from a new model of river basin evolution *Earth Surf. Proc. Landforms* **16** 237

Willmott C J, Robeson S M and Feddema J J 1994 Estimating continental and terrestrial precipitation averages from rain-gauge networks *Int. J. Climatol.* **14** 403

Xiangjiang H and Niemann J D 2006 Modelling the potential impacts of groundwater hydrology on long-term drainage basin evolution *Earth Surf. Proc. Landforms* **31** 1802

Young R A and Spamer E E 2001 *The Colorado River: Origin and Evolution* (Grand Canyon, AZ: Grand Canyon Association Monograph)

Yu F, Faybishenko B, Hunt A G and Ghanbarian B 2017 A simple model of the variability of soil depth *Water* **9** 460

Yu F and Hunt A G 2017a An examination of the steady-state assumption in soil development models with application to landscape evolution *Earth Surf. Process. Landforms* **42** 2599

Yu F and Hunt A G 2017b Damköhler number input to transport-limited chemical weathering calculations *ACS Earth Space Chem.* **1** 30

Yu F, Hunt A G, Egli M and Raab G 2019 Comparison and contrast in soil depth evolution for steady state and stochastic erosion processes: Possible implications for landslide prediction *Geochem. Geophys. Geosyst.* **20** 2886

Zhang L, Peng M, Chang D and Xu Y 2016 *Dam Failure Mechanisms and Risk Assessment* (New York: Wiley)

IOP Publishing

Networks on Networks (Second Edition)
Role of connectivity in physics of geobiology and geochemistry
Allen G Hunt and Muhammad Sahimi

Chapter 11

Ecohydrological applications: watershed hydrology and water balance

11.1 Background

Watershed hydrology addresses the intersection of ecology, geochemistry, and hydrology within a specific area on Earth's surface. To understand why this is the case, consider a systems approach to problem. A watershed is a geographic region defined in terms of water collection over an area and delivery to a point. In practice, it sheds water, though in sufficiently arid regions, only rarely. There are other inputs and outputs to a watershed and other systems that it interacts with, besides water cycling.

In geomorphology, the naturally occurring watershed is referred to as a drainage basin, and a key goal is to understand the relationship between its organization, in terms of stream networks, hillslopes, and divides, and its processes, i.e., collecting and shedding water and the water-borne transport of loose material that are usually referred to as sediments. The 'shape' or 'form' of the drainage basin depends on the history of the sediment transport, but since the transport depends on the ability of the basin to collect water, the basin form guides future transport. Such a circular linkage makes it difficult to disentangle form and process. Nevertheless, this is only the beginning of the complexity. Looking at it more carefully, we see that the sediment is not produced without weathering, and evidence is accumulating that physical weathering is limited by the process of chemical weathering. Considering the results of the previous two chapters, without water flowing into the soil, as well as the fractured bedrock below, there will ultimately be no sediment to erode. Consequently, the geomorphological concept of a drainage basin cannot be considered in the absence of geochemistry and weathering, nor wholly without an accounting for the third dimension, i.e., soil depth and development.

In hydrologic terms, the two most important system inputs are the precipitation and the solar energy, which affect the partitioning of water into its solid, liquid, and

doi:10.1088/978-0-7503-5698-5ch11

gaseous forms The particular state water controls its subsequent transport and/or storage. In the present analysis, we neglect effects due to snow and ice, which would require a number of adaptations to the present theory. The obvious hydrologic outputs are run-off and change in subsurface storage. The water flowing into the soil engages the geochemical system. Its flowing out through the streams is typically the chief agent for removing sediment, as well as weathering reaction products and, thus, engages the geomorphological and geochemical systems simultaneously. The water leaving a drainage basin in the subsurface also contributes to subsurface run-off and carries dissolved products of the weathering reaction, but little in the way of suspended sediments.

Water also leaves a basin in the form of water vapor, as driven by the physics of the phase change of water through input solar radiation. The source of the water vapor flux is evaporation off bare ground or plants, or through plant transpiration in which plants draw water from soil through the process of evaporation off the stomata. The fundamental influence of vegetation on drainage basin evolution is in its extraction of water from the hydrologic system by providing a pathway for return of water to the atmosphere, which represents transpiration. Clearly, the process also engages the ecological system and couples to the carbon cycle through photosynthesis. Such a conjunction of all the four process types makes the watershed the focus of a wide range of current studies. Moreover, since the combined effects of evapotranspiration, E, represent the fate of almost 2/3 of global precipitation, it has, on average, a dominating influence in the hydrologic system. Recycling of the moisture through reprecipitation is also of fundamental relevance to the development of ecostytems, as well as flooding.

While the hydrologic processes considered already link the solid earth, biosphere, and hydrosphere, the atmosphere was considered only insofar as its ability to deliver water and allow the passage of solar energy to the surface. Another critical linkage is in the carbon content of the atmosphere, mostly in the form of CO_2. The geochemical cycle intersected at the watershed is responsible for the largest global carbon sink, namely, the deposition and lithification of carbonates in carbonate rocks, on account of such silicate weathering products as bicarbonate. Likewise, the carbon drawn down from the atmosphere, which is deposited in vegetation, accounts for the most rapidly changing carbon sink. These two carbon pathways link, respectively, to the subsurface component of the run-off and the transpiration component of E. They also, evidently, have a direct impact on plant growth and productivity, but also indirectly influence vegetation growth through the process of climate change, due to changes in atmospheric carbon content. A particularly important impact of climate change is on the water cycle. The tendency to accelerate the hydrological cycle puts more water in the atmosphere, which is not, however, evenly distributed. Thus, the zeroth-order effect of a warmer climate is to make wet (dry) areas wetter (drier), due to the greater moisture fluxes and higher values of moisture convergence and divergence. Second-order effects likely disrupt oceanic heat storage and currents, together with changing storm frequencies, hurricane patterns, and atmospheric wind characteristics, including the jet stream. Therefore, climate change may introduce unforeseen spatial, as well as temporal variability in the atmosphere. Another point worth noting is that some calculations of the time scales of the rock cycle are based on the rate of delivery of sediments to the ocean basin (Garrels and McKenzie 1972). If this perspective is valid, then the rate of chemical weathering at the surface may be the most important single limiting factor in determining the rate of the sedimentary rock cycle.

11.2 The water balance

The water balance describes the partitioning of precipitation into evapotranspiration E, run-off Q, and changes in storage S. Thus,

$$P = E + Q + \Delta S, \tag{11.1}$$

where ΔS represents the rate of change of the subsurface water storage S. Each term in equation (11.1) is typically represented as water depths accumulated or lost per year. For Q, a depth lost per year is obtained from, for example, the flux of water exiting a catchment in a river by dividing by the area of the drainage basin. Although water balance, equation (11.1), may appear trivial, predicting E and its variability in terms of the principal input climatological variables, P, and potential evapotranspiration, PET, as well as catchment storage characteristics S involves great complexity of the process and couplings across many length scales, which explains why solving equation (11.1) is considered the central problem of the hydrological sciences for about a century (Horton 1931, as reported in Dooge 1988, NAS 1991, NSF 2020), and, in fact, its central importance to water resources was realized well before Horton's address (1931) to the American Geophysical Union (AGU), having been addressed early in the 20th Century by Schreiber (1904) and Oldekop (1911). The solution connects at a nexus of critical zone processes, thereby linking hydrology, soil science, and ecology (Eagleson 1982, NAS 1991, Rodríguez-Iturbe and Porporato 2007, Hunt 2021, Hunt et al 2021, Nijzink and Schymanski 2022, Hunt et al 2023a, 2023b), establishing its significance in ecology, geomorphology, climate, and the history and future of life on Earth (Bonan and Doney 2018, Nordt and Driese 2013, Milly et al 2005, Oki and Kanae 2006, Vörösmarty et al 2010, Zhang et al 2022). One measure of the significance of a solution to equation (11.1) is, for example, the prediction that the strongest single predictor of the net primary productivity (NPP) of plant ecosystems is the value of E (Budyko 1974, Rosenzweig 1968).

Whereas P, E, and usually Q as well, all have the same sign every year, the sign of ΔS can change from year to year. Thus, in a temporal average of equation (11.1), the magnitude of ΔS tends to decrease as the time interval of the averaging increases, whereas, in the absence of climate change, the other quantities tend to converge to non-zero values. It has, therefore, been proposed, and mostly agreed to, that averaging equation (11.1) can be performed for eliminating the term ΔS, yielding,

$$P = Q + E \tag{11.2}$$

The length of time over which such an average is performed in order to justify the simplification is not universally agreed upon, but has been assumed to be 30–50 years (Gentine *et al* 2012). Understanding the partitioning of water at Earth's surface into run-off, storage change, and evapotranspiration is, thus, important to many disciplines, but also likely not understandable without a combined understanding of these disciplines and the interaction of their fields of study with the water balance, particularly at the scale of the watershed. On average, transpiration is almost 2/3 of evapotranspiration, and evapotranspiration almost 2/3 of the precipitation, making transpiration roughly 45% of precipitation. In all these areas, being able to predict the water cycle and its variability, as well as its variation along climate gradients, is important.

From an ecosystem perspective, the single most important output is productivity. In hydrology, which incorporates directly and to some extent the field of geomorphology in its attention to drainage basin form, water resources—i.e., total run-off, and erosion—mostly surface run-off, have been historically the key interests. Typically, change in storage is neglected for long-term averages of the water partitioning. We predict the water balance by optimizing ecosystem productivity as a function of the geochemistry of soil depth and the plant growth dependence on evapotranspiration. Our results are within 2% of the global average.

11.2.1 Optimality and percolation model of water balance

Solving equation (11.1) is simplified significantly if ΔS can be neglected, but its omission is justified at best over multi-decadal time scales (Gentine *et al* 2012), and not at the annual scales over which data are typically reported. It was postulated by Budyko (1958) that the solution of equation (11.1) for $E(P)$ with $\Delta S = 0$ can be represented as, $E/P = f(\text{PET}/P)$, where f is a function that must approach zero in the limit $E/P \to 0$, and unity in the opposite limit, $P/\text{PET} \to 0$. Several phenomenological models based on the Budyko hypothesis have been proposed in order to describe the water balance. Analytical results include those of Pike (1964), Fu (1981), Milly (1993), Choudhury (1999), Zhang *et al* (2004), Yang *et al* (2008), and Zhang *et al* (2022). Alternatively, for a specific set of conditions, the water balance may be solved using distributed process models (Arnold *et al* 1998, Henley *et al* 2011, Lan *et al* 2018, Martinez and Gupta 2010, Westra *et al* 2014), although their nature is not to develop an analytical solution to the water balance. The analytical models include non-parametric ones (see, e.g., Budyko 1958, Pike 1964, Turc 1954), which only describe trends in the data. They could also be parametrized models that address the variability in, for example, E/P, for a given PET/P. No consensus has, however, emerged regarding factors that control such model parameters (Greve *et al* 2015, Mianabadi *et al* 2020, Zhou *et al* 2015a), suggesting that parameters of ad hoc equations may not facilitate direct physical interpretation (Hunt *et al* 2023a, 2023b, Reaver *et al* 2022).

Distributed models also have their difficulties. Due to the complexity of processes and spatial variations among global catchments, they generally require many parameters and wide ranges of observational data for calibration, restricting their power for making accurate predictions at larger scales (Arnold *et al* 1998, Beven 1993, Moretti and Montanari 2007). In any case, an accurate, physically based, general solution that links critical zone processes explicitly to a functional form $f(\text{PET}/P)$ is lacking. Historically, a third approach to the water balance has been proposed to relate to ecological (or other) optimization principles (e.g., Eagleson 1982, NAS 1991, Rodriguez-Iturbe and Porporate 2007). It has been shown by two groups, Hunt and co-workers (Hunt *et al* 2020a, Hunt *et al* 2021a, Hunt *et al* 2023a, 2023b), and Nijzink and Schymanski (2022) that it is possible to predict essentially the same Budyko curve based on the principle of ecological optimality. In this chapter, ecological optimality is combined with principles of percolation theory to develop a direct calculation of the variability of streamflow with changing climatological variables, known as the streamflow elasticity (Němec and Schaake 1982).

The fundamental hypothesis (Hunt *et al* 2020a) is that the maximization of the ecosystem net primary productivity, NPP, with respect to the partitioning of P into Q and E provides the basis for predicting E, which is distinct from other optimization schemes that do not relate directly to the hydrologic fluxes (Eagleson and Tellers 1982, Guswa 2008, 2010, Milne and Gupta 2017). To use a flux-based hypothesis, however, it is necessary first to be able to predict $\text{NPP}(Q, E)$, a function built on scaling equations that were developed to predict soil production and vegetation growth rates (Hunt 2017), and were, themselves proposed to address more immediate goals, rather than to produce an analytical equation for E in terms of the Budyko variables. In that context, however, it was possible then to use a framework that had already been tested in its individual parts and demonstrated as predictive (Hunt 2017, Egli *et al* 2018, Hunt *et al* 2020b).

As discussed in chapter 10, soil formation is limited by chemical weathering— disintegration of soil by chemical reactions—since plants and vegetables cannot use bare rock to grow, but need the cations in the minerals that are transported by flowing water. In turn, chemical weathering is limited by the rate of transport of the cations—the solute in flowing water. Thus, solute transport plays a fundamental role in the processes that we consider here. Sahimi (1987) argued that in porous media that are sufficiently disordered that water flow is dominated by the critical paths (see chapter 4), properties of solute transport follow the percolation power laws. In particular, Lee *et al* (1999), who studied solute transport in porous media, showed that

$$t \sim x^{D_{bb}}, \tag{11.3}$$

which was already discussed in the previous chapters, where D_{bb} is the fractal dimension of the percolation backbone, the flow-carrying part of the percolation clusters. Here, t is the time, and x is the transport distance. In the context of soil formation, our focus, D_{bb} is associated with downward water fluxes, and for three-dimensional (3D) connectivity it should equal either 1.87 for saturated conditions, or 1.86, for wetting conditions (Sheppard *et al* 1999), a negligible distinction. The fact that $D_{bb} > 1$ implies retardation in the solute transport, a result of the topological complexity of the preferred flow paths near the percolation threshold that define the water flow rate, and also have orders of magnitude lower cumulative resistance than non-preferred paths. For dimensional consistency, we rewrite equation (11.3) as

$$x = x_0 \left(\frac{t}{t_0} \right)^{D_{bb}},$$

(11.4)

with x_0 being the fundamental spatial scaling factor, as a network node separation, or the median particle size, d_{50}, and t_0 the ratio of d_{50} and the mean annual vertical flux at the pore scale, $(P - E)/\phi$, where ϕ is the porosity (Yu and Hunt 2017c). When erosion is negligible, equation (11.4) yields the soil layer depth (Hunt 2017). In the case of surface reactions in porous media, such as weathering, solute transport is the limiting factor, and the product of the solute velocity, dx/dt, and the molar density of the weathering products provides the rate of weathering. Provided that chemical weathering limits soil formation (Yu and Hunt 2017b), dx/dt also gives the soil production rate. Suppose that z is the soil depth. Then, $dz/dt = dx/dt - q_d$, where q_d is the denudation rate of soil, since dx/dt also represents soil formation rate (per unit area). Given that, $x_0/t_0 = (P - E)/\phi$, using equation (11.3), one finds the steady-state soil depth z to be given by (Yu and Hunt 2017c);

$$z = x_0 \left(\frac{P - E}{D_{bb} \phi q_d} \right)^{1/(D_{bb} - 1)}.$$

(11.5)

The predictions of equations (11.4) and (11.5) for z were verified for durations of time t such that $10 \, \mathrm{yr} < t < 10^8 \, \mathrm{yr}$ (Yu and Hunt 2017a), namely, $1 \, \mathrm{m}/10^6 \, \mathrm{yr} < q_d < 1000 \, \mathrm{m}/10^6 \, \mathrm{yr}$ (Yu and Hunt 2017c) and ranges of climate such that $2 \, \mathrm{mm} \, \mathrm{yr}^{-1} < P < 10 \, \mathrm{m} \, \mathrm{yr}^{-1}$ (Hunt and Ghanbarian 2016), as well as for slope angles from 0 to 45° (input to q_d) (Yu et al 2019), while, for typical values of all the parameters, equation (11.5) yields the typical soil depth of about $1 \, \mathrm{m}$ (Yu et al 2017). Most important for the water balance is that, $z \approx (P - E)^{1.15} \, [1.15 = 1/(D_{bb} - 1)$, using $D_{bb} = 1.87]$.

To access soil nutrients carried along by flowing water pulled in by root potentials, trees' roots tend to grow horizontally along the optimal paths of their clusters—the percolation clusters—with the distance reached from the central stem being a power-law function of time, even when root tip extension rates are constant. As discussed in chapter 4, the optimal paths in heterogeneous media also form fractal structures with a fractal dimension D_{op}. It was shown (Hunt et al 2020b) that the root lateral extent (RLS) R_L is given by,

$$R_L = E_g \left(\frac{t}{t_g} \right)^{1/D_{op}},$$

(11.6)

which predicts both tree height and R_L over a time period between 11 and 200 years, with E_g being the evapotranspiration of a growing season of duration t_g. Equation (11.6) is simply a rewrite of equation (9.2) for longer time and spatial scales in such a way that does not alter predictions of root lateral spread, and represents a temporally-based upscaling of a velocity that diminishes as a power law in time

Next, we determine the net primary productivity, NPP. Since the root lateral extent equals transpiration after one growing season, and the cumulative transpiration after multiple seasons, and given that the root structure is fractal (Lynch 1995, Levang-Brilz and Biondini 2003, Dannowski 2005, Yang *et al* 2022), its mass equals transpiration raised to the power of its fractal dimension D_f and, thus, NPP $\propto E_g^{D_f}$. We assume that $E_g \propto E$, which was verified in a meta-study of 20 individual investigations (Hunt 2017), and is also consistent with spatiotemporal variability of tree growth. Since NPP is also propotional to the soil depth z, then,

$$\text{NPP} \propto z E^{D_f} \propto (P - E)^{1/(D_{bb} - 1)} E^{D_f}. \qquad (11.7)$$

As mentioned above, the prevalent view in ecology is that, to access and utilize resources for growth and reproduction, plant ecosystems exploit diversity optimally. Since soil depths and vegetation growth are typically formulated in terms of the principal hydrologic fluxes Q and E, we optimize NPP in terms of E, since it is not available to contribute to chemical weathering and deepening the soil. Thus, optimizing NPP, i.e., solving $\partial \text{NPP}/\partial E = 0$, yields

$$E = \frac{D_f(D_{bb} - 1)}{1 + D_f(D_{bb} - 1)} P \equiv \alpha(d) P, \qquad (11.8)$$

where d is the spatial dimension. One now has to characterize the root network and its fractal properties.

Levang-Brilz and Biondini (2003) grew 55 species of plants common in the Great Plains ecosystem in pots and then extracted them to measure their below-ground biomass (BGB) and root lateral spread R_L. The data for R_L as a function of root mass, which yields $1/D_f$, were collected under conditions that minimized both light and water limitations. The species chosen were typical of Great Plains, USA ecosystems, for which grasses make up over 90% of the biomass (Barker and Whitman 1988). Levang-Brilz and Biondini (2003) noted that, taken in aggregates, the fractal dimensions of the roots of the grasses separated into two groups, one of which was characterized by $D_f \simeq 2.65$ and dubbed Grasses 1, and a second one for which $D_f \simeq 1.79$, values that differ from the predictions of percolation in 2D, $D_f \simeq 1.9$ and 3D, $D_f \simeq 2.5$ by only about 6%. The aggregate fractal dimensionality of the forb species was reported to be 2.5, although there was large variability among these species as well. Detailed discussion of this point was presented by Hunt *et al* (2021), to which we refer the interested reader.

Since plant roots expand mostly horizontally near the surface, with their vertical extent being approximately equal to the soil depth, and due to the experimental data of Levang-Brilz and Biondini (2003), we assumed that the fractal dimension D_f of their network is the same as that of the 2D sample-spanning percolation clusters (see chapter 4) (Sheppard *et al* 1999), $D_f \simeq 1.9$, which together with $D_{bb} \simeq 1.87$, yields, $E = 0.623P$. This prediction agrees with the data for the global water balance, as eight studies reported (Hunt *et al* 2021) that, $0.59 \leqslant E/P \leqslant 0.67$, while a separate meta-study reported (Hunt 2017) a constant value of $\alpha = 0.64$. We will return to comparison with the experimental data shortly.

It was proposed (Hunt 2021, Hunt *et al* 2020a, 2020b) that when the root systems are deeper due to critical requirements for water (Fan *et al* 2017), the appropriate value of the optimization constant should be determined for the 3D percolation value of D_f, i.e., $D_f \simeq 2.5$. Thus, let us consider how the results would change, if the relevant root fractal dimensionality were, $D_r = D_f = 2.5$. For consistency in units, the power $(P\text{–}E)$ from the soil depth factor would have to be replaced with $(P - E)^{(3-D_f)/(D_{bb} - 1)}$. Then, the coefficient α for $d = 3$ would change to $\alpha \simeq 0.813$.

How does the equation for E vary with climate, expressed in terms of the aridity index? We define aridity index AI by $\text{AI} = PET/P$. Hunt (2021) suggested that the optimization scheme could apply to vegetation present, even for $\text{AI} \gg 1$, but that such vegetation could, at most, occupy the surface area equivalent to a fraction P/PET, while evaporation is 100% from the remaining surface. The result is

$$\frac{E}{P} = 1 - (1 - \alpha)\frac{P}{\text{PET}}, \tag{11.9}$$

$$\frac{Q}{P} = (1 - \alpha)\frac{P}{\text{PET}}, \tag{11.10}$$

where the second equation follows from, $Q + E = P$. If, $D_f = 3$, $\alpha = 1$, and for all $\text{AI} > 1$, we obtain, $E/P = 1$, the Budyko limit. Otherwise, there is an important distinction with the Budyko function. Budyko phenomenology requires, $1 - E/P \approx (\text{AI})^{-3}$, which we refer to as cubic asymptotics. Therefore, our equations are distinct from Budyko's phenomenology.

11.2.2 Extension of the theory and comparison with other models

Since models of plant root in the context of percolation theory can extend from 2D to 3D, we hypothesize that practical long-term, steady-state bounds on the water balance components for $\text{AI} \geqslant 1$ are provided by the 2D and 3D results given in the above equations. Thus, let us estimate the corresponding limits in the regime $\text{AI} \leqslant 1$, the energy-limited regime. The simplest theoretical alternative for $\text{AI} < 1$ is based on assuming that $(P - PET)$ simply runs off, unaccounted for, and that the optimization applies to the remaining P equal to PET. Hence, one has, $E = \alpha \, PET$. Then

$$\frac{E}{P} = \alpha\frac{\text{PET}}{P} \tag{11.11}$$

$$\frac{Q}{P} = 1 - \alpha\frac{\text{PET}}{P}. \tag{11.12}$$

For the actual plant (or ecosystem) value, $D_f = 3$, we have, $\alpha = 1$, and the equation, $E/P = PET/P$ is recovered, implying that the known limits for E on the Budyko plot are recovered for all AI. In general, at $\text{AI} = 1$, equations (11.9) and (11.11) both yield, $E/P = \alpha$, while equations (11.10) and (11.12) both predict, $Q/P = 1 - \alpha$. The theory thus produces continuous behavior for AI < 1 and AI > 1 for any value of α. Three points must be made, however. One is that the derivative of E/P with respect to AI is not continuous, except when, $\alpha = 0.5$. Furthermore, there is no guarantee that any particular trajectory across Budyko space in natural systems will require consistency of α across the entire range of AI values. As water becomes scarcer with increasing AI, root depths increase (Fan *et al* 2017), which we should expect to lead to less anisotropy of the root system and a higher fractal dimensionality. Thus, we may expect a crossover to higher values of α with increasing AI, although there is no reason to expect that this crossover will occur at any particular value of AI. Finally, although systems for which PET/P < 1 are classified as energy-limited, if the precipitation arrives in winter and the bulk of the solar irradiance in summer, they may still experience significant water limitations. Therefore, the conceptualization of a rigid distinction between energy- and water-limited systems is oversimplified and might be improved by a composite treatment, allowing separate limitations for distinct fractions of a year. Nevertheless, for the sets of data that we are aware of, experimental scatter is of sufficient magnitude that such a theoretical limitation is not of significance.

For later comparison with experiment (see below) in the water-limited (AI > 1) regime, equation (11.9) is rewritten as,

$$\frac{E}{\text{PET}} = \left(\frac{P}{\text{PET}}\right)\left[1 - (1 - \alpha)\frac{P}{\text{PET}}\right]. \tag{11.13}$$

Up until now, Budyko's theory has mostly inspired multiple mathematical developments, but a limited number of papers that presented new physical insights. For example, Sposito (2017a, 2017b) incorporated into Budyko's framework the changes in vadose zone water storage in a manner that is both parsimonious in hypotheses and broad in scope. As pointed out by Budyko (1958), clear limits on E/P exist in the limits when PET/P approaches zero and infinity. In the former case, E cannot exceed PET, whereas in the latter P provides a similar upper bound. Mathematical functions have been sought out that conform to such general behavior. For parametric functions, which provide a continuous range of predictions in terms of a parameter, these equations may be obtained exactly from applying a limiting value of that parameter. A range of proposed functions is presented in table 11.1. The results are expressed traditionally in terms of $\varphi = \text{PET}/P$. Schreiber's and Oldekop's formulations were considered lower and upper bounds for E/P, while that of Budyko, typically superior in the mean, was formulated as the geometric mean of the bounds (Choudhury 1999). More recent phenomenological models (Choudhury 1999, Fu 1981, Zhang *et al* 2004, 2008) are most easily conceptualized in terms of a generalization of the Pike model. Choudhury (1999), for example, replaced Pike's powers 2 and 1/2 by arbitrary reciprocal fractions n and $1/n$, with $n > 1$. The original phenomenological model of Pike is scarcely distinguishable from Budyko (Choudhury 1999); however, Choudhury's (1999) generalization allows it to generate a much wider range of E values for any value of AI than that predicted by

Table 11.1. Various models relating evapotranspiration E to precipitation P.

Model	Reference
$f(\varphi) = 1 - \exp(-\varphi)$	Schreiber (1904)
$f(\varphi) = \varphi \tanh(1/\varphi)$	Oldekop (1911)
$f(\varphi) = 1/\sqrt{1 + 1/\varphi^2}$	Pike (1964)
$f(\varphi) = \{\varphi \tanh(1/\varphi)[1 - exp(-\varphi)]\}^{1/2}$	Budyko (1958, 1974)
$f(\varphi) = 1/[1 + (1/\varphi)^n]^{1/n}$	Choudhury (1999)
$f(\varphi) = 1 + \varphi - (1 - \varphi^\omega)^{1/\omega}$	Zhang et al (2004), Fu (1981)
$E/P = (1 + w\varphi)/(1 + w\varphi + 1/\varphi)$	Zhang et al (2001)

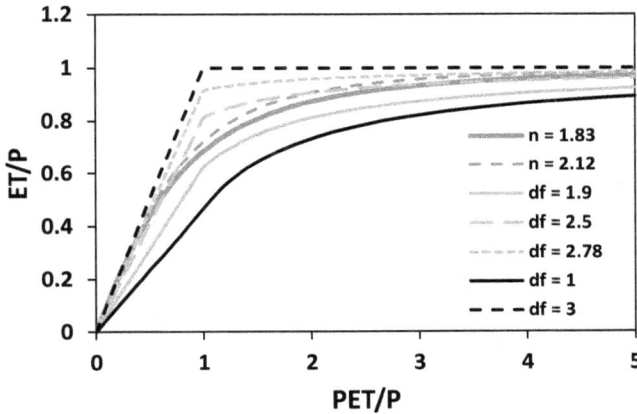

Figure 11.1. Comparison of the Budyko function derived from the optimization of net primary productivity with respect to E with the Choudhury (1999) function for various values of n for the Choudhury function, and the percolation model with a fractal dimension d_f. Hunt et al (2021) John Wiley & Sons. Copyright 2021. American Geophysical Union. All Rights Reserved.

Oldekop's and Schreiber's functions together. A significant, unanswered, question is: how much variability should be contained within a single, universal, function for E/P? (Berghuijs et al 2020).

Figure 11.1 presents a comparison of the predictions of the percolation model with one of the parametric models in table 11.1, that of Choudhury (1999), which shows the predictions of Choudhury's model for two values of its parameter n that are $n = 2.12$ for forested catchments, and $n = 1.83$ for grasslands, and compares the predictions for $D_f \simeq 1.9$ (2D percolation), 2.5 (3D percolation), and 2.78 (one standard deviation above the mean for Grass Group 1 from experiments reported by Levang-Brilz and Biondini (2003); see below, for a discussion quantifying variability of D_f). Reflecting the physical nature of the percolation model, the Budyko limit on E is recovered for $D_f = 3$, in contrast to Choudhury's model for which the Budyko limits are recovered in the limit of $n \to \infty$. Recall in this context that the limiting value of $D_f = 3$ arises from a (physical) dimensional constraint; the horizontal (RLS) and vertical (depth) dimensions of the root system cannot together account for more than three dimensions of root mass. A mathematically less rigorous, but still physically significant bound, is provided by $D_f = 1$, a linear relationship between R_L and below-ground biomass (BGB). Note that, for AI > 2.5, curves in figure 11.1 that are predicted by percolation theory are essentially indistinguishable from those of the Choudhury (1999) phenomenological model. In the range AI < 1, Choudhury's curves resemble closely percolation predictions for $3 - D_f \ll 1$, i.e., $D_f \gg 2$. The percolation model tends to generate a broader range of possible values of PET/P than the Choudhury function for any given E/P when AI < 1. This distinction is due to the fact that, in contrast to other Budyko-type formulations, only for $D_f = 3$ does the percolation model predict a slope of 1 in the limit AI $\to 0$. Otherwise, it predicts a smaller slope, which is α.

We also note in figure 11.1 that the physical constraint $1 \leqslant D_f \leqslant 3$ provides a predicted bound for data that, except in the region AI < 0.5, is in reasonable agreement with the model of Choudhury for the range $1 \leqslant n \leqslant \infty$. Moreover, the percolation prediction, $D_f \simeq 2.5$ for root fractal dimensionality is in rather close correspondence to Choudhury's model prediction for $n = 2.12$ over the entire range of PET/P, considered valid for forests. A third important result in figure 11.1 concerns sensitivity. The distinction between the predictions for $D_f = 1$ and $D_f = 2.1$ is considerably smaller than the corresponding distinction between the predictions for $D_f = 2.1$ and $D_f = 2.9$, which is not surprising, as the limit of $D_f \to 3$ in the percolation model corresponds to the Choudhury's limit of $n \to \infty$. Thus, sensitivity to input root measurements increases with increasing D_f, a measure of the space-filling ability of the plant roots.

11.2.3 Bias and uncertainty in measurements at large evapotranspiration or aridity index

Before the predictions of percolation theory are compared with the experimental data, an important aspect of such comparisons must be pointed out. Gentine *et al* (2012) suggested that the explicit form of Budyko's $E(P)$ relationship may account for all the long-term average data, if one assumes that the variability in the data are attributable to random experimental error. Thus, in order to generate reasonable uncertainty in the prediction, Gentine *et al* (2012) used ±10% adjustment to the original Budyko function (see also Potter *et al* 2005). This approach would, however, give rise to long-term average E/P as high as nearly 1.1 for AI = 5, which in this picture can only be acceptable as a random error. An analysis of figure 1(b) of Gentine *et al* (2012) indicated that at least 80% of the data for long-term averages exceed the Budyko prediction for AI > 2.1, implying that, unless there is a systematic experimental error at large AI, the Budyko function also underestimates E in the same range. The percolation model described in this chapter attributes significant variability to E due to plant root architecture, but it cannot also argue that E should exceed the Budyko limit, $E/P > 1$. Since both the percolation model and Budyko formulation underestimate E for large AI, Hunt *et al* (2021) investigated potential systematic biases that could lead to model overestimation of E at high AI. For this purpose, they considered two types of bias: (1) diminution in groundwater storage due to either climate change effects on, or exploitation of, aquifers in agriculture, and (2) underestimation of precipitation at high AI.

Worldwide estimates of groundwater depletion vary significantly, and are of interest for their potential contributions to sea level rise. The pre-2000 estimates vary from 0.075 mm yr^{-1} to 0.3 mm yr^{-1} (Wada *et al* 2017). More recent studies (Wada *et al* 2017) of the global water budget suggested mean storage losses as high as 0.71 mm yr^{-1}. The data reported by Gentine *et al* (2012) are exclusively from drainage basins in the United States, and most authors agree that there has been sufficient water demand from agriculture in the U.S. to deplete water resources significantly. Specific numbers on this vary. Liu *et al* (2018) reported a yearly mean depletion of almost 4% of P (20 mm yr^{-1}) in the Columbia basin, approximately 0.4% of P (3 mm yr^{-1}) in the Mississippi drainage, and 0.5% of P (7 mm yr^{-1}) in the Pearl River basin. After (areal) averaging, the best estimate for the three basins is, annually, 0.8% of P storage loss, with the (areal) average AI being 1.41. The depletion to storage tends, however, to increase with increasing AI, and is largest in the Columbia River basin.

For the second potential confounding input, precipitation measurements, which are typically less accurate than often assumed, were analyzed. Rainfall during intense events may be underestimated by 5%-40% using tipping buckets. Light rainfall, fog, and dew are difficult to capture accurately by any method. Moratiel *et al* (2016) stated, 'If the canopy is wet due to fog, dew, or light rainfall, however, energy contribution to surface evaporation will reduce transpiration and hence soil water losses. When surface evaporation occurs, the E overestimates the soil water depletion by an amount approximately equal to the surface water evaporation.' Thus, the ratio E/P may be overestimated by a fraction similar to the contribution to P of fog or dew. In order to estimate a magnitude of this effect, Hunt *et al* (2021) carried out a survey of the literature, whose results are presented in table 11.2. Daily contributions of fog or dew to P were typically in the tenths of millimeters, annual values measured in centimeters. Such a contribution to P is obviously of greater significance in arid regions, where annual precipitation is typically less than 20 cm, and often less than 10 cm. The mean and standard deviation of the fractional underestimation, 0.11 ± 0.08, calculated based on table 11.2, do not include the extreme results of 0.75 of Cape Mountain and the Atacama Desert, but use the upper and lower bounds separately, when ranges of values were given. The difference between 0.11 and 0.08 is 0.03, or 3% of the precipitation. Only one of the data in table 11.2, the Pilarcito Creek watershed of Chung *et al* (2017), suggests a contribution to P of fog and dew that is as small as the up to 4% impact on the water balance from changes in storage. One may, therefore, conclude that, although the statistical summary may be biased toward investigations of sites with particularly large contributions to P, it is unlikely that the effect of fog and dew on E is, on the average, smaller than that of changes in storage. Moreover, both effects are of the same sign and typically increase in a relative sense with increasing AI.

Table 11.2. Fog and/or dew contributions to annual mean precipitation P.

Fraction of mean P	Region	References
0.75	Cape Mountain, S. Africa	Matimati (2009)
0.05–0.1	Arid Andes valley	Kalthoff et al (2006)
0.15	Balsa Blanca Spain (4-year mean)	Uclés et al (2014)
0.19	Semi-arid coastal south-western Madagascar	Hanisch et al (2015)
0.27	Sand dune areas of India (maximum)	Subramanian and Rao (1983)
0.049–0.102	Continental semi-arid grassland	Aguirre-Gutiérrez et al (2019)
0.13	Chinese loess plateau	F.-L. Yang et al (2015)
0.111	Rambla Honda, Spain	Moro et al (2007)
0.1–0.25	Negev, Israel	Kidron (1999)
0.045	Netherlands grassland	Jacobs et al (2006)
0.055–0.069	Northern Germany	Xiao et al (2009)
0.035–0.15	New Zealand snow tussock grasslands	Fahey et al (2011)[†]
0.01–0.03	Upper Pilarcito Creek watershed, N. Calif.	Chung et al (2017)
0.75	Atacama Desert ($P = 0.8$ mm yr^{-1})	Westbeld et al (2009)
0.021–0.27	California coastal islands	Fischer et al (2009)

[†]The lower limit was within uncertainty of mean annual precipitation, while the upper limit was influenced by other factors. Thus, both limits are possibly overestimations.

11.2.4 Comparison with data

We now compare the predictions of the percolation model with experimental data. For later context, figure 11.2 shows a comparison between the predicted range of E by the percolation model and the ca. 18 000 data points from the aforementioned figure 1(a) of Gentine et al (2012). The 2D and 3D predictions use the theoretical estimates of percolation theory for root fractal dimension D_f, with $D_f = 3$ representing its maximum maximum value for a completely space-filling system, for which roots can extract all water entering the soil. $D_f = 1$ is, on the other hand, a physical minimum that represents a purely linear relationship between below-ground biomass and root lateral spread. As explained above, $D_f = 2.78$ is an upper bound established by plants growing in pots (Levang-Brilz and Biondini 2003) that was determined by assuming that plant root fractal dimensions are distributed according to a Gaussian distribution. Although, on the whole, the theoretical upper and lower values of D_f do constrain the observed data very well, the percolation model does appear to somewhat underestimate the observations at high AI. In figure 11.2 the ability of the percolation model for predicting E/P as a function of PET/P is tested to see if its theoretical parametric values are in agreement with a wide range of data, as well as whether its upper and lower bounds constrain the large majority of the data. As the figure indicates, it appears as though the predicted values of D_f underestimate E/P somewhat at higher PET/P values.

Data of Duan et al (2006): figure 11.3 compares the predictions of the percolation model for E/PET with the Budyko phenomenology with the data of figure 3 of Duan

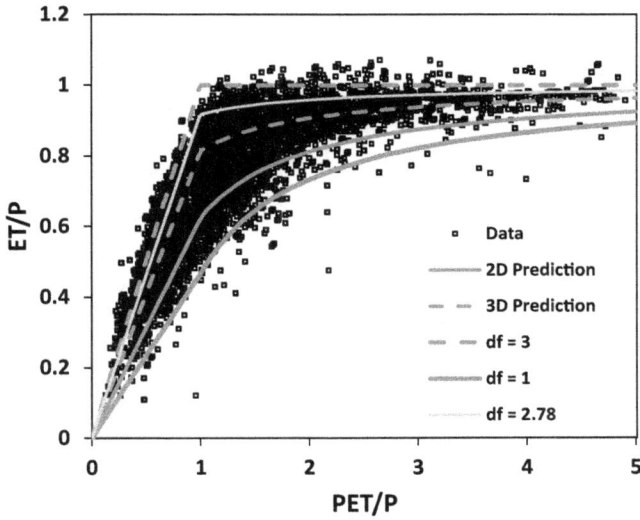

Figure 11.2. Predictions of the percolation model for E/P as a function of PET/P. Approximately 18 000 data points shown on the figure were digitized from figure 1(a) of Gentine *et al* (2012). Hunt *et al* (2021) John Wiley & Sons. Copyright 2021. American Geophysical Union. All Rights Reserved.

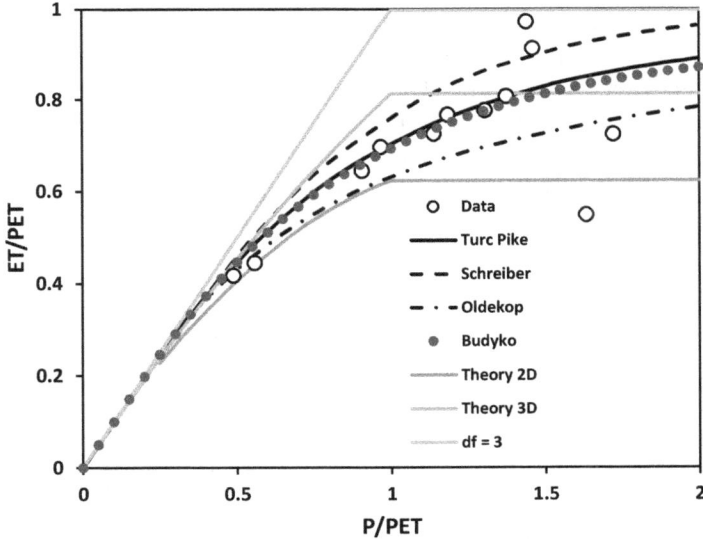

Figure 11.3. Comparison of the predictions of the percolation model for 2D and 3D systems in the (red and blue lines) with experimental Duan *et al* (2006) data and traditional phenomenological results. The maximum for E/PET results when the plant root fractal dimensionality, $d_f = 3$. Whereas eight data points fall between the phenomenological predictions of Schreiber and Oldekop, 9 of the 12 data points fall between the predictions of the percolation model for 2D and 3D values of the root fractal dimensionality. Hunt *et al* (2021) John Wiley & Sons. Copyright 2021. American Geophysical Union. All Rights Reserved.

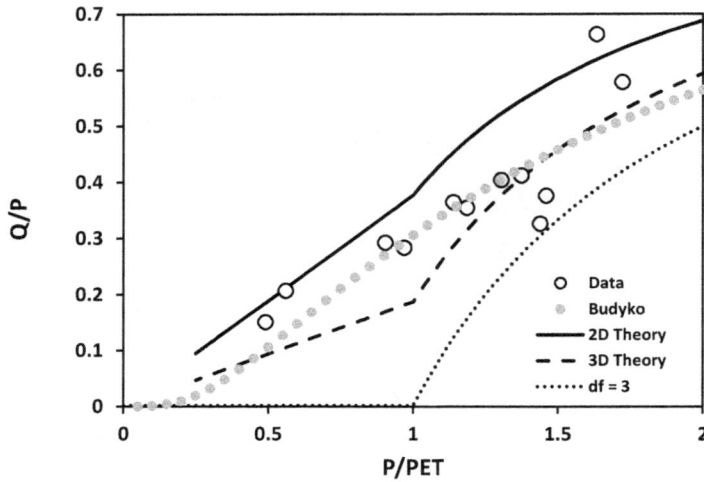

Figure 11.4. Predictions of the percolation model for Q/P as a function of $P/$PET and their comparison with the Budyko model and data reported by Duan *et al* (2006). As required by the results shown in figure 11.3, nine of the 12 data points again fall between the 2D and 3D predictions of the percolation model. Hunt *et al* (2021) John Wiley & Sons. Copyright 2021. American Geophysical Union. All Rights Reserved.

et al (2006). Despite the artificial appearance of the discontinuous change in slope, the theoretical predictions with $D_f = 1.9$ ($\alpha = 0.623$) and $D_f = 2.5$ ($\alpha = 0.813$) constrain the data slightly better than the Schreiber and Oldekop functions taken together. The conclusion would not change, if we used instead the D_f values, 2.65 and 1.79, reported by Levang-Brilz and Biondini (2003) mentioned above. Figure 11.4 compares the percolation predictions for Q/P with existing phenomenological models with the data of Duan *et al* (2006). A distinct discrepancy between the predictions of the percolation model and Budyko's model at high values of PET is also clear, which is due to the cubic asymptotic form of the Budyko function, AI^{-3}, described above.

Data of Gentine et al (2012): figure 11.5 compares predictions for E/P as a function of PET/P with the annual E data reported by Gentine *et al* (2012). Since there was some indication in figure 11.1 that theory underestimated experimental data at high AI, an adjustment was made to the data in figure 11.5 that is in general accord with overall studies on sea level rise (Wada *et al* 2017), which necessitate an input from depletion of groundwater reserves, and approximately tied to the specific hydrology of western United States, as exemplified by the Columbia river, which indicates that groundwater storage has been declining by as much as 4% of P annually (Liu *et al* 2018). To make the adjustment, the 4% figure was used as a limit for the case AI $\rightarrow\infty$ and a linear model based on the inverse of AI was used, with effects of storage change that vanish at AI = 1. With this adjustment, the results predicts changes of only -1.5% for the Columbia River with its AI of 1.6, as well as -1.3% for the Mississippi, and 0% for the Pearl River drainages, for which storage changes are -0.4% and -0.5%, respectively.

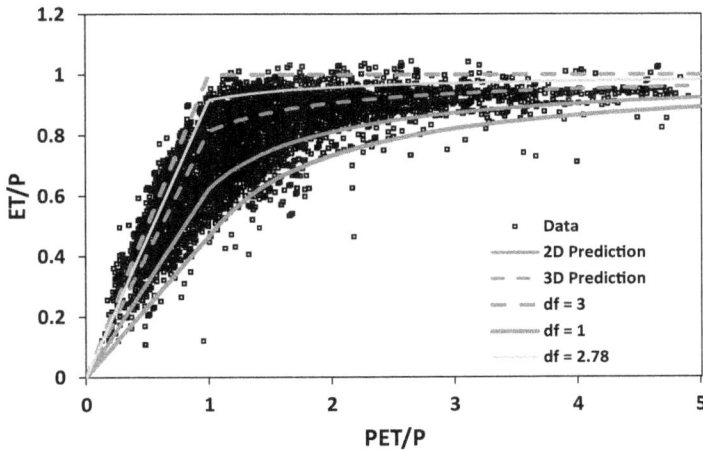

Figure 11.5. Comparison of the predictions of the percolation model with data, adjusted to reflect the tendency of groundwater storage to diminish in the United States stream basins, with an effect assumed to be linear in AI^{-1} between $AI = 1$ and $AI = \infty$. In this comparison, the maximum storage change in the limit of infinite AI is assumed to be 4%, approximately equal to the losses from the Columbia River (at $AI = 1.64$). Thus, the model underestimates storage losses in the Columbia and Pearl River basins, but overestimates such losses in the Mississippi basin. Note that such a correction places the large majority of the data points between the 2D and 3D percolation model at large AI, giving the distribution a symmetric appearance with respect to theoretical predictions. Hunt *et al* (2021) John Wiley & Sons. Copyright 2021. American Geophysical Union. All Rights Reserved.

The comparison of the predictions of the percolation model, using also the upper value of D_f from Grasses 1 of Levang-Brilz and Biondini (2003) in figure 11.5 prompts several important remarks. One is that the tendency to underestimate E at high AI in figure 11.1 can be removed by accounting for drawdown of groundwater storage. The second is that the lower limit of the predicted E is rather sharp, i.e., it is not plastic. Thus, values of E below the lower limit are likely due to important characteristics that have been left out, such as the effects of snowmelt on the moisture content and groundwater recharge. The third remark is that, even when systematic trends in storage are removed, yearly summaries of E/P still incorporate the effects of weather-related fluctuations in groundwater storage. However, such fluctuations (typically under about 5%) in figure 11.5 appear equally spaced around the percolation predictions. In 26 large river basins globally (with an overall mean value of 0.618 for E/P), yearly fluctuations in storage appear to average about $\pm 3\%$ of E (Liu *et al* 2018). While subsurface storage changes can be positive or negative, such changes in regions with intensive agriculture are predominantly negative. Note also that at large AI, extreme desert conditions can restrict dominant vegetation types to a few species, such as those in the Mojave Desert, namely, creosote bush (*Larrea tridentata*) and bursage (*Ambrosia dumosa*) (Hunt and Wu 2004). With increasing aridity, the diversity of the ecosystem may suddenly change, if one of such species is unable to adapt. Evidence of this kind of

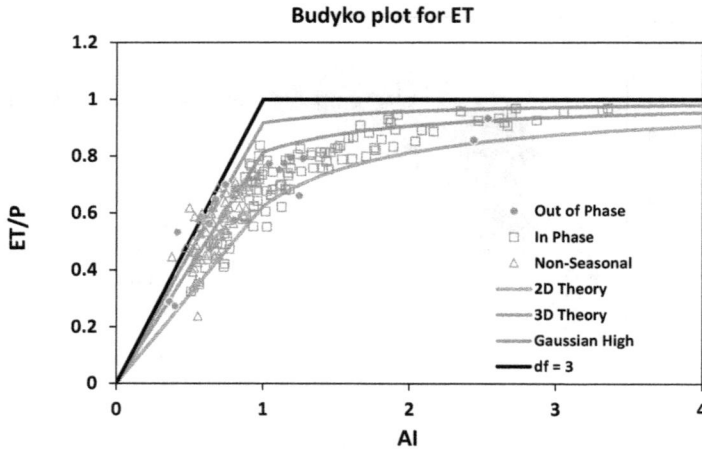

Figure 11.6. Comparison of long-term data for E reported by Gentine *et al* (2012) with the predictions of the percolation model in 2D ($d_f = 1.9$), 3D ($d_f = 2.5$), the upper limit based on the Gaussian model of observed Grass 1 d_f variability ('Gaussian High'), and the limit resulting from the maximum value of the root biomass fractal dimension of $d_f = 3$. Hunt *et al* (2021) John Wiley & Sons. Copyright 2021. American Geophysical Union. All Rights Reserved.

discrete behavior may be present for AI > 3.5, where approximately horizontal rows of data, each terminating at a higher AI value, are present.

In figure 11.6, the long-term—50-year average—results, obtained by Gentine *et al* (2012) are replotted and compared with predictions of the percolation model with for $D_f = 1.9$ (2D), 2.5 (3D), and 2.78 (one standard deviation above the mean for Grass 1). In this representation, the yearly average storage loss (between 0% at AI = 1 and 4% in the limit AI $\rightarrow\infty$), as a fraction of P was used to reduce the ratio E/P predicted by the models. As can be seen, for a wide range of AI, the data are mostly between the 2D and 3D percolation predictions, but they tend to cross from an upper constraint of the 3D prediction to $D_f = 2.78$ at high AI. A sharp distinction between the out-of phase and in-phase data is noted, particularly for AI < 1. In particular, the in-phase data exhibit a crossover from constraints between 2D and 3D, to between 3D and $E = P$ as AI exceeds 1.6. Two possible reasons for this result are, (1) under highly arid conditions, the dominant plant property of relevance is the ability to extract the maximum water possible (corresponding to large D_f), and (2) precipitation is underestimated for large AI. While the 2D ($D_f = 1.9$) predictions provide a good lower bound for nearly all the data in figure 11.6, the 3D predictions do not provide an upper bound, except for a relatively narrow range of AI between 1 and about 1.8. Nearly all the data fall above the 3D prediction for Al > 1.8. For AI < 1, the 3D value of D_f almost perfectly distinguishes between measurements of E for in-phase and out-of-phase precipitation, with nearly all the in-phase values below the 3D predictions, and nearly all the out-of-phase results above it. Since the statistics involve hundreds of data points, the probability that this prediction is due to random variability is essentially zero. One interpretation is that precipitation that arrives out-of-phase with the maximum in irradiance is not conducive to an optimal use of water by plants, but with increased water waste, here noted as evaporative losses. If this interpretation is correct, it is a result that is not in accord with current thinking (Berghuijs *et al* 2014) regarding the role of seasonality of precipitation, though it is consistent in a more general sense with the study of Madany *et al* (2021), as discussed by Thompson (2021), in Eos, published by the American Geophysical Union that showed a diminution of water use efficiency (WUE) when resources are scarce. Further support of this diagnosis is provided by the fact that non-seasonal drainages fall equally on both sides of the predictions of the 3D percolation model. It should also be pointed out that typical phenomenological models that are currently used would be unlikely to discover such a correlation, since they condense the variability in prediction curves dramatically in the low AI range, relative to the large AI range.

Figure 11.7 uses 1/AI, the inverse of the aridity index, and the variable $1 - E/P = Q/P$ to clarify the distinction between in-phase and out-of-phase vegetation in the low AI regime. The general appearance of the data can support linear dependence of Q/P on 1/AI, but requires a threshold value of P/PET to be exceeded before run-off is measured. It could also support a quadratic or cubic dependence near 1/AI = 0. In figure 11.8 whether alternative explanations could be relevant were checked, which are, (1) based on the hypothesis that

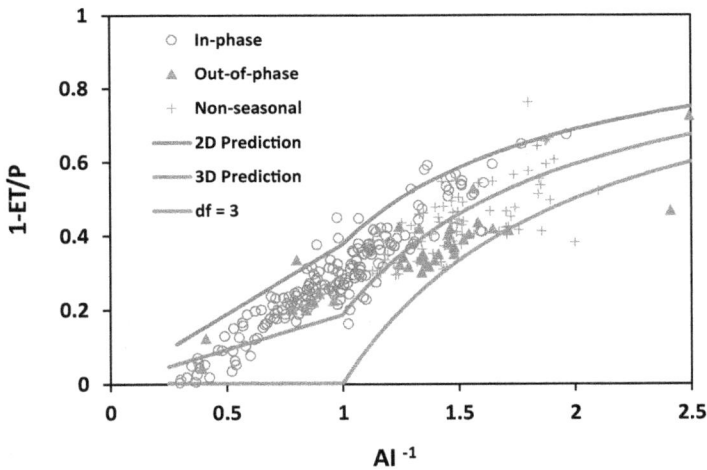

Figure 11.7. Comparison of the predictions of the percolation model with the full dataset of Gentine *et al* (2012). In the range $1 \leqslant AI^{-1} \leqslant 2$, it is noteworthy that nearly all the out-of-phase data fall between the prediction of the 3D percolation model and those with $d_f = 3$, whereas nearly all the in-phase data fall between the 2D and 3D predictions of percolation model. Also, the non-seasonal data are spread approximately evenly between the 2D predictions and $d_f = 3$ predictions. Note the discrepancy at low P/PET, which may be better represented by a traditional model, e.g., Budyko (1958), which generates a cubic dependence on 1/AI of $1 - E/P$. In figure 11.8 this discrepancy is addressed. Hunt *et al* (2021) John Wiley & Sons. Copyright 2021. American Geophysical Union. All Rights Reserved.

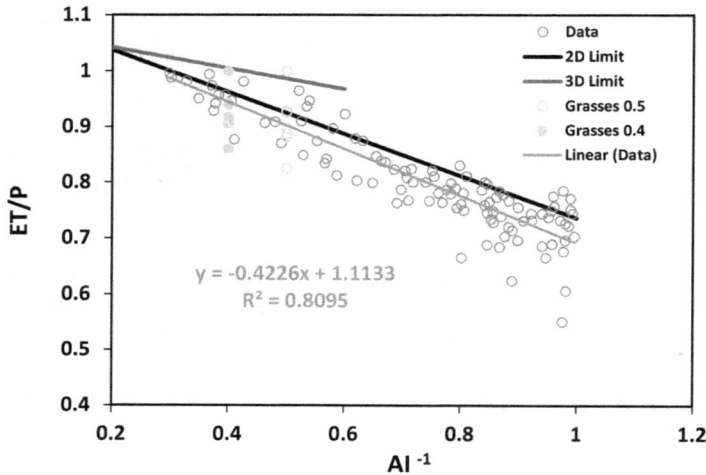

Figure 11.8. Plot of E/P versus P/PET for the Gentine *et al* dataset for AI $\geqslant$ 1. Hunt *et al* (2021) John Wiley & Sons. Copyright 2021. American Geophysical Union. All Rights Reserved.

underestimation of precipitation leads to an overestimation of E/P, and (2) whether there is evidence that vegetation in arid environments adapts so that the fractal dimension D_f of root system is increasing. The interpretation that underestimation of precipitation leads to the discrepancy between theory and prediction at large AI is supported in that the intercept (extrapolated value of E/P as AI $\rightarrow \infty$ approaches) is equal to 1.11, which is a discrepancy equal to the mean underestimation of precipitation shown in table 11.2. However, the selection of individual grasses from Grass Group 1 sweeps out the range of E values shown, which crosses over from the 2D limit to the 3D one, calculated using an 11% P underestimation assumption. Figure 11.8 also investigates the behavior of $1 - E/P$ in the asymptotic limit, AI $\rightarrow \infty$. Grasses 0.5 and 0.4 refer to the predicted E values at $\text{AI}^{-1} = 0.5$ and 0.4 of seven of the individual grass species identified as being in the group (Grasses 1) with an aggregate fractal dimension of 2.65 (Levang-Brilz and Biondini 2003). The reason for excluding five outlier values of D_f is that some of them, such as $D_f = 142$, 5, and 1.5 for, respectively, *Agropyron cristatum*, *Panicum virgatum*, and *Hordeum jubatum*, do not appear to be reasonable estimates for values of ecosystem D_f. Including these species in the determination of the standard deviation of the mean value of $s = 1/D_f$ does not, however, change the statistics substantially. Overall, any of the possible interpretations mentioned can be supported. While a systematic error of 11% in precipitation may seem large, it should be noted that nearly 4% of P annual drawdown of storage is documented by Liu *et al* (2018) for the Columbia River—a semi-arid region—with basin average AI of 1.64. Thus, even a significant exaggeration of the underestimation of precipitation is at least partially compensated by a lack of simultaneous accounting for systematic changes in storage.

11.3 Arid lands

The fundamental water balance equation described above is merely a statement of the conservation of water mass. For increasing spatial or temporal scales, the equation becomes increasingly accurate. But, at smaller length scales, a range of caveats and, hence, certain complications arise that include, (1) a non-zero horizontal divergence of a subsurface flow field with water transport towards the surface, and (2) spatial boundaries of the flow in the subsurface that differ from those at the surface, the catchment boundaries. Nevertheless, application to large spatial scales does not eliminate potential complications. In the limit of large aridity, AI $\gg$ 1, several factors complicate comparisons of a water balance theory with observations. Most prominently, perhaps, is the tendency for the contributions of fog, dew, and events with very light rainfall to constitute a larger fraction of the total precipitation. A compilation of evidence from many sources (Aguirre-Gutiérrez 2019, Chung *et al* 2017 Fahey *et al* 2011, Fischer *et al* 2009, Hanisch *et al* 2015, Kalthoff *et al* 2006, Kidron 1999, Matimati 2009, Moratiel *et al* 2016, Moro *et al* 2007, Uclés *et al* 2014) indicates that typical magnitudes of such contributions are on the order of a centimeter or two, a value that can even exceed the measured rainfall at sites in the world's arid lands. Moreover, agriculture in arid parts of, for example, western United States requires irrigation from groundwater sources that is as much as 8% of the precipitation of an entire river basin (Chen *et al* 2020). Given that predicted E in such regions can be up to 90% of P, an underestimation of moisture input into the system equal to 10% of P would constitute a prediction of streamflow that is half of what should be determined. It has been pointed out that all of such complicating factors tend to produce underestimates of the observed E/P ratio.

Other complications in arid lands include the facts that vegetation tends to collect dust beneath it, both from the direct collection in the foliage that is subsequently washed to the ground (Field *et al* 2010), and due to the effects of rainsplash in moving loose surface dirt around (Planchon and Mouche 2010). Since vegetation tends to intercept the fall of raindrops, those that strike the ground underneath vegetation tend to do so with fewer changes in momentum after having already lost considerable momentum through impacts with the plant. Thus, splashing of dirt towards plants is enhanced with respect to the opposite. The production of soil through dust collection is an alternate path and tends, particularly underneath desert plants, to vitiate the need to allow some water to bypass plant use in the process of weathering and soil formation. Thus, the parameters of the optimization described above are changed. Finally, the assumption for deriving the percolation theory of water balance has been that water falling on land between bushes in arid regions would evaporate fully, which is an oversimplification. Plants tend to occupy zones where surface water converges, particularly when there is an easy pathway for the water to follow into the soil, such as shallow fractures or cracks (Stothoff *et al* 1999). On the other hand, in areas where plants do not grow well, the surface tends to become armored and impermeable (Wells *et al* 2014). The net result is a tendency for water to flow towards plants in desert regions.

11.4 Forests and grasslands

Among terrestrial ecosystems, grasslands cover the largest area (Akash *et al* 2021), but forests store the most carbon (Pregitzer and Euskirchen 2004). Thus, due to their heightened importance, studies on evapotranspiration of forests and grasslands are relatively common, affording accentuated opportunity for comparisons, while in global studies across ecosystem types, such categories have also been addressed separately. In order to improve our understanding of the relative importance of factors that influence the water balance and carbon cycle, we study and analyze data for the two types of ecosystem, forests and grasslands, under a range of climatic conditions.

Environmental conditions conducive to forest growth are more likely to be energy-limited, i.e., $\text{PET} < P$, than water-limited, the limit in which $\text{PET} > P$. Data for evapotranspiration E for forests may, however, extend into water-limited conditions, making it necessary to enable us to apply the models derived above to this aspect of ecosystems (Hunt 2021).

11.4.1 Water limitations

As discussed earlier in this chapter, effects of water limitation are accounted for by assuming that E is optimized by vegetation over a fraction of ground equal to P/PET with $E = 100\%$, otherwise. That the fractional surface coverage of vegetation is at least proportional to $\varphi = P/\text{PET}$ was shown by Yang *et al* (2009). Thus, wherever vegetation exists, the above optimization is carried out. The fundamental assumption about water limits is generally consistent with the tendency of plants in arid regions to grow in zones of water convergence, such as land surface depressions, preferential flow along fractures (Stothoff *et al* 1999), or between regions of surface armoring by clasts, which inhibit infiltration (Wells *et al* 2014). The result is given by equation (11.9). Because $E/P = 1$ is a limiting value of E, equation (11.9) may tend to overestimate E on bare ground (Hunt 2021), while we suggest that it might underestimate E under plants. The reasons for such underestimation include soil formation through accretion by diffusion of particles towards plants during rainstorms, and the deposition of aeolian sediments on plants and subsequent deposition on the soil through interception.

11.4.2 Energy limitations

Next, we consider the opposite case, the limit of small AI ($P > \text{PET}$). In analogy to the case of large AI, where Q was assumed to be negligible in areas without vegetation, we assume that, to the lowest order, infiltration fluxes are negligible for that portion of P that exceeds PET, and that the excess precipitation $P - \text{PET}$ simply runs off and is removed from the system. The optimization is then performed on a fraction of P equal to PET with the result being equation (11.11). One might add a bit more complexity to this analysis by allowing a fraction β of the excess precipitation, i.e., $P - \text{PET}$, to infiltrate into the soil and contribute to soil production. The factor β will tend to decrease with precipitation intensity, but will also depend on subsurface characteristics. Instead of $Q \leqslant$ PET as an optimization parameter, we now have $Q = P - E$, together with an additional term $\beta(P - \text{PET})$. Then, the optimization proceeds as described above, starting from

$$Q_s = \text{PET} - E + \beta(P - \text{PET}), \tag{11.14}$$

with Q_s being the soil producing subsurface component of the run-off. The optimization follows exactly as before, but with the substitution PET $\rightarrow \beta P + (1 - \beta)\text{PET}$, yielding

$$\frac{E}{P} = \alpha(d)[\beta + (1 - \beta)\text{AI}]. \tag{11.15}$$

When $\beta = 0$, equation (11.11) is recovered. Moreover, at AI $= 1$, the predictions conform to those of equation (11.11) for any value of β. Otherwise, E is increased relative to what equation (11.11) predicts. So long as the value of E predicted by equation (11.15) is less than PET and under the condition that PET $< P$, it could be used. If equation (11.15) predicts $E >$ PET, then, PET should be substituted for E. For later comparison, the predictions of equation (11.15) can easily accommodate the variables E/PET and P/PET, which is used by some authors because PET varies less from year to year than does P, reducing variation of the dependent variable when annual variability in the water balance is addressed. Such a transformation yields,

$$\frac{E}{\text{PET}} = \alpha(d)\left[\beta\frac{P}{\text{PET}} + (1 - \beta)\right]. \tag{11.16}$$

11.5 Comparison with data

The predictions of the percolation model have been compared with six sets of data (Hunt *et al* 2023a). They are from: (1) FLUXNET (Baldocchi *et al* 2001), a global network of micrometeorological tower sites that use eddy covariance methods to measure the exchanges of carbon dioxide, water vapor, and energy between the biosphere and atmosphere; (2) The Model Parameter Estimation Experiment (MOPEX), an international project aimed at developing enhanced techniques for the *a priori* estimation of parameters in hydrologic models and in land surface parameterization schemes of atmospheric models (Spieler *et al* 2020); and (3) various studies by Wang *et al* (2009) on Nebraska Sand Hills, by Ning *et al* (2017) on Loess Plateau in China, by Donohue *et al* (2007) on Upper Cotter in Australia, by Palmroth *et al* (2010) on Upper Neuse River Basin in North Carolina, and by Benyon *et al* (2015) on Coranderrk, North Maroondah and Crotty Creek in Australia.

Data of Williams et al *(2012)*: The data consists of 167 sites from the FLUXNET. When net radiation was not available (for six sites), its value was approximated as 80% of global radiation. For ten sites, the mean seasonal precipitation was missing for one or two months for which monthly global precipitation climatology product data in the nearest neighbor was used. The total annual precipitation arriving in a frozen form—snow, sleet, hail, graupel, etc—at each site was approximated from the sum of precipitation in months with an average air temperature below 0 °C. The climate of each site and vegetation type were determined, respectively, based on the Köppen–Geiger climate classification and the International Geosphere–Biosphere Program land cover type classification. Maximum leaf area index and local precipitation were acquired from the site-specific ancillary data. Williams *et al* (2012) determined the best-fit curve through-mean annual data for all the sites to describe the mean tendency of E/P as it varied with PET/P by adopting the functional form suggested by Pike (1964), and calculated site-specific departures of E/P from the expected value for each site's PET/P based on the fitted curve. The 'grassland' data were extracted from Williams *et al* (2012) and their data are compared with those reported by two other studies, those of Ning *et al* (2017) for the Chinese Loess Plateau, and Nebraska Sand Hills, reported by Wang *et al* (2009).

A comparison of the predictions of the percolation model with the data of Williams *et al* (2012) is shown in figure 11.9, which depicts interannual variability of the water balance. The original Budyko curve is represented by the dashed curve, the

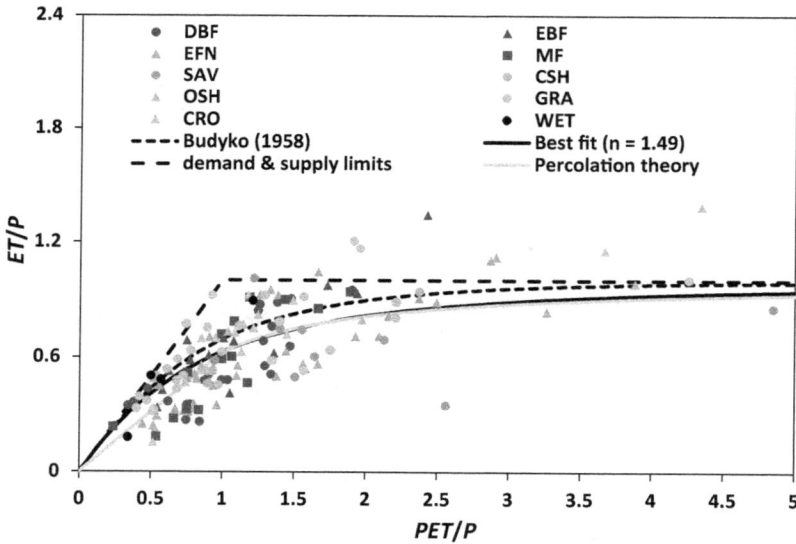

Figure 11.9. E/P versus AI for the biome data from the FLUXNET sites based on annual climatologies [data were obtained by digitizing figure 4(b) of Williams *et al* 2012]. Available energy for calculating PET was estimated from net radiation. DBF is deciduous broadleaf, EBF denotes evergreen broadleaf, ENF indicates evergreen needleleaf, MF means mixed forest, SAV represents savanna including woody savanna, CSH is closed shrubland, OSH denotes open shrubland, GRA indicates grassland, CRO is cropland, and WET represents wetland. Best fit represents the Choudhury (1999) model with $n = 1.49$, reported by Williams *et al* (2012), and percolation theory denotes the predictions for for AI > 1 with $\beta = 0$ for AI < 1 in the case that the 2D exponent, $d_f = 1.9$ is appropriate for the root mass. Reprinted from Hunt *et al* (2023a), copyright (2023), with permission from Elsevier.

best-fit through the data is given by the solid curve, and the 2D prediction of the percolation model by the green curve. Note that, at AI = 1, the percolation model yields, $E/P \simeq 0.623$, essentially, the global mean value, and is indistinguishable from the best fit value at that point. In the range $1 < AI < 3$, the predictions of the 2D percolation model are nearly coincident with the 'best fit.' Outside the range, the predictions of the 2D model are below the best fit. This may not be a drawback for AI < 1, as conventional models (see table 11.1) require an asymptotic approach to the demand limit, which leads to a weaker correspondence than for the percolation predictions. Note that the mean value of E/P over climate types (except arid that contribute the least to the global water balance) reported by Williams *et al* (2012) is 0.634 ± 0.027, whereas the 2D percolation model predicts, $E/P \simeq 0.623$. Schlesinger and Jasechko (2014) reported a global value, $E/P \simeq 0.639$. This correspondence may be of greater significance than meets the eye: only in the percolation description does a characteristic E value at global scales appear in the functional form of the E/P. Thus, the correspondence to both the best fit, the Choudhury-type function (table 11.1), and the global mean supports the view that the functional form proposed here is correct. As Williams *et al* (2012) noted, however, their own data compilation lies below other such compilations, while a large fraction of the data also lie below the prediction of the percolation model, as would be expected.

Williams *et al* (2012) stated that, 'Grasslands on average have a higher evaporative index than forested landscapes, with 9% more annual precipitation consumed by annual evapotranspiration compared to forests,' which is in general accord with the hypothesis presented here that, at least the forests with plentiful water supplies, but nutrients concentrated near the surface, should have shallow root systems (2D characterization, consistent with Fan *et al* (2017), as documented by Hunt and Manzoni (2016), Hunt (2017), Hunt *et al* (2021)). Since, as discussed above, explicit data for root fractal dimensions of grassland species have been presented, the interannual variability of E for grasslands is examined in more details by using the data of Williams *et al* (2012). Because their dataset covers a relatively narrow range of AI, the data reported by two other studies, one from the Chinese Loess Plateau (Ning *et al* 2017), and a second one for the Great Plains (Wang *et al* 2009) are also compared with the predictions. The comparison is shown in figure 11.10. Regarding the site Loess Plateau in China, Yamanaka *et al* (2013) stated, 'Dominant vegetation is grassland or shrubland established under drought stress,' while Ning *et al* (2017) stated, 'The impacts of vegetation changes on E have been widely studied with the Budyko framework by assuming surface conditions can be represented by the controlling parameter [ω] (see table 11.1). However, according to the developed relationships in our study, the controlling parameter is not only related to surface condition change, but also to climate seasonality.' It is, however, known that climate seasonality can also affect rooting depths, which should be correlated with root fractal dimension.

In the study of the Nebraska grasslands in and near the Sand Hills, Wang *et al* (2009) stated, 'Vegetation cover is predominantly composed of native warm C4 grasses,' which are more adapted to warm or hot seasonal conditions under moist or dry environments. Basins for the 9 points below Grass 1 prediction (see above) are

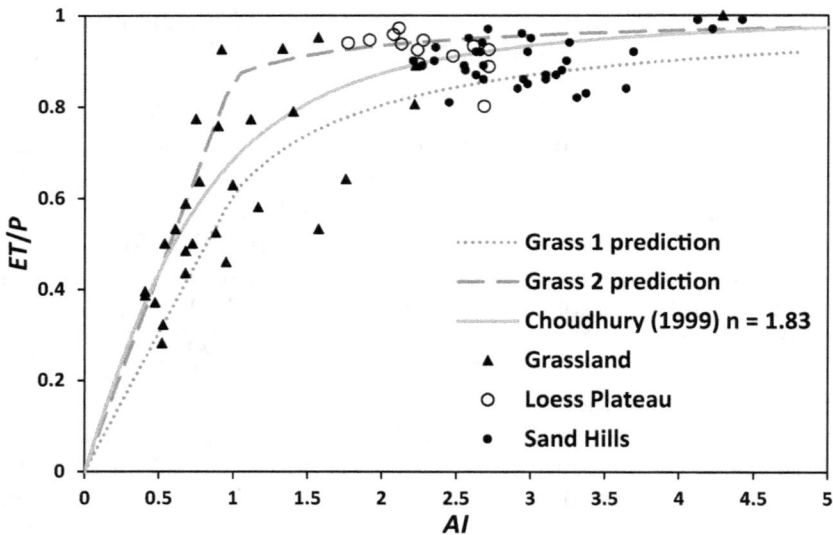

Figure 11.10. Representation of three models for *E* of grasslands within the Budyko model and its comparison with the Choudhury function with *n* = 1.83, considered a characteristic of grasslands. The two predictions, 'Grass 1' and 'Grass 2' use the experimentally determined fractal dimensionalities of groups of grasses by Levang-Brilz and Biondini (2003). 'Grassland' represents the grassland sites in Williams *et al* (2012), while 'Loess Plateau' data are from Ning *et al* (2017) and 'Sand Hills' data were reported by Wang *et al* (2009). Reprinted from Hunt *et al* (2023a), copyright (2023), with permission from Elsevier.

almost entirely within the Sand Hills with very high hydraulic conductivities and abnormally high infiltration, leading to an unusually high subsurface contribution to run-off 'under a similar range of *P*, the basins in the Sand Hills exhibit much higher [run-off] than those outside of the Sand Hills' (Wang *et al* 2009).

Data of Carmona et al (2014): The data were derived from MOPEX (Duan *et al* 2006), and focused on nine watersheds from a range of climate zones along the Budyko curve. The study explored the annual variability in their water balance, meaning that each range of the Budyko curve was explored in a distinct watershed. Although the overall appearance still follows typical Budyko phenomenology fairly closely, individual basins appear to generate linear behavior over their more limited ranges of climatic variables, PET/*P*. The comparison between the predictions of the percolation model and the annual water balance data in terms of *Q*/PET, as compared with *P*/PET, reported by Carmona *et al* (2014) is shown in figure 11.11 for nine catchments across the USA. Thus, the individual datasets reflect variations in the water balance for which the effects of variability in vegetation are reduced. Nevertheless, the predictions for the water balance that assume long-term averages are still used and any changes in storage are neglected. The reported root fractal dimensions of two distinct groups of grasses from the Great Plains are also used in the model, partly because their values (see above), 1.79 and 2.65, are rather close to the theoretical values of percolation in 2D and 3D that are, respectively, 1.9 and 2.5. As the comparison indicates, except for Florida (orange), the data from every state

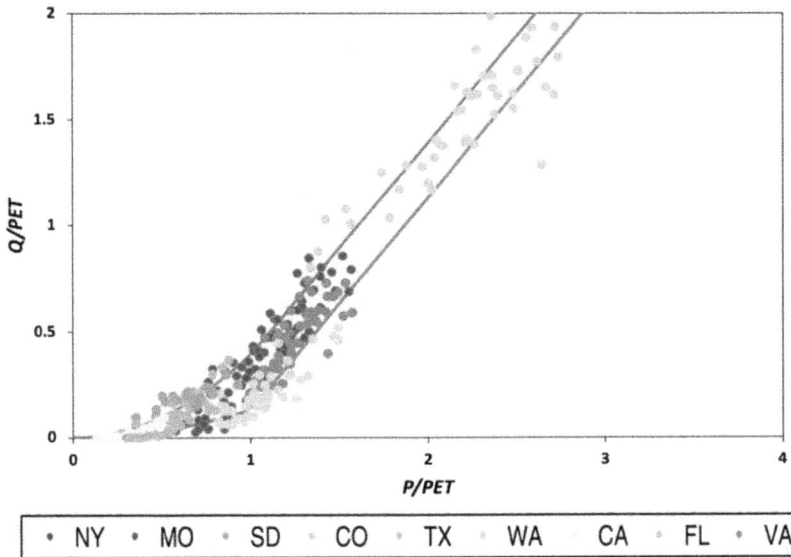

Figure 11.11. Comparison of the predictions of the percolation model with annual water balance data compiled by Carmona *et al* (2014) for nine catchments across the United States. The red line represents the prediction using Levang-Brilz and Biondini's (2003) data for the fractal dimension of one group of grass root systems, $d_f = 2.65$, corresponding roughly to the 3D value of percolation, $d_f = 2.5$. The blue line represents the predictions using Levang-Brilz and Biondini's (2003) fractal dimension for the second group of grasses, $d_f = 1.79$, corresponding roughly to the 2D value of percolation theory, $d_f = 1.9$. Each color dot represents a different state. Note that it is possible during some seasons to also have excess precipitation in climates for which $P < $ PET, so that the region of figure 11.14 (see below) where constant slope is predicted could extend down into $P/\text{PET} < 1$, a possibility that might explain some of the (relatively minor) discrepancies with the predictions in this region. Reprinted from Hunt *et al* (2023a), copyright (2023), with permission from Elsevier.

represented is reasonably well constrained to lie between these the two predictions. One might also conclude, however, that there is a discrepancy with the Washington site (light green) at the high end of Q/PET versus P/PET, which is discussed further below.

Next, consider the data for E/PET as a function of P/PET, shown in figure 11.12. We note the rather singular case of Florida (orange), with E/PET larger than the predicted bound, which might support the suggestion (Williams *et al* 2012) that wet climates with precipitation in-phase with solar irradiance tend to have higher E than temperate climates with similar precipitation and AI, but out-of-phase precipitation, since in the latter case winter precipitation is more likely to run off. Such interpretation is, however, the opposite of most data. One should also note that several Florida data points violate the condition restricting E to values less than PET, permitting an alternate interpretation based on data reliability, and that the discrepancy between prediction and data in the case of Washington state is accentuated compared with the Q representation.

The discrepancy in the case of Washington state is somewhat reduced by modifying the theory (see above), including the modeled fraction of $P - \text{PET}$ that

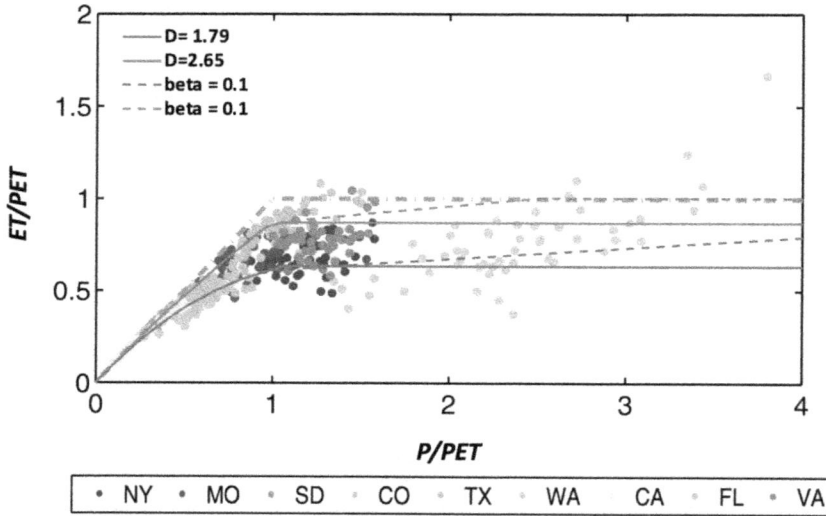

Figure 11.12. Predictions of the percolation model, using observed (Levang-Brilz and Biondini 2003) fractal dimensions, $d_f = 2.65$ (red solid), $d_f = 1.79$ (blue solid), bound reasonably well observed annual values of E/PET (Carmona *et al* 2014) for drainage basins in nine states. Allowing 10% of the excess precipitation (P–PET) to infiltrate and contribute to soil formation outside the optimization calculation for NPP results in the associated blue and red dashed line predictions. The grey dashed line represents the Budyko limits based on the assumption that the change in water storage is zero. Reprinted from Hunt *et al* (2023a), copyright (2023), with permission from Elsevier.

contributes to soil formation. Note that other authors (Cheng *et al* 2011) have asserted that vegetation growth rates depend linearly on precipitation, with the proposed consequence that E could still increase linearly with increasing P for $P>$ PET, and suggest that statistical tests on individual catchments may favor the linear model over the Budyko curve for annual variability of E. The linear increase comes about because the diversion of excess precipitation to soil formation frees up additional water for use by the plants; see the modified optimization given by equation (11.15). Nevertheless, the linear prediction does not appear to capture the discrepancy in the small P/PET range. For higher P/PET, the violation of the condition that $E \leqslant$ PET may be a consequence of the drawdown of water storage: if that limit is not enforced, the equation would be more nearly compatible with the data. Others, such as Williams *et al* (2012), simply remove data from their analyses, but not from their figures, for which $E>$ PET or $E > P$. The data for Florida, while outside predicted range by the percolation model, do not stand out to the same extent in the E representation. Other states have similar discrepancies.

Data of Donohue et al (2007): Donohue *et al* (2007) addressed the annual variability of E of a single drainage, namely, the Upper Cotter drainage of the Australian Capital territory, which is known to vary from dry sclerophyll to wet sclerophyll predominantly *Eucalyptus* forests, in terms of the Budyko variables. On the drier side of the Great Dividing Range, the catchment is a mixed *Eucalyptus* forest. The data for *Eycalyptus* growth rates, reported by Roberts *et al* (2001) and

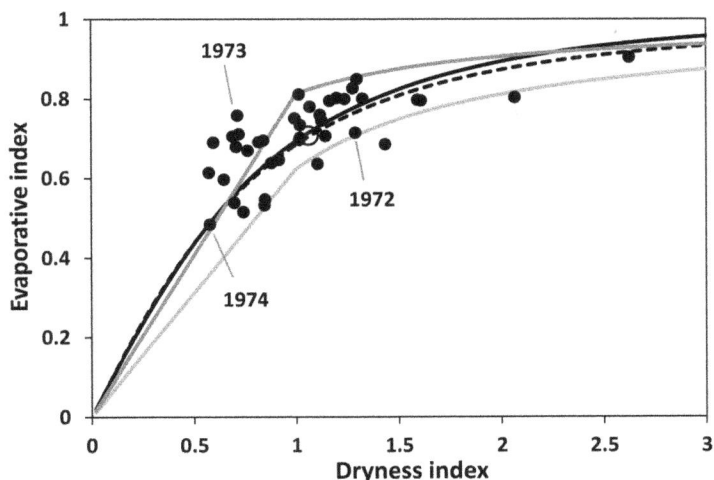

Figure 11.13. Comparison of the predictions of percolation model for *E/P* as a function of AI with annual variability of the same for the Upper Cotter Creek Catchment (Donohue *et al* 2007). Most of the data lie between the 2D and 3D predictions of the percolation model, but there is a cluster of points near '1973' which lie above the 3D predictions and, indeed partially above the limit *E*= PET for AI < 1. Reprinted from Hunt *et al* (2023a), copyright (2023), with permission from Elsevier.

Vertessy *et al* (2001), was analyzed by Hunt *et al* (2022) in terms of a 3D root model. Thus, here, the possibility that either 2D or 3D percolation model is relevant is considered. The comparison between the data and the predictions is shown in figure 11.13. The comparison with the Choudhury-type function (table 11.1) with $n = 1.8$ is also depicted. As discussed above, $n = 1.83$ corresponds to a grassland (Yang *et al* 2008).

Vertessy *et al* (1998) present a summary of data for the evapotranspiration E of grasslands and forests in southeastern Australia, where PET ranges from about 1000 mm yr^{-1} to 1400 mm yr^{-1}. The data and their comparison with the percolation model are shown in figure 11.14, which suggests the interpretation that, for Australian catchments, grasslands correspond better with the 2D percolation prediction for E, whereas forests are more in line with the 3D prediction, with a significant contribution to the variability arising from the (undocumented) variability in PET.

Data of Gentine et al (2012): The data reported by Gentine *et al* (2012) for *E/P* as a function of AI are widely distributed across climate zones, and were also derived from the complete MOPEX dataset (Duan *et al* 2006). In this case, however, the authors restricted their selection to watersheds where it was possible to take 50-year averages, in order to explicitly minimize the effects of change in water storage. Each site was labeled according to whether the precipitation peaked during winter (out-of-phase), summer (in-phase), or was non-seasonally distributed throughout the year.

The comparison between the predictions of the percolation model with the data is shown in figure 11.15. In both figures 11.15(a) and (b), the lower bound is well expressed by $E = 0.623$ PET, which is what the 2D percolation model predicts,

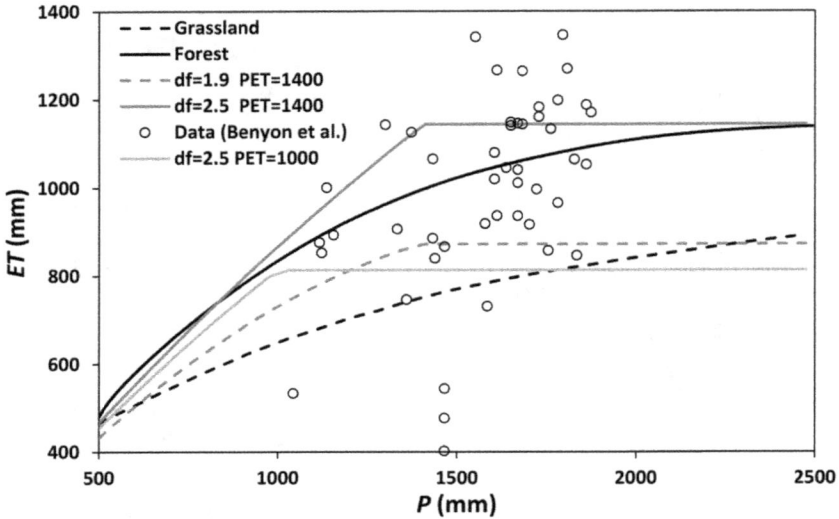

Figure 11.14. Summary of the data compiled by Vertessy *et al* (1998), which had been reported by a large number of contributing authors, including Benyon *et al* (2015), whose data are also included specifically, but were analyzed separately (see below). The curves grassland and forest are Vertessy *et al*'s (1998) qualitative summaries of the data for E as a function of P, but for varying values of PET. The 2D and 3D percolation values, $d_f = 1.9$ and 2.5 at PET = 1400, appear to be in agreement with the qualitative observations for grasslands and forests, respectively. Benyon *et al* (2015) data are for *Eucalyptus*-dominated catchments, but without specific values of PET. Most of the forest data are bounded between the two predictions with PET = 1000 and 1400, and $d_f = 2.5$. Reprinted from Hunt *et al* (2023a), copyright (2023), with permission from Elsevier.

while the upper bound is commensurate with $E =$ PET, though a few watersheds do appear to yield higher values of E. Such data for long-term averages cannot be readily interpreted, unless groundwater mining is occurring (Hunt *et al* 2021). There is no evidence in figure 11.15(a) for the relevance to non-seasonal ecosystems of the separate upper bound of $E = 0.813$ PET, the prediction of the 3D percolation model, as obtained by optimizing the use of E in NPP (see above). Figure 11.15(b) is, however, quite different. Throughout the range of AI values, the in-phase data are reasonably constrained between the predicted 2D and 3D functions. Conversely, at least for AI < 1, the large majority of the out-of-phase watersheds fall between the prediction of the 3D percolation model and the maximum E possible, although out-of-phase watersheds, similar to in-phase watersheds, fall near the upper end of the range between the predictions of the 2D and 3D percolation model for AI > 1. Plants in climates with opposite phase between energy and water availability (e.g., Mediterranean climate) tend to exhibit a much deeper rooting structure in order to cope with water stress (Schenk and Jackson 2002, Gentine *et al* 2012), an observation compatible with the data. Nevertheless, Williams *et al* (2012) observed that, in general, forests in Mediterranean climates have greater run-off than expected from the ratio PET/P alone, the opposite of what is observed here.

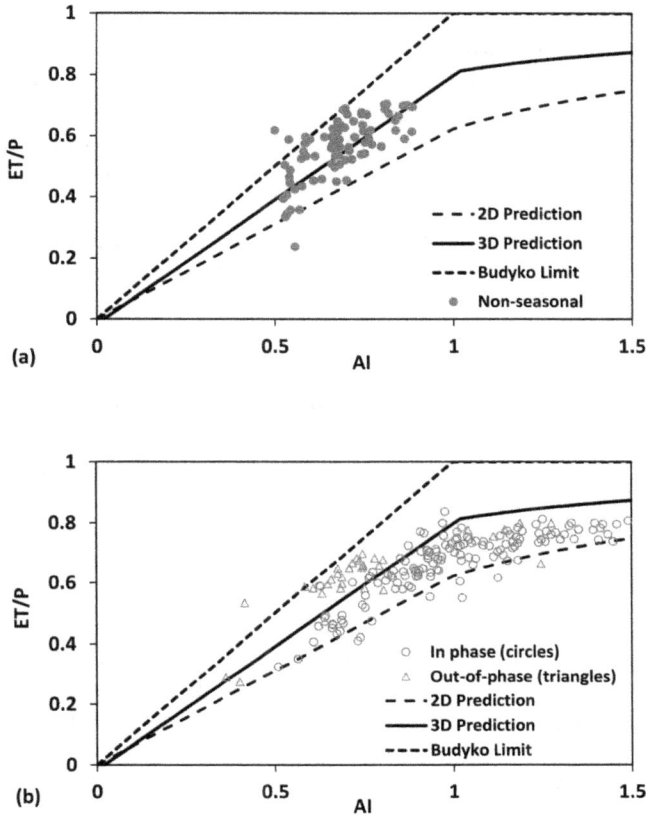

Figure 11.15. (a) Data for E/P as a function of PET/P= AI (Gentine *et al* 2012) for watersheds where precipitation is non-seasonal. (b) The same graphical representation of E/P as a function of PET/P (Gentine *et al* 2012) in watersheds where precipitation is either in-phase or out of phase with maximum solar irradiance. Reprinted from Hunt *et al* (2023a), copyright (2023), with permission from Elsevier.

Data of Palmroth et al (2010): The data represent long-term trends in streamflow through forest and agricultural watersheds of about 60 catchments in the Piedmont region of the Neuse River in North Carolina. The comparison is made with their data, presented in terms of the function $Q(P)$. Although there are no PET data readily available for the Palmroth *et al* (2010) study site, the relevant data for the same physiograph province, located only about 100 km south of the study site in South Carolina's piedmont region (Amatya *et al* 2018), are used in the comparison. To make proper predictions, a regional value for PET is required. Although a comprehensive data source for North Carolina was not found, the same was located for South Carolina. Amatya *et al* (2018) reported values of PET for 58 stations in South Carolina by two distinct methods. The mean values of PET for the 24 Piedmont stations were 1137mm yr^{-1} (the Priestly–Taylor method, which generates actual E values based on a coefficient empirically derived from PET) and 1202mm yr^{-1} (the Hargreaves–Samani method, a temperature-based evapotranspiration method based on an empirical relationship where reference E is regressed with solar radiation and air temperature data), each with standard deviation of about 10 mm yr^{-1}. The mean value of the two is 1170 mm yr^{-1}. The mean values for coastal plain were about 30 mm yr^{-1} higher, and mountain mean values about 70 mm yr^{-1} lower. Moreover, the same authors did a comparative study on North and South Carolina (Amatya *et al* 2016) PET values, with a specific comparison between the coastal plain sites, Carteret in North Carolina (1146 mm yr^{-1}) and Santee, South Carolina (1136 mm yr^{-1}). The difference between the coastal plain sites is within the margin of error. Using the single site-specific parameter of the Piedmont PET, predictions of the percolation model for upper and lower bounds of streamflow using 2D and 3D values of root fractal dimensions, provided by percolation theory, are shown in figure 11.16(a).

Data of Benyon et al (2015): These data are for streamflow Q as a function of precipitation P, and were collected for the humid climate of southeastern Australia for *Eucalyptus* forests, and were obtained through at long-term hydrological research sites at Coranderrk, North Maroondah, and Crotty Creek around the margin of the Yarra Ranges National Park, 60–100 km northeast of Melbourne. The Yarra Ranges National Park region is mostly within the point potential evapotranspiration contours 1000 mm yr^{-1} < PET < 1200 mm yr^{-1} (Chiew *et al* 2002). Outside the boundaries of the national park, point PET is typically between 1200 mm yr^{-1} and 1400 mm yr^{-1} (Chiew *et al* 2002). Benyon *et al* (2015) determined regional losses (evapotranspiration) to be 1113 mm yr^{-1}, though 15 of their 48 watersheds exhibited E values that clearly exceeded this value. But they determined that in clear-cut areas, where nearly all the *Eucalyptus* had been removed, E ranged from around 400 mm yr^{-1} in the year after clear cutting, to over 1300 mm yr^{-1}, with the average losses in the year after clear cutting being 467 mm^{-1}, but site-specific PET values were missing, adding uncertainty to the analysis and preventing unified graphical representation in terms of AI. Since the actual losses closely resemble the known PET, the actual losses are used as a substitute for PET also in the clear-cut cases.

To compare predictions with the data of Benyon *et al* (2015), since $Q/P = 1 - E/P = 1 - [1 - \alpha(d)]P/\text{PET}$, we obtain

$$Q = \begin{cases} 0.377\dfrac{P^2}{\text{PET}} & 2D, \\[2mm] 0.187\dfrac{P^2}{\text{PET}} & 3D, \end{cases} \tag{11.17}$$

if we use $\alpha \simeq 0.623$ for $d = 2$ and $\alpha \simeq 0.813$ for $d = 3$. It should be pointed out that for the relatively wet southeastern Australia, the data of Benyon *et al* (2015) are presented referenced from the low P limit (large AI). Since the PET is relatively consistent across the watersheds, it is then possible to use the value quoted by the authors, 1113 mm yr^{-1}, as water 'lost' to evapotranspiration for making concrete predictions for the limits of streamflow Q for various root models, which is indicated figure 11.16(b). Note that the 467 mm yr^{-1} losses to E are those experienced one year

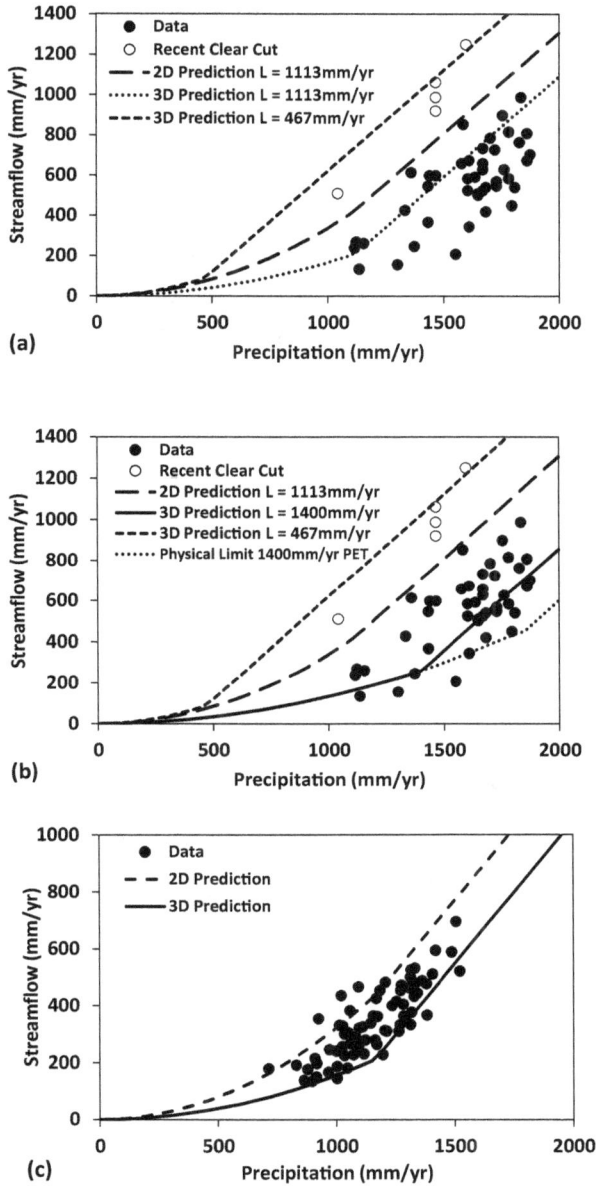

Figure 11.16. (a) Comparison of the data for streamflow values reported by Palmroth *et al* (2010) with the upper bound (2D percolation prediction, $d_f = 1.9$) and lower bound (3D model, $d_f = 2.5$) predicted by the percolation model. Each uses the mean PET of Piedmont stations in neighboring South Carolina. (b) Streamflow as a function of precipitation for the site studied by Benyon *et al* (2015). L is the regional value of 'losses' to evaporative processes, given by (Benyon *et al* 2015) as 1113 mm yr^{-1}; its use here is effectively as the PET. As noted in the text, Chiew *et al* (2002) give contours for point PET in the higher elevations of the Yarra ranges of 1000 mm yr^{-1} and 1200 mm yr^{-1}. However, the surrounding area point PET values, including parts of the study area, fall predominantly between the 1200 mm yr^{-1} and 1400 mm yr^{-1} contours. (c) Change in the

predictions from assuming 1400 mm yr^{-1} for the regional PET. Here, no values imply an ET value outside the physical limit. A significant fraction of the data fall, however, between the 3D optimization and the maximum, $E = $ PET. Reprinted from Hunt *et al* (2023a), copyright (2023), with permission from Elsevier.

after clear cutting, which is represented by the open circles. Note also that there is a significant representation of data that fall below the predictions of the 3D percolation model. This is addressed using the Budyko variables in figure 11.16(c) (see below). Comparison of the data of Palmroth *et al* (2010), shown in figure 11.16 (a), and those of Benyon *et al* (2015), presented figure 11.16(b), indicates that the North Carolina streamflow data are better constrained between the predictions of the 2D and 3D models than are the data from Southeastern Australia.

Figure 11.16(c) presents the results of using 1400 mm yr^{-1} for PET, which, according to the map of Chiew *et al* (2002), is the largest likely PET value within the area of the study sites. Some data still fall outside the predicted range obtained by using the 2 and 3D values of the fractal dimension D_f, but in this case all the run-off data are within the allowed range. The dotted line in figure 11.16(c) represents the standard Budyko curve with maximum $E = P$ for PET $>P$ and $E = $ PET for PET $<P$.

It is useful to consider the representation of the same Benyon *et al* (2015) data in customary Budyko variables. In the absence of site-specific values of PET, the same value PET $= 1113$ mm yr^{-1}, equal to 'regional losses,' was utilized for all sites. What becomes clear is that a large portion of the data is above the physical limit of $E = $ PET. Thus, use of regional losses as a regional estimate of PET underestimates actual values in some cases. Although Benyon *et al* (2015) did not explicitly report E, since their data only provided Q and P, E is easily extracted through $E = P - Q$. In 15 of 48 cases, E so extracted was larger than 1113 mm yr^{-1}, with the maximum value of E being 1346 mm yr^{-1}, which is 20% larger than the given PET. To evaluate the likelihood that these sites had PET values that were larger than 1113 mm yr^{-1}, for all E values that exceeded 1100 mm yr^{-1}, a PET of 1400 mm yr^{-1} was attributed, which is larger than the largest inferred E and equal to the largest PET contoured by Chiew *et al* (2002). For watersheds with $E < 1100$ mm yr^{-1}, the estimated losses of PET $= 1113$ mm yr^{-1} was left. Such a 'binary' model of PET values is perhaps more realistic than no distinction at all, while, in the absence of any other evidence, a more complex model would be difficult to justify. The result is '*Eucalypts* modified' shown in figure 11.17. The adjustment amounts to displacing values of $E >$ PET to a slightly higher AI with $E \leqslant$ PET, which, while it does not alter the fact that most of the data are above the 3D optimum, it does move the data below the physical limit of $E = $ PET.

In the Otway and Yarra ranges of southeastern Australia near Melbourne and the study sites used by Benyon *et al* (2015), precipitation is, indeed, maximum in winter, though somewhat later in the winter for the Yarra ranges than in the Otway. Thus, one possible interpretation of the large E values in this region is that *Eucalyptus* trees, with their deep taproots and, thus, close to having 3D root topology, have adapted to the lack of phasing of temperature and precipitation by accessing deeper

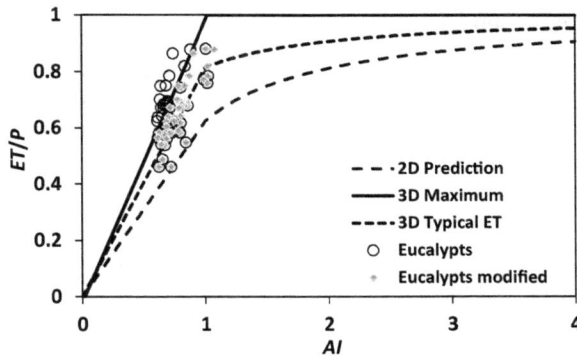

Figure 11.17. The data for E/P, as computed through $(1 - Q/P)$, are plotted first as a function of AI by assuming that PET is 1113 mm yr^{-1}, the regional value for losses to E reported by Benyon *et al* (2015). As discussed in the text, however, E can be as large as 1346 mm yr^{-1}, leading to violation of the condition, $E \leqslant$ PET. A procedure was developed to account for the implied higher PET values in the estimation of the AI, as discussed in the text and called 'Eucalypts modified' in the figure. Elimination of known violations of the condition that, $E \leqslant$ PET translated all the data points out of the forbidden region. Reprinted from Hunt *et al* (2023a), copyright (2023), with permission from Elsevier.

water, which is in general accord with the conclusion of other authors, such as Vertessy *et al* (2001).

11.6 Net primary productivity

In a meta-study of twenty individual investigations, including those of Rosenzweig (1968) and Webb *et al* (1986), Hunt (2017) showed that the first equation in (11.7), i.e., NPP $= cE_{D_f}$ with an exponent, $D_f \simeq 1.9$—the fractal dimension of the sample-spanning percolation cluster acting as surrogate for the root network—provides accurate description for NPP for up to about 3500 gC m^{-2} yr^{-1}, which is nearly equal to the existing published theoretical limits of NPP for Earth-based solar energy inputs (De Wit 1965, Krause-Jensen and Sand-Jensen 1998). Thus, a physical limit in energy is what produces a sharp cut-off in NPP. But wherever solar energy is significantly lower than in the tropics and subtropics, the cut-off should occur at smaller energy values, which in fact should be the case whenever the solar energy supply is insufficient to evaporate all the annual precipitation, or when $P >$ PET, a condition that may also be formulated as PET/$P \equiv$ AI < 1, with AI being the aridity index, as before. This particular strategy allows the use of a function of two variables, with distinct inputs of solar energy and water, and a clear distinction between energy-limited environments with AI < 1 and water-limited environments with AI > 1. While the energy input for evapotranspiration derives from the Sun and the atmosphere, the water is ultimately drawn by the roots from the soil through the suction induced by evaporation from the stomata. Optimization of NPP with respect to the hydrologic fluxes concerns the two chief pathways that water takes in the soil, namely, infiltration and evapotranspiration.

11.6.1 The percolation model

Recall that equations (11.9) and (11.11) represent the predictions of the percolation model for, respectively, AI > 1, and AI < 1. Thus, if we substitute the two equations for E into (11.7), assuming, as shown in Yu *et al* (2017), that predicted soil depths from percolation theory tend to a universal value of ca. 1 m

$$
\text{NPP} = \begin{cases} c[\alpha(d)\text{PET}]^{D_f} & \text{AI} < 1 \\ cP\left\{P - [1 - \alpha(d=2)]\dfrac{P^2}{\text{PET}}\right\}^{D_f} & \text{AI} > 1 \end{cases}
\tag{11.18}
$$

where for $d = 2$, $\alpha = 0.623$, and $D_f = 1.9$, and c is the proportionality constant defined above. Note that for AI = 1, the two equations are identical. Equations (11.18) represent the predictions of the percolation model for net primary productivity, NPP. In the next section, the predictions of equations (11.18) are compared with the data (Hunt *et al* 2024).

11.6.2 Comparison with data

Two sets of data are compared with the predictions of equation (11.18). One was compiled by Budyko (1974) who summarized all the data for NPP available at the time, organized by latitude, with corresponding solar energy inputs that varied over a factor of 7, compatible with a variation of, for example, PET from 1750 to 250 mm yr^{-1}. He represented a synthesis of the data graphically in units of solar energy and the amount of energy that was required to evaporate the precipitation that falls. While unfamiliar, such units are generally (though not precisely) equivalent to PET and P. The data were digitized by Vijay Gupta in 2002 and preserved for two decades in the usual units by German Poveda.

The second dataset was reported by Scurlock and Olson (2002) who presented the variability of NPP with P, but with no explicit representation of solar energy as an input variable. They used, for comparison, calculations of the Miami model of Lieth (1975), which addresses precipitation and temperature as independent variables. Scurlock and Olson (2002) did, however, comment that it was important, particularly in the boreal forests, to account for the temperature dependence in the Miami model, while omitting that part of the model led to significant overprediction of NPP. In this way, Scurlock and Olson (2002) implied that values of NPP lower than those predicted on the basis of P alone were most likely due to smaller values of solar energy than assumed. Of several possibilities considered by these authors, the Osnabrueck global dataset is selected here, whose source is in compilations by Esser (1991) and Esser *et al* (1997). Their figure 5 presents the Osnabrueck database (189 sites) for above-ground NPP (ANPP), as well as total NPP (119 sites). The maximum in NPP of about 3500 gC m^{-2} yr^{-1} in their datasets is at $P = 1300$ mm yr^{-1}. To predict the total NPP, we adjusted the constant multiplier c to take on a value that would result in NPP = 3500 gC m^{-2} yr^{-1} at $P = 1300$ mm yr^{-1}. Thus, in order to obtain the predicted ANPP, all the predictions of equations (11.18) were multiplied by 0.58.

Data of Byduko (1974): figure 11.18 presents the direct comparison of the prediction of NPP by the percolation model with the data summary of Budyko (1974). Figure 11.18(a) demonstrates the ability of the model for capturing the chief

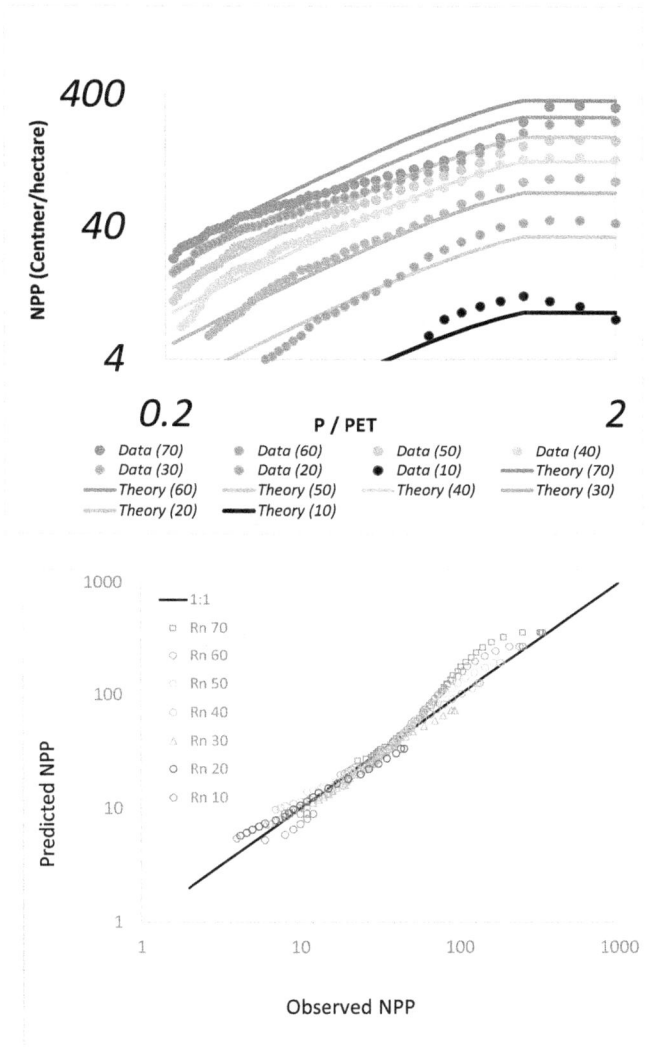

Figure 11.18. (a) Data digitized by Gupta *et al* in 2002, originally published by Budyko (1974). Solar energy varies over a factor of $70/10 = 7$, while the aridity index AI varies over a factor of 10. NPP, given in dry weight units of Centner/hectare, varies over a factor of 100. (b) Point-by-point comparison of the predicted and observed values of NPP with a single unknown, but adjusted parameter describing a universal efficiency in converting solar energy to plant mass, as well as a value of the crossover AI chosen to be 0.8, instead of the predicted value of 1. *Rn* denotes radiation, in accord with Budyko's discussion, but was interpreted by Gupta *et al* (2002) as PET. The overall R^2 value associated with this comparison is 0.97. Hunt *et al* (2024) John Wiley & Sons. Copyright 2024 The Authors.

variability of NPP with aridity index AI, as well as with input solar energy. In order to make this set of predictions, the constant c had to be adjusted, but the adjustment was employed only for the top curve at the largest solar energy, and then the same constant was used in all the remaining curves. The implication of the relevance of a universal factor converting solar energy to plant matter has its own significance. The value of AI dividing the energy- and water-limited conditions that provides the best fit is not the predicted value of 1, but rather 0.8, which represents a small, but significant discrepancy between predictions and data. Figure 11.18(b) compares, point-by-point, the predicted and measured values of NPP as a function of P/PET and PET simultaneously. Despite some systematic errors as evinced by changes in the curvature, the high R^2 value is interpreted in the comparison of figure 11.19 as indicating that the fundamental characteristics of the scaling relationship are in accord with Budyko's data synthesis, making it capable of predicting the latitudinal trends he observed for NPP. One may conclude that the results translate to variability in NPP from the variables P and PET generally. Since the prediction of the optimization of the product of root productivity and soil depth for the maximum NPP predicts E to be proportional to P, the predictions for water balance also predict a soil depth that is in accord with the discussion of section 10.3 on the 'Mystery of 1 m deep soils.' Thus, to predict NPP, we used a constant soil depth, assumed to be 1 m. Aside from steep topography, however, there is one kind of climate for which soils are systematically and consistently less than 1 m, namely, regions with AI > 1, and heavy monsoonal precipitation, for which the surface fraction of run-off in very heavy precipitation is very high and also carries a great amount of sediment from particularly high erosion (V Singh, personal communication 2024). In such regions, one should expect the percolation model to overpredict soil depth, which it does, representing its largest systematic deviation.

Data of Scurlock and Olson (2002): figure 11.19 compares the exact predictions of equations (11.18) with the data for (a) total NPP, and (b) above-ground NPP in the Osnabrueck database, reported by Scurlock and Olson (2002) in their figure 5. In figure 11.19(b), the numerical value c was chosen such that the model would yield NPP = 3550 gC m^{-1} yr^{-1} with P increasing beyond PET = 1300 yr^{-1}. This results, in this pair of comparisons, in a maximum in predicted NPP of 3550 gC m^{-2} yr^{-1}, already at PET of 1300 mm yr^{-1}, which may be smaller than the maximum PET value represented by Budyko. Figure 11.19 demonstrates that the predictions of the percolation model using a universal constant that predicts the appropriate variability of NPP over Earth's climate zones are consistent with nearly the entire cloud of data points reported by Scurlock and Olson (2002) in their figure 5(a), in contrast to the linear fit employed by them, which had an R^2 value of 0.15. A change in the lowest chosen value of PET, from 500 to 300 mm yr^{-1}, would lower the associated prediction for ANPP by a factor nearly 3 to about 100 gC m^{-2} yr^{-1}. Since the appearance of the distribution in Scurlock and Olson (2002) is essentially identical to those reported by Churkina *et al* (1999) in their figure 2, the comparison between predicted and reported NPP ranges as a function of P in figure 11.19 appears likely to be relevant to other datasets, also. However, examination of figure 2 of Churkina *et al* (1999) reveals its incompatibility with the process of digitization of the data.

(a)

(b)

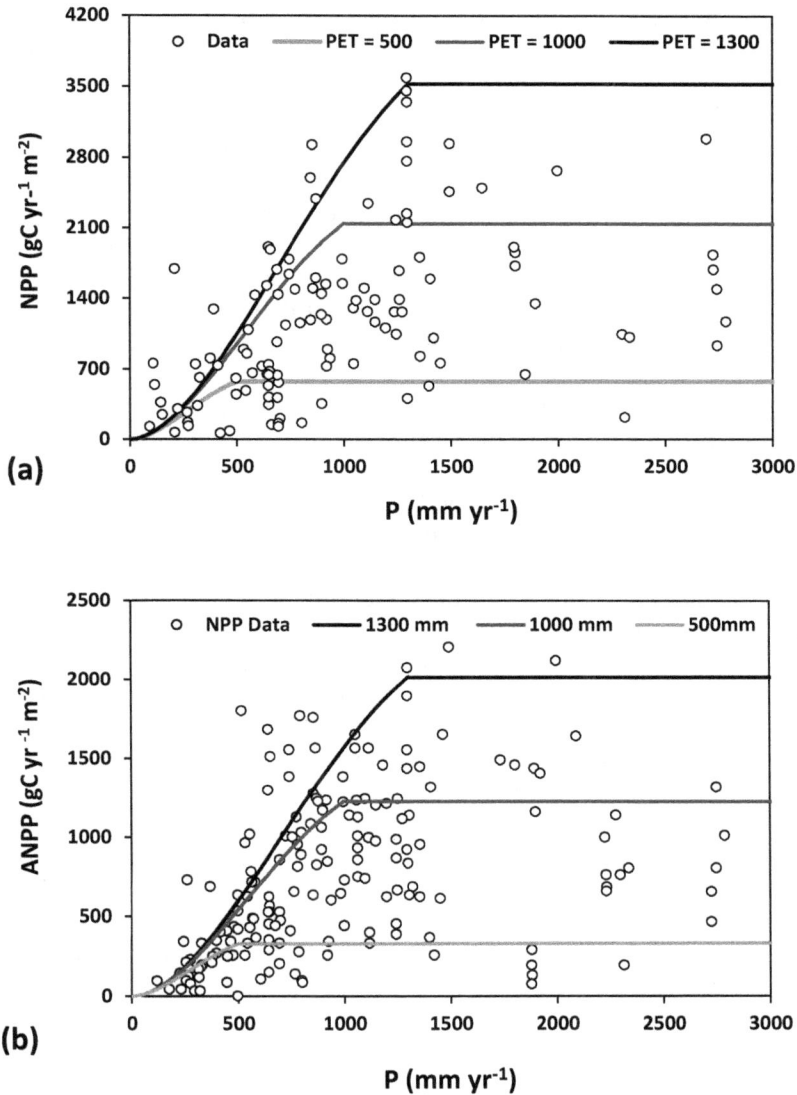

Figure 11.19. (a) Comparison of the predicted total NPP, P, by the percolation with the data (Scurlock and Olson 2002). (b) Comparison of predicted NPP, P, with data for above-ground NPP, or ANPP (Scurlock and Olson, 2002). Use of two separate numerical factors is consistent with predicting a maximum value of total NPP of 3550 gC m^{-2} yr^{-1} at a maximum PET of 1300 mm yr^{-1}1, and above-ground ANPP equal to 58% of the value for total productivity, approximately consistent with observations. Hunt *et al* (2024) John Wiley & Sons. Copyright 2024 The Authors.

11.6.3 Discussion

Rosenzweig (1968) argued that knowledge of E is the single best predictor of NPP, because it accounts for limitations of solar energy and water simultaneously. His investigation showed that NPP was likely related to E by the power law, $NPP = cE^s$, which is of the same form as equation (11.7), except that Rosenzweig (1968) estimated empirically the exponent to be, $s \simeq 1.69$, through comparison of the data across the world's biomes. Webb *et al* (1986) noted a tendency for a Rosenzweig type relationship to break down at large E, suggesting instead that an exponential function would be more suitable in supporting an asymptotic approach to maximum value of NPP. As mentioned above and demonstrated in the last section, in a compilation of 20 studies, including those of Rosenzweig (1968) and Webb *et al* (1986), Hunt (2017) showed that the theoretical value $s = D_f = 1.9$ was an excellent predictor of the data, though a sharp cut-off in NPP at a light-limited value of about 3500 gC m^{-2} yr^{-1} (de Wit 1965, Krause-Jensen and Sand-Jensen 1998) was observed.

On the other hand, Dynamic process models (see, e.g., Adams *et al* 2004, Cramer *et al* 1999) with a large number of parameters have gained ascendancy because, as Raich *et al* (1991) stated, 'We can more successfully predict rates of NPP in terrestrial ecosystems, if we model the basic processes controlling productivity and how they are influenced by environmental factors.' The state of research into the water balance has similarly been unhelpful in providing useful predictions, although Manabe (1969) used the Budyko formula to check the accuracy of his climate model. Specifically, none of the analyses of the water balance reviewed by Mianabadi *et al* (2020) has a strong physical basis, as they all have adjustable parameters with unknown relationships to real soil-vegetation systems (Dooge 1988, Greve *et al* 2015, Horton 1931, Klemes 1988, Reaver *et al* 2022, Zhou *et al* 2015).

The percolation model presented in this section for the net primary productivity NPP of ecosystems, which is based on the percolation model for the water balance that partitions precipitation P into run-off and evapotranspiration E and ecological optimization, explains about 97% of the variability in the data summary of Budyko (1974). The optimal dependence of E on climate variables P and PET yields the largest possible NPP. Although the ecological principle of optimal NPP was originally suggested by Odum (1959) and the power-law form for NPP (E) by Rosenzweig (1968), the percolation model differs from Odum's in that it proposes that soil depth limits NPP, not plant susceptibility to moisture deficits, and from Rosenzweig's in that the exponent of the power law is not an empirical parameter, but the universal fractal dimension D_f in percolation theory.

11.7 Elasticity of streamflow

The relative changes in streamflow $\Delta Q/Q$ can be calculated as (Dooge 1992)

$$\frac{\Delta Q}{Q} = \epsilon_p \frac{\Delta P}{P} + (1 - \epsilon_p)\frac{\Delta PET}{PET}, \tag{11.19}$$

where $\epsilon_p = (P/Q)(\partial Q/\partial P)$ and $(1 - \epsilon_p) = (P/PET)\partial(PET)/\partial P$, is the streamflow elasticity (Němec and Schaake 1982). Equation (11.19) is valid as long as catchment conditions do not change too severely, and the climate forcing factors $\Delta P/P$ and $\Delta PET/PET$ are not too large. A valid solution to equation (11.1) will provide a valid solution of equation (11.19), though it may not be clear that such a solution can describe the actual behavior of catchments (Reaver *et al* 2022). In order to calculate ϵ_p, the percolation model developed in the previous sections is developed further to take into account explicitly the effect of ΔS, but without a specific process model for the storage (changes). The solution provides guidance for estimating the magnitude of storage variability in a given catchment, if one hypothesizes that the analytical formulation produces the appropriate zero storage change solution to the water balance and precipitation elasticity simultaneously. By direct comparison of the predictions for the variable storage water balance with the annual MOPEX data (that consist of 18 000 data pairs), it is then possible to estimate magnitudes of typical annual variability in storage driving the variability in $E(\text{PET}, P)$, which are then used in the predictions of streamflow elasticity. As usual, comparisons between the variability of elasticity ϵ_p with a number of studies of continental-to-global extent are made, for which ϵ_p is variously represented as a function of aridity index, AI, its inverse, the humidity index, or of run-off coefficient, Q/P, as well as its inverse.

In addition, the percolation model is shown to be consistent with the conclusion that any reduction (increase) in storage will decrease (increase) streamflow elasticity. While the approach is self-consistent, as well as generally consistent with other ideas in the current literature, it is also distinct. For example, Anderson *et al* (2023) address expected effects of catchment storage properties on streamflow elasticity as follows: 'Low flows [...] in natural rivers [depend on] inflow from catchment storage sources, such as groundwater, lakes, or wetlands (Smakhtin 2001),' and tend to be associated with low elasticity. 'High streamflow magnitudes are controlled [...] by precipitation events and antecedent soil moisture conditions (Ivancic and Shaw 2015, Slater and Villarini 2006),' increasing streamflow response to precipitation. The authors point out that snow, with its tendency to reduce the magnitude of streamflow response, is a second example of the buffering effects of water storage. Thus, in their view, storage in different reservoirs can have widely differing effects on streamflow response.

11.7.1 The percolation model

The following derivation is due to Hunt *et al* (2023b). Starting with the equation, $Q = [1 - \alpha(d)]P^2/\text{PET}$, which was derived above, the precipitation and energy elasticities are determined by differentiating the equation to obtain,

$$dQ = [1 - \alpha(d)]\left(2\frac{PdP}{\text{PET}} - \frac{P^2 d(\text{PET})}{\text{PET}^2}\right), \quad \text{AI} > 1 \tag{11.20}$$

which, given the expression for Q, is equivalent to,

$$\frac{dQ}{Q} = 2\frac{dP}{P} - \frac{d(\text{PET})}{\text{PET}}. \tag{11.21}$$

This implies that in the Dooge's formulation, one obtains a logarithmic derivative of Q with respect to P and, therefore, $\epsilon_p = 2$, i.e., a precipitation elasticity of 2 and energy elasticity of -1. One can also use the expression for Q for AI <1 to obtain

$$\frac{dQ}{Q} = \frac{dP - \alpha(d)d(\text{PET})}{P - \alpha(d)\text{PET}} = \left[\frac{1}{1 - \alpha(d)\text{AI}}\right]\left(\frac{dP}{P} - \alpha(d)\text{AI}\frac{d(\text{PET})}{\text{PET}}\right), \quad \text{AI} < 1 \quad (11.22)$$

which also satisfies the Dooge's complementarity, but not with simple coefficients that are independent of AI.

Note that since α is explicitly dependent upon vegetation strategy, it can be used to represent the effect of vegetation changes on the water balance. Since a more isotropic root system is likely to be deeper, an increase in α would also be consistent with an increase in the rooting depth, or active soil layer. Such a procedure would add the terms

$$-\frac{(P^2/\text{PET})d\alpha}{(1 - \alpha)P^2/\text{PET}} = -\frac{d\alpha}{1 - \alpha}, \quad \text{AI} > 1 \quad (11.23)$$

and

$$-\frac{\text{PET}d\alpha}{P - \alpha\text{PET}} = -\frac{\text{AI}d\alpha}{1 - \alpha\text{AI}}, \quad \text{AI} < 1 \quad (11.24)$$

As discussed above, the parameter α is sensitive to rooting depth, such that increasing it tends to increase the value of α. Consider that, for AI >1, increasing aridity due to either a decrease in P or an increase in PET should cause rooting depth to increase, which would result in a vegetation contribution that enhances the contribution due to either P or PET alone and, hence, a value $\epsilon_p > 2$. Consider, for example, a situation in which P increases in a dry climate, adding grasses to a landscape that had none. In such a case, it would be conceivable to reduce $\alpha(d)$ from 0.813, its value for $d = 3$, to 0.623, which is for 2D media (see above), with the result being an increase in $\Delta Q/Q$ (about equal to) dQ/Q of $0.187/0.19 \approx 1$, in addition to the direct contribution of a factor of 2. Conversely, the loss of grass due to overgrazing and its replacement by woody shrubs with very deep roots could cut streamflow in half, i.e., $-0.187/0.377 \approx -0.5$, even in the absence of climate change.

Although uncertainty regarding the distinction between evapotranspiration of forests versus grasslands exists, woody plant encroachment (WPE) in regions historically associated with grasslands has often been cited as a reason for diminished streamflow. Furthermore, many factors may influence the actual change in streamflow (Huxman *et al* 2005), yet the mean of the run-off coefficients from nine studies of WPE is 70% higher with 'low woody plant' concentration than with 'high woody plant' concentration, rather close to our estimate of 100%, if the change of root fractal dimensionality is from 3D to 2D in all plants in the region. It should be pointed out that it would be difficult for the present analysis to account for a separate reduction in vegetation coverage due to increasing aridity, since an increase in AI is already assumed to decrease vegetation coverage, represented as 1/AI.

Evidence reported by Yang *et al* (2009) suggests that such an onset of vegetation loss occurs for AI >1. Nevertheless, although an increase in AI at a given location for AI >1 does not produce an explicit contribution to the precipitation elasticity, it would seem to be consistent with an additional contribution to streamflow decrease.

In the remainder of this section, only the 2D value of α, $\alpha(d = 2) \simeq 0.623$, are used, partly because of its close agreement with the global mean value of E/P (see above), and partly because the agreement with experiment on the streamflow elasticity is not enhanced by considering potential variability in $\alpha(d)$ as well. Moreover, the predicted median value of ϵ_p is the same for AI >1 for either choice of $\alpha(d)$. To understand better the data for the annual streamflow precipitation elasticity, the effect of the potential change ΔS in storage S in the 2D systems is examined. Writing

$$\frac{Q}{P} = (1 - 0.623)\frac{P}{\text{PET}} - \frac{\Delta S}{P}, \tag{11.25}$$

and differentiatng it yields,

$$dQ = 0.377\frac{2PdP}{\text{PET}} - 0.377\frac{P^2 d(\text{PET})}{\text{PET}^2} - d\Delta S, \quad \text{AI} > 1 \tag{11.26}$$

which means that

$$\frac{dQ}{Q} = \left(\frac{1}{1 - 2.65\Delta S\ \text{PET}/P^2}\right)\left(2\frac{dP}{P} - \frac{d(\text{PET})}{\text{PET}}\right) + \frac{1}{1 - 2.65P^2/(\text{PET}\Delta S)}\frac{d\Delta S}{\Delta S}$$
$$= \left(\frac{1}{1 - 2.65\Delta S \text{AI}/P}\right)\left(2\frac{dP}{P} - \frac{d\text{PET}}{\text{PET}}\right) + \frac{1}{1 - 2.65P/(\Delta S \text{AI})}\frac{d\Delta S}{S}, \tag{11.27}$$

(where $2.65 \approx 1/0.377$) which does not follow Dooge's theorem, but approaches the equality in the limit $\Delta S \to 0$. The final term, a storage elasticity, will be neglected in the following, but was included for completeness.

A similar procedure is used to derive the corresponding equation for AI $<!$, with the result given by

$$\frac{dQ}{Q} = \frac{dP - 0.623d(\text{PET}) - d\Delta S}{(P - 0.623\text{PET}) - \Delta S} = \left(\frac{1}{1 - 0.623\text{AI} - \Delta S \text{AI}/\text{PET}}\right)\left(\frac{dP}{P} - \frac{d(\text{PET})}{\text{PET}} - d\Delta S\right), \quad \text{AI} < 1 \tag{11.28}$$

Since, it is assumed that for AI <1 energy limits the productivity, while water does so for AI >1, we make an analogous argument for storage. Thus, the maximum $|\Delta S|$ for energy-limited systems is a small fraction of PET, while for water-limited systems, it is the same fraction of P. For a reasonable estimate of year-to-year changes in storage, we estimate the fraction to be 10%. In the Great Lakes watershed values of ΔS of about 7% have been reported (Niu *et al* 2014), while in western USA values as high as 30%–40% have been cited (Garcia and Tague 2015), while Liu *et al* (2018) estimate annual storage changes among the world's largest river basins to be ca. 3% of P. Focusing on the precipitation elasticity, and keeping in mind that the sign of ΔS may be positive or negative, we find using changes of storage that are 10% of P or PET that

$$\epsilon_p = \begin{cases} \dfrac{1}{1 - 0.623\text{AI} \pm 0.1\text{AI}} & \text{AI} < 1 \\[2ex] \dfrac{2}{1 \mp 0.1 \times 2.65\text{AI}} & \text{AI} > 1 \end{cases} \tag{11.29}$$

Note that the contrasting assumptions regarding the relevance of PET and P to storage in humid and arid climates, respectively, leads nevertheless to equations that predict dependence of ϵ_p on AI, which is increasing for either $\text{AI} < 1$ or $\text{AI} > 1$. In the crossover from the humid to the arid regime, however, there is a sharp drop in ϵ_p.

If positive changes in storage are as likely as negative ones, the equations that neglect storage should be interpreted as predicting the median values for the precipitation elasticity. For the same absolute value of storage change, however, positive changes in storage predict a larger negative modification of ϵ_p than do negative changes in storage in their reduction of ϵ_p. Thus, the median is not the mean value, and the latter expected to be somewhat larger than the former. Some authors use the inverse of the aridity index, P/PET, called the humidity index, as the independent variable in presentation of data for ϵ_p. Some authors also use the run-off coefficient, Q/P (or its inverse) to present data for ϵ_p. Note that the result for $\epsilon_p = 1/[1 - \alpha(d)\text{AI}]$ for $\text{AI} < 1$ is also equal to P/Q, as is, in general, any equation for E that is a linear function of P and PET.

11.7.2 The data

The predictions of the percolation model for ϵ_p are compared with data obtained non-parametrically, as well as data from the largest compilations currently available. In addition to the large sources, information from the relatively small database of Zhou *et al* (2015b) with values of non-parametrically obtained streamflow elasticities for mainland China, which is also used for comparison. It is known that ϵ_p may depend on the details of the precipitation anomaly, as well as on the time period associated with measurement (Zhang *et al* 2022). With one exception (Deusdara-Leal *et al* 2022), only datasets for which the reported ϵ_p are obtained over annual periods are considered. Note that Deusdara-Leal *et al* (2022) made no distinction between positive and negative anomalies in P. The climatic data of Sankarasubramanian *et al* (2001) were generated using PRISM—Parameter-Elevation Regressions on Independent Slopes Model—(see Daly *et al* 1994), while the streamflow data derived from the Hydroclimatic Data Network (Slack *et al* 1993). Note that data for only 5 of the 17 United States regions mentioned were explicitly given by Sankarasubramanian *et al* (2001). Thus, the digitized data included only the published figures, covering regions 1 (New England), 3 (the southeast), 10 (the Missouri River drainage), 12 (the western Gulf of Mexico), and 17 (the Pacific Northwest), with median (mean) $\epsilon_p = 1.73$ (1.82), and (excluding a single data point for $\text{AI} > 9$) a maximum AI of 4.5. The entire dataset included 1291 catchments with at least 20 years of records and areas exceeding 129 km^2. For each catchment, the reported ϵ_p, independent of precipitation, was the median of annual measurements, defined by

$$\epsilon_p = \text{median}\left[\left(\frac{\langle P \rangle}{\langle Q \rangle}\right)\left(\frac{Q - \langle Q \rangle}{P - \langle P \rangle}\right)\right], \tag{11.30}$$

where $\langle \cdot \rangle$ denote the average value. Sankarasubramanian and Vogel (2003) also presented data for ϵ_p as a function of inverse run-off ratio P/Q, which were also used for the comparison. The data of Harman *et al* (2011) included 405 catchments across the United States, obtained from MOPEX (Duan *et al* 2006). Catchments of sizes 80 km^2 to more than 10 000 km^2, for which streamflow data were available for at least 25 years, were used. The data that were taken from figure 1 of this source (see top panel for total run-off elasticity) had a median elasticity of 2 and a mean elasticity of 2.12, although the authors reported variously a mean of 2.10 and 2.08, and were indicative of a maximum AI of 4.2, although the authors reported an absolute maximum AI of 8. Harman *et al* (2011) split the run-off into fast and slow components, but reported simultaneously a combined elasticity, which is what we used.

To make further comparison, the figure on page 12 of Chiew *et al* (2006) was digitized that reported ϵ_p as a function of Q/P with sites on every continent except Antarctica, as well as on islands, such as Ceylon, New Zealand, and Iceland. The data had a median $\epsilon_p = 1.68$ and a mean value of 1.74. The criteria for selection was possession of concurrent sources for monthly streamflow (Peel *et al* 2000), precipitation, and temperature (Global Historical Climatology Network, GHCN). This stipulation allowed calculation of the elasticity non-parametrically for 532 catchments with time periods from 23 to 64 years and areas ranging from 100 km^2 to 76 000 km^2. For each catchment, the median of the annual elasticity values was determined according to the procedure of Sankarasubramanian *et al* (2001) described above. The Australian data (Chiew 2006) for ϵ_p as a function of AI, as well as as a function of Q/P have a median value of 2.43 and a mean value of 2.57, and a maximum AI of 6, with the remaining data confined to AI < 5. The catchments ranged in size from 50 km^2 to 2000 km^2. The major river basins compiled in the supplement to Tang and Lettenmaier (2012) included 25 catchment areas between 80 000 and 100 000 km^2, 26 catchment areas that exceed 1 000 000 km^2, and the remaining of about 140 with areas between 100 000 and 1 000 000 km^2, covering roughly half the world's continental area. Tang and Lettenmaier (2012) thus provided a dataset complementary to all the remaining data. In this calculation/compilation of ϵ_p, the 194 largest river basins were chosen [referred to as Simulated Topological Network (STN-30p); see Vörösmarty *et al* 2000]. The authors did not report the aridity index for the basins. Since Liu *et al* (2018) reported AI data for 28 major river basins, one of the co-authors, Yongqiang Zhang, provided 400 additional AI values that had been obtained using the methods described by Lan *et al* (2018), from which 125 data pairs (ϵ_p, AI) were extracted after accounting for different names of the same rivers. Zhou *et al* (2015b) calculated ϵ for basins within China by non-parametric methods, as well as by the Budyko method, but represented ϵ_p as a function of AI for the non-parametric methods. According to Zhou *et al* (2015b), the three sites with very low values of ϵ_p are cold sites. Overall, the regions that produced the smallest values of ϵ_p were either in the very humid southeast, or cold and dry climates in the far northwest.

When assessing the accuracy of the predictions by comparing them with actual data, it is illustrative to also use an existing phenomenological model for comparison, of which Budyko's (1958) function was used in this chapter, since it is venerable and still often used, although its lack of parametrization restricts is versatility. Gentine *et al* (2012) carefully compared Budyko's model with MOPEX data (Duan *et al* 2006) for 50-year averages of E/P, showing that it provides good accounting of E/P as a function of PET/P throughout the range of climates, and argued that any discrepancy between the Budyko curve and the long-term average data can be attributed to experimental error of about 10%.

To summarize the essence of the data, Sankarasubramanian *et al* (2001) found that median ϵ_p values of individual catchments vary from 1 to 2.5 over the entire United States with the exception of a small (cold) portion of Montana and North Dakota that have lower values. These values compare well with the prediction of the percolation model, namely, that over the range $0 < $ AI 1, the median elasticity should vary from 1 to 2.65 and for AI 1, it should be exactly 2. In only 0.7% of the range of AI values, from 0.964 to 1, is the predicted median elasticity greater than 2.5, while its absolute maximum is 2.65. Regarding the data for North Dakota and eastern Montana, studies of streamflow elasticity in catchments with significant snow cover often reveal ϵ_p values near 1 and even smaller (e.g., Berghuijs *et al* 2016, Sankarasubramanian *et al* 2001, Zhou *et al* 2015b, Khan *et al* 2022), which has been interpreted as a buffering effect of snow (Sankarasubramanian *et al* 2001). Clearly, time lags between P and run-off may be considerable, while, on shorter time scales, run-off may most strongly be positively correlated with solar energy, rather than P. Xing *et al* (2018) investigated the precipitation and energy elasticities of 35 drainage basins in China. They found mean values of P and PET elasticities to be 1.972 and -0.972, respectively, satisfying the Dooge (1992) relationship. The range of P elasticities in their catchments was from 1.24 to 2.45; the PET elasticity values again satisfying the Dooge (1992) relationship. The geographical distribution of their elasticities produces the smallest values in high altitudes with drought conditions, where PET is small, and in the far southeast where P is largest.

For the Kuye River basin, Zheng *et al* (2021) calculated pre-anthropogenic and post-anthropogenic values for the precipitation and energy elasticities, using three different phenomenologies for the water balance, rather than non-parametric methods. Only in the pre-anthropogenic period (1956–66) do the values obtained using all the distinct phenomenologies agree closely. In that case, $\epsilon_p = 1.95 \pm 0.04$. Others (e.g., Chiew *et al* 2006, Zheng *et al* 2009) discussed (or, Zhou *et al* (2015b), reported data consistent with) the tendency for the precipitation elasticity to increase with increasing aridity for AI <1 and then, particularly the median value of its spatial variability, to tend towards independence of AI when AI >1. The predictions of the percolation model for AI >1 when $\Delta S = 0$, is $\epsilon_p = 2$, which appears to be the most commonly observed overall value of the P elasticity. The prevalence of precipitation elasticity values near 2 is confirmed in a study of 521 catchments worldwide by Chiew *et al* (2006), by presenting a table, table 11.3. Note that the

Table 11.3. Climate zones, number of catchments, median values of ϵ, and its range (after Chiew *et al* 2006).

Climate zone (Köppen[†])	Number of catchments	median ϵ_p	range of ϵ_p
Tropical	79	1.7	0.8–3.1
Very wet	20	1.2	0.8–1.9
Moderately	59	2.0	0.9–3.3
Arid	45	1.8	0.4–2.9
Cold arid	32	1.6	0.4–3.1
Warm arid	13	2.0	0.5–2.5
Temperature	262	1.9	0.9–3.1
Wet winter	32	2.0	0.9–3.4
Wet summer	35	1.8	0.8–2.8
No seasonality	195	1.9	1.0–3.1
Cold	135	1.1	0.5–1.9

[†]The Köppen climate classification is one of the most widely used climate classification systems.

range of ϵ_p is defined by Chiew *et al* (2006) as lying between the 10th and the 90th percentile, while the mean ϵ_p at these two percentiles is greater than or equal to the median in every climate category, except one.

11.7.3 Inferences about storage effects

A number of authors (see e.g., Berghuijs *et al* 2016, and references therein) have noted that annual changes in storage should have significant effects on the water balance, as well as on precipitation elasticity of streamflow. Drawdown of storage reflects, in part, extra water available for streamflow, which reduces the effect of negative fluctuations in precipitation over the same time period. While an increase in storage reduces immediate run-off, it can also make streams more sensitive to precipitation fluctuations over longer times. Since the present study uses data from other sources that did not provide water storage changes, it is not possible to use the storage variable predictively, but only diagnostically. Thus, guidance was sought from two sources, namely, (1) references to storage changes as a function of precipitation for specific catchments with known climates, and (2) a data summary for the water balance that allows one to estimate storage variability based on the percolation model. For the former, we refer to Niu *et al* (2014) who found that, in Great Lakes watersheds with AI near 1, annual variability of ΔS is about $0.07P$, while Garcia and Tague (2015) gave specific estimates of storage change in the western United States From Oregon and California to Colorado, with the annual variability of ΔS varying from about 20% to 30%-40% of P. For the latter estimates, comparison of the predictions of the percolation model of water balance, described above, with the MOPEX annual water balance data published by Gentine *et al* (2012) produces specific values of ΔS. Hunt (2021) and Hunt *et al* (2021) used over 18 000 points from the annual water balance extracted from MOPEX (Duan *et al* 2006) by Gentine *et al* (2012), which, in figure 11.20 are compared with E/P, as obtained from $1 - Q/P$ and invoking the effects of storage.

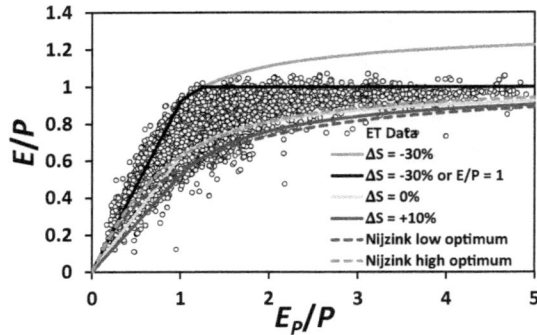

Figure 11.20. A demonstration that the prediction of percolation theory in 2D together with annual storage changes of 10% of P (PET, for AI < 1) and -30% of P (PET for AI < 1) bounds the MOPEX annual water balance data accurately. For AI > 1, the data scarcely approach these bounds; moreover, -30% of P exceeds significantly the usual Budyko limit imposed for longer time frames. MOPEX data from Gentine *et al* (2012) were digitized and used here. Note the very close correspondence between the predictions for optimal carbon assimilation (Hunt *et al* 2021) and the optimal carbon profit model of Nijzink and Schymanski (2022), both of which predict a water balance that is very close to the empirical model of Choudhury with $n = 1.49$, although Nijzink and Schymanski (2022) include also a 'low optimum' that corresponds to $n = 1$, as well as the predictions presented here with a 10% storage increase. Reproduced from Hunt *et al* (2023b). CC BY 4.0.

As figure 11.20 indicates, storage changes ΔS with values $-30\% < \Delta S < 10\%$ (blue and red solid curves), as applied to a median water balance obtained from percolation model in 2D appears to be sufficient to constrain the water balance data over the entire range of AI. In fact, for AI > 1, this range of storage changes overestimates the water balance variability, both above and below the $\epsilon_p = 2$ solution. Thus, the same limits on ΔS should, in most cases (particularly within the conterminous USA) serve to constrain the elasticity data, and is referenced to PET (P) for AI < 1 (AI > 1). For large AI, it is likely that storage changes proposed to be as large as -30% are unrealistically high, since total shallow subsurface storage may be a small fraction of P. Thus, there is a cut-off at $E/P = 1$ when its predicted value exceeds 1.

11.7.4 Comparison with data

Data of Tang and Lettenmaier (2012) *and Zhou et al* (2015a): figure 11.21 compares the values of ϵ_p reported by Tang and Lettenmaier (2012) for the world's major river basins, and Zhou *et al* (2015a) for selected rivers in China, with the predicted dependence using storage changes of only $+8\%$ and -10% in equations (11.29). The results for ϵ_p for major river basins (with typical sizes in the hundreds of thousands of square kilometers) fall comfortably within the predicted limits, but the data from China exhibit a noticeably wider scatter than is predicted by the storage changes used. While a decrease in relative storage magnitudes with increasing basin size is expected based on the general arguments regarding the summation of, for example, uncorrelated random variables (at length scales longer than typical atmospheric Rossby wavelengths, i.e., 500 km, one often finds negative correlations), it may also

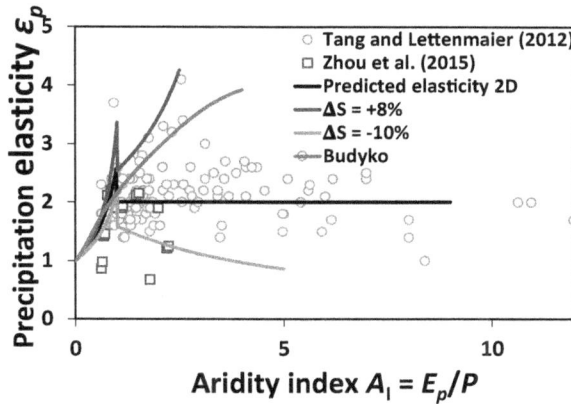

Figure 11.21. Comparison of the predictions of the percolation model with the data of Tang and Lettenmaier (2012) and Zhou *et al* (2015a), using the storage changes mentioned. The 'predicted elasticity 2D' curve refers to zero storage change, interpreted in the text to be a median value. For reference to traditional theoretical descriptions, the phenomenological model of Budyko (1958) is included. Reproduced from Hunt *et al* (2023b). CC BY 4.0.

be a consequence of the different means Tang and Lettenmaier (2012) used to generate ϵ_p. Clearly, the phenomenology of Budyko (1958) is not well adapted to the data shown, although incorporation of ΔS variability would allow a more realistic appearance (yet without bringing the predicted median value into closer agreement). The prediction of a median value of $\epsilon_p = 2$ for AI >1 looks to be in accord with the Tang and Lettenmaier data, and is in quantitative agreement with the conclusions of Chiew *et al* (2006) mentioned above.

Data of Sankarasubramanian et al (2001): figure 11.22 presents a comparison of predictions of equation (11.29) for ϵ_p with the corresponding data in three regions, presented by Sankarasubramanian *et al* (2001). The three regions, 1, 3, and 12, correspond to New England, the southeast, and western Gulf of Mexico, respectively, which document the kind of response expected with increasing AI. Note the increase in the median value of ϵ_p with increasing aridity index for AI <1, but no apparent increase for AI >1 (where the data are noticeably more symmetrically distributed about $\epsilon_p = 2$), in agreement with the predictions. Thus, what appears at first glance to be purely an artifact of data collection provides additional evidence for distinct dependence of elasticity on AI for AI >1 and AI <1.

Figure 11.23 presents a comparison of the predictions of the percolation model for ϵ_p with data for the Missouri River basin and the Pacific Northwest. The former dataset is concentrated mainly in the region AI >1, but, owing to the strong rainshadow effects of the Cascade mountain range, the latter dataset spreads across both humid and arid regions. Again, the predicted tendency for the median value of ϵ_p to stay constant at AI >1, but to increase with increasing AI for AI <1 seems to be borne out, although here, the crossover may occur at a lower AI (higher humidity index) than predicted.

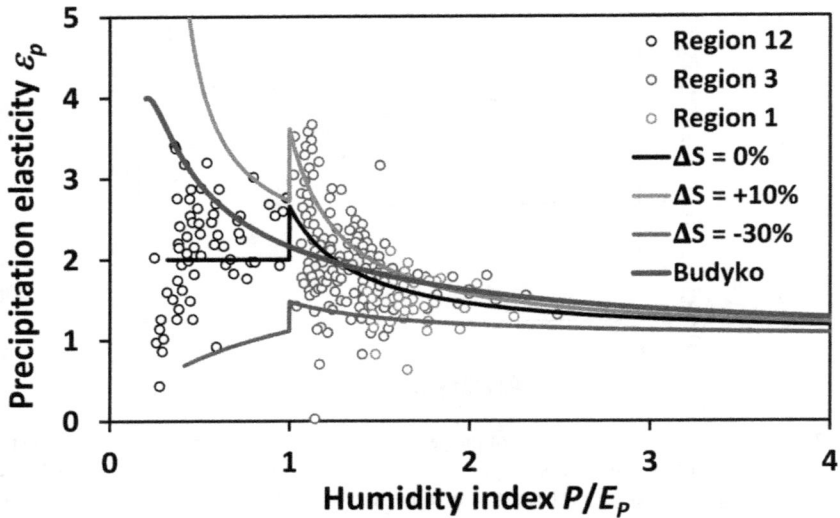

Figure 11.22. Comparison of the predictions of the percolation model with observed ϵp from Sankarasubramanian *et al* (2001) for regions 1 (New England), 3 (the Southeast), and 12 (west Gulf coast). Here, we used the changes in storage found to be suited to the MOPEX water balance data reported by Gentine *et al* (2012) as shown in figure 11.21. For comparison, the predictons of the phenomenological model of Budyko (1958) is again provided. Reproduced from Hunt *et al* (2023b). CC BY 4.0.

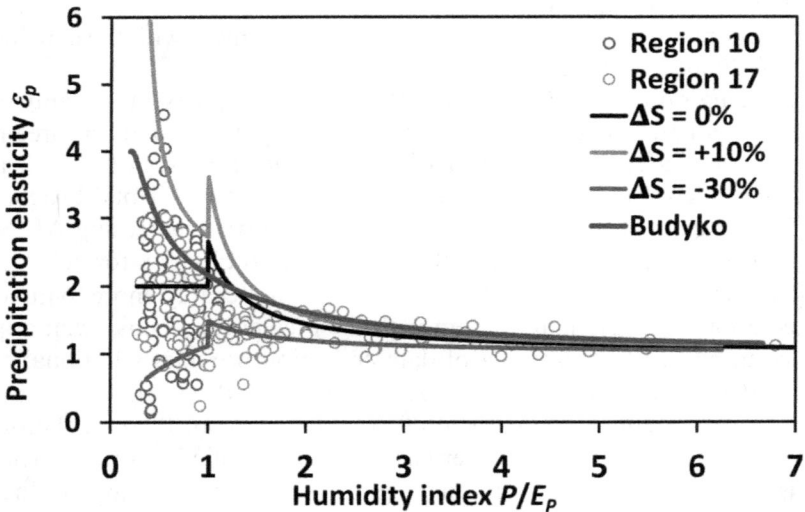

Figure 11.23. Comparison of predicted and values of ϵ_p with the data as a function of aridity index. Region 10 corresponds to the Missouri River basin, while Region 17 represent the Pacific Northwest (Sankarasubramanian *et al* 2001). As in figure 11.22, the limits on ΔS implied by the comparison of the water balance prediction with the MOPEX data, summarized by Gentine *et al* (2012) are used. The Budyko prediction is again used for comparison. Reproduced from Hunt *et al* (2023b). CC BY 4.0.

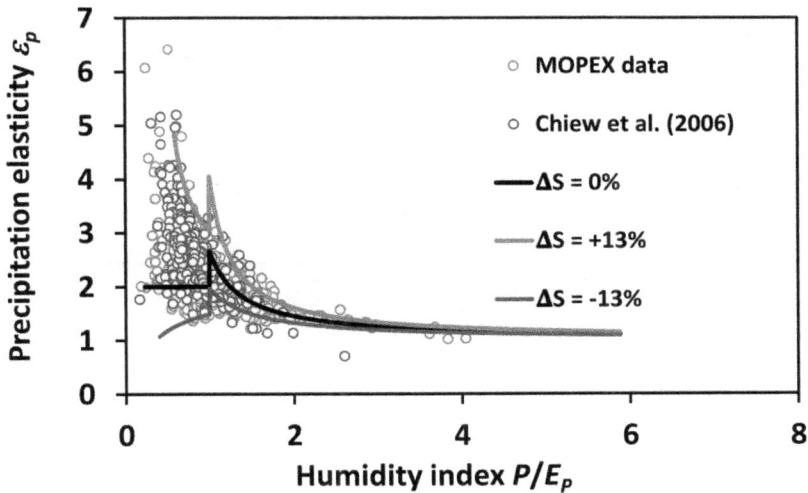

Figure 11.24. Comparison of the Australian data of Chiew (2006) and of MOPEX data reported by Harman *et al* (2011) with predictions of the percolation model. For this combination of datasets, a smaller range of storage changes, $\Delta S = \pm 13\%$, suffices to constrain the large majority of data. Reproduced from Hunt *et al* (2023b). CC BY 4.0.

Data of Harman et al (2011) *and Chiew et al* (2006): In figure 11.24, the MOPEX data of Harman *et al* (2011) and the Australian data of Chiew (2006) for ϵ_p, as a function of humidity index HI=1/AI, are compared with the predictions of the percolation model. Equation (11.29), with storage changes $\Delta S = \pm 13\%$, provides reasonable upper and lower bounds for the observed data, significantly smaller than the range of storage changes chosen to constrain the original MOPEX water balance data reported by Gentine *et al* (2012). While there is no additional confirmation of a peak in ϵ_p at AI $= 1$, the predicted changes in slope and curvature are verified. The median value of ϵ_p at large AI appears to be larger than 2, although the median ϵ_p for one of the (entire) datasets is reported to be slightly less than 2 Harman *et al* (2011). Chiew (2006) and Chiew *et al* (2006) do report a median elasticity larger than 1 for Australia, however—the only continent with such a feature. Although not explicitly shown, the Budyko (1958) prediction would perform somewhat better here.

Data of Sankarasubramanian et al (2001): figure 11.25 inspects data reported by Sankarasubramanian *et al* (2001) for ϵ_p that were reported as a function of inverse run-off ratio, P/Q, rather than as aridity index, AI, which originate from the same study, Sankarasubramanian *et al* (2001), as presented in figures 11.22 and 11.23, however. Thus, one might expect a similar validity of prediction here. However, a somewhat wider range of storage changes needed to be used, in order to predict a comparable agreement with the data. A possible explanation for the apparent added variability is the sensitivity of the run-off ratio itself to variables that are unaccounted for. The vertical line in the curve appropriate for $\Delta S = 0$ is predicted for AI $= 1$ to fall at $P/Q = 1/0.377 = 2.65$. The current representation in terms of P/Q, thus, obscures the particular relevance of AI $= 1$ by distributing the predicted sharp crossover in terms of AI over different P/Q values from each distinct value of ΔS.

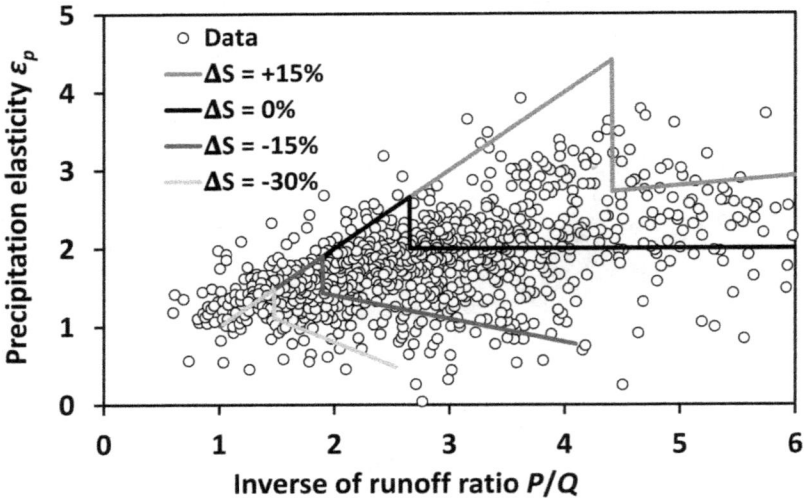

Figure 11.25. Comparison of predictions for ϵ_p as a function of inverse run-off ratio, P/Q, with the data reported by Sankarasubramanian and Vogel (2003), which are, however, derived from the same dataset that was published in Sankarsubramanian *et al* (2001). Here, it appears that a somewhat larger positive ΔS value of 15% (rather than 10%) was needed to constrain the large majority of the data. Note that, as observed by Sankarasubramanian and Vogel (2003) and implied as well in the publications of Chiew (2006) and Chiew *et al* (2006), the percolation model predicts that all data should fall below the line given by the equality of inverse run-off ratio, P/Q, and ϵ_p. Clearly, this prediction is violated. However, the implied prediction that Q not exceed P is also violated, and the discrepancies with the predictions of ϵ_p are largest where $Q > P$. In constructing these predictions, the same change in storage was entered into both the run-off ratio and the calculation of ϵ_p. Reproduced from Hunt *et al* (2023b). CC BY 4.0.

Data of Chiew et al (2006): In figure 11.26, the dependence of ϵ_p on the run-off ratio, Q/P, as reported by Chiew *et al* (2006) and Chiew (2006) from a global study as well as an Australian study, are compared with the predictions of the percolation model. Each branch of the curve is cut off either at the margin of the figure, or at $AI = 5$, the largest AI reported by the authors. Predictions for each ϵ_p change sharply at $AI = 1$, but at different values of Q/P. Several features of the data are picked up by the theoretical framework, including, (1) if the change in storage ΔS could take on any value, the predictions would ultimately trace out the entire curve of $\epsilon_p = 1/(Q/P)$, the predictions for small AI (a result considered an upper limit by Sankarasubramanian *et al* 2003). For the given data, however, this bound becomes irrelevant for ΔS greater than about 10%, even though such large storage changes can still be relevant for $AI > 1$. (2) Cutting off the predictions at $AI = 5$ is in rather close agreement with the data. (3) The sharp change in behavior of ϵ_p at $AI = 1$ shows up in the data near $\Delta S = 8\%$. (4) For small run-off ratio, changes in storage can be important in extending the range of run-off ratio values, for which the percolation model applies, to near zero. It should be noted that the Budyko approximation, without addition of changes in storage, does not particularly capture the trends of the data.

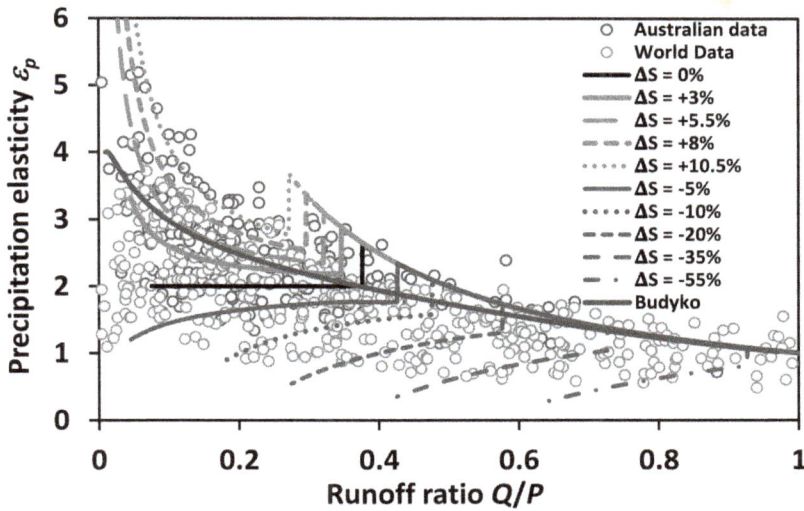

Figure 11.26. The annual data for ϵ_p as a function of the run-off coefficient Q/P, reported by Chiew (2006) and Chiew *et al* (2006), together with the Budyko (1958) curve and predictions of the percolation model. In this comparison it was necessary to include ΔS values as large in magnitude as (-) 55%, much larger than in any other study. Each branching curve is cut off either at the margin of the figure or at the largest AI value found in the data, which is 5, as given in the text. The Budyko (1958) prediction is given for comparison. Reproduced from Hunt *et al* (2023b). CC BY 4.0.

Simultaneous comparison with all the major datasets: Figure 11.27 compares the predictions of equation (11.29) for ϵ_p as a function of humidity index with all the major datasets simultaneously. The bounds used on ΔS were found to be compatible with the MOPEX data obtained reported by Gentine *et al* (2012), which are +10% and −30%. Of all the datasets, only that of Sankarasubramanian *et al* (2001) has a significant number of points (about 25) lying significantly outside the boundaries associated with the storage changes given. The zero-storage change prediction, unlike the Budyko curve, remains near the middle of the range of data points, consistent with a relatively universal result from the Chiew (2006) data summary, giving values of nearly 2 for the median elasticity in almost every climate class. The predictions are much less sensitive to changes in storage at large values of the humidity index, in agreement with the data. By following separately each of the two predicted peaks in elasticity at AI = 1 and large AI, each of the four datasets exhibits evidence of the relevance of the crossover in functional form of ϵ_p predicted at AI = 1. The Tang and Lettenmaier (2012) data exhibit, as expected, an overall narrower range of ϵ_p, particularly at very low values of the humidity index, than all other sources. Thus, a much smaller range of relative storage change values is sufficient to account for the variations in elasticity for very large river basins than for the smaller catchments typically chosen for investigation in large studies, in general accord with the study of Liu *et al* (2018), which finds typical changes in storage values for the world's largest river basins on the order of 3% of P, a factor ca. 3 smaller than values quoted earlier for smaller basins quoted earlier (ca. 10% in the Great Lakes region, for example).

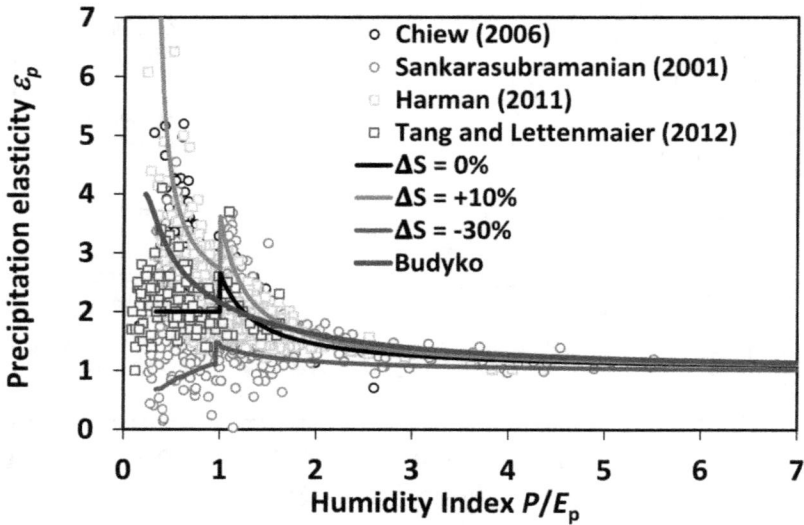

Figure 11.27. Comparison of all the major datasets for ϵ_p as a function of humidity index. Bounds on the storage change are taken to be $\Delta S = 10\%$ and -30%, the values compatible with the MOPEX data for the water balance predictions. The Budyko (1958) curve is shown once again for comparison. Reproduced from Hunt *et al* (2023b). CC BY 4.0.

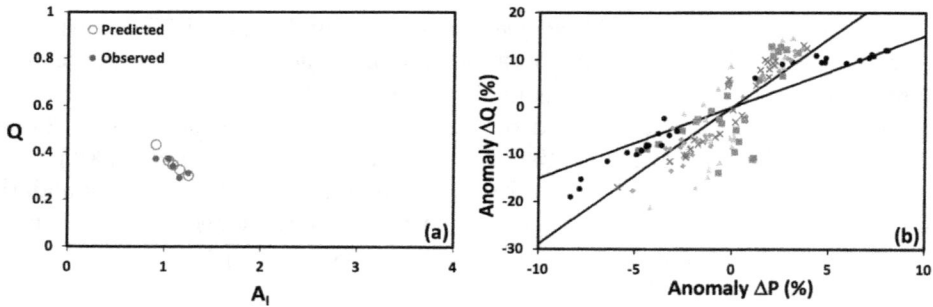

Figure 11.28. (a) Comparison of the predicted dependence of Q on the aridity index AI with the data. (b) The anomaly ΔQ versus ΔP. Reproduced from Hunt *et al* (2023b). CC BY 4.0.

Data of Deusdara-Leal et al (2022): A study by Deusdara-Leal *et al* (2022) on streamflow in southeastern Brazil reported data for Q, AI, and the streamflow elasticity ϵ_p. The predictions of the 2D model percolation for the water balance is accurate for their streamflow indices, as the comparison shown in figure 11.28(a) indicates. Figure 11.28(b) presents the percolation prediction for ϵ_p (lines with specified slopes), as calculated for zero storage change, and compares with annual data for the $\Delta Q(\Delta P)$, which correspond to the expected streamflow precipitation sensitivity for the published aridity index values. A clear conclusion from the comparison is that the percolation predictions for the elasticity underestimate the actual values, since a significant fraction of the data fall outside the predictions (between the red and blue lines in the figure). Interestingly, however, a large share of the data comes from multiyear correlations. The precipitation was also anomalously low for almost all the regions for the entire period. Thus, cumulative effects of continued drying may accentuate the values of the streamflow elasticity over time. Notably, however, the one river basin that follows the minimum theoretical slope calculation has undergone forest restoration, helping to stabilize soils as well as water storage.

As the above comparisons indicate, there may be advantages to representing precipitation elasticity as a function of the run-off ratio, since that allows simultaneous assessment of the relevance of storage changes and aridity index and, thus, a measure of the diminishing relevance of large magnitude storage changes at large HI values. This finding is in accord with the understanding that storage in arid regions is typically small. In this interpretation, ϵ_p has a universal median value of 2 for AI > 1, which is a consequence of the relevance of the case of zero storage change to the median value of the precipitation elasticity. The relatively small range of variation of ϵ_p for any particular value of the aridity index in the world's great river basins is due to the rarity of having the same sign of subsurface storage changes over the very large areas of such basins, particularly when they span multiple climate regimes. In a related matter, Zhang et al (2021) found that catchment elasticity values tended to increase with time over sub-annual time scales, but at times exceeding one year, tended to saturate at values between 1.3 and 2.3 with a median value near 1.8, not very different from the summary of Chiew et al (2006). Clearly, the Australian catchments studied by Chiew (2006) have a greater sensitivity to precipitation anomalies (larger ϵ_p) than the worldwide selection of catchments in Chiew et al (2006). If the percolation model is a guide, then, this is associated with a greater importance of positive storage changes in Australia, as compared with negative storage changes globally. This appears to be counterintuitive, but it is in accord with the study of Deusdara-Leal et al (2022) for southeastern Brazil. Perhaps when drought is persistent, a minimum storage is reached and storage changes are mostly positive, a negative correlation based on antecedent anomalies. It should be pointed out that the failure to capture the magnitude of the southeastern Brazil elasticity values is then likely due to a neglect of potential storage changes.

The small values of precipitation elasticity in snowy catchments is related to the phenomenon of snowmelt, which has important implications. For example, absorption of solar energy through warming the snow and the phase change of melting should tend to reduce effective value of AI. It is known (Dooge 1988) that the precipitation and temperature elasticities sum to 1. On sub-annual time scales, snowmelt, and particularly ice melt, tend to make the streamflow response to solar energy positive, with the consequence that the response to precipitation must trend smaller and more nearly negative. As other authors have also noted (e.g., Sankarasubramanian et al 2001), snow can be regarded as buffering streamflow response or simply as another form of water storage.

It should also be pointed out that the data reported by Zhang et al (2022) imply that accounting for storage results in nearly constant $E(\text{PET}/P)$ for AI > 1, and very little difference from the original Budyko function for AI > 0.5, whereas predictions of the percolatiuon model imply strong dependence of $E(\text{PET}/P)$ for small AI, as well as strong dependence of ϵ_p on storage changes, except for small AI. The two perspectives are, of course, very different. The two perspectives of different root fractal dimensionalities, and different storage changes, while giving very similar variability in the water balance, should betested further including comparisons with data, in order to establish whether there is a quantitative relationship between their predictions. Ideally, these comparisons should also involve compilations of data that address streamflow elasticity and the direct water balance at the same sites.

References

Adams B, White A and Lenton T M 2004 An analysis of some diverse approaches to modelling terrestrial net primary productivity *Ecol. Model.* **177** 353

Aguirre-Gutiérrez C A, Holwerda F, Goldsmith G R, Yepes E, Carbajal N, Escoto-Rodríguez M and Arredondo J T 2019 The importance of dew in the water balance of a continental semi-arid grassland *J. Arid Environ.* **168** 26

Akash M H, Mondal S and Bisht D 2021 Sustainable livestock production and biodiversity *Emerging Issues in Climate Smart Livestock Production, Biological Tools and Techniques* ed S Modal and R L Singh (New York: Academic) p 91

Amatya D M, Tian S, Dai Z and Sun G 2016 Long-term potential and actual evapotranspiration of two different forests on the Atlantic coastal plain *Trans. ASABE* **59** 647

Amatya D M, Muwamba A, Panda S, Callahan T, Herder S and Pellett C A 2018 Assessment of spatial and temporal variation of potential evapotranspiration estimated by 4 methods for S. Carlina *J. South Carolina Water Resour.* **5** 3

Anderson B J, Brunner M I, Slater L J and Dadson S J 2023 Elasticity curves describe streamflow sensitivity to precipitation across the entire flow distribution *Hydrol. Earth Syst. Sci. Discuss.* **1** https://hess.copernicus.org/:http://arXiv.org/abs/s/hess-2022-407/

Arnold J G, Srinivasan R, Muttiah R S and Williams J R 1998 Large area hydrological modeling and assessment part I: Model development *J. Am. Water Resour. Assoc.* **34** 73–89

Baldocchi D *et al* 2001 FLUXNET: a new tool to study the temporal and spatial variability of ecosystem-scale carbon dioxide, water vapor, and energy flux densities *Bull. Am. Meteorol. Soc.* **82** 2415

Barker W T and Whitman W C 1988 Vegetation of the northern Great Plains *Rangelands* **10** 266

Benyon R G, Lane P N J J, Jaskierniak D, Kuczera G and Hayden S R 2015 Use of a forest sapwood area index to explain long-term variability in mean annual evapotranspiration and streamflow in moist eucalypt forests *Water Resour. Res.* **51** 5318

Berghuijs W R, Gnann S J and Woods R A 2020 Unanswered questions on the Budyko framework *Hydrol. Process.* **34** 5699

Berghuijs W R, A Hartmann A and Woods R A 2016 Streamflow sensitivity to water storage changes across Europe *Geophys. Res. Lett.* **43** 1980

Berghuijs W R, Woods R A and Hrachowitz W 2014 A precipitation shift from snow towards rain leads to a decrease in streamflow *Nat. Clim. Change* **4** 583

Beven K 1993 Prophecy, reality and uncertainty in distributed hydrological modelling *Adv. Water Resour.* **16** 41–51

Bonan G B and Doney S C 2018 Climate, ecosystems, and planetary futures: The challenge to predict life in Earth system models *Science* **359** eaam8328

Brilli L, Bechini L, Bindi M, Carozzi M, Cavalli D, Conant R and Bellocchi G 2017 Review and analysis of strengths and weaknesses of agro-ecosystem models for simulating C and N fluxes *Sci. Total Environ.* **598** 445

Budyko M I 1958 The heat balance of the Earth's surface (in Russian) ed N A Stepanova (United States Department of Commerce, Weather Bureau)

Budyko M I 1974 *Climate and Life* (New York: Academic) p 508

Bundt M, Widmer F, Pesaro M, Zeyer J and Blaser P 2001 Preferential flow paths: Biological 'hot spots' in soils *Soil Biol. Biogeochem.* **33** 729

Carmona A M, Sivapalan M, Yaeger M A and Poveda G 2014 Regional patterns of interannual variability of catchment water balances across the continental US: a Budyko framework *Water Resour. Res.* **50** 9177

Chen H, Huo Z, Zhang L and White I 2020 New perspective about application of extended Budyko formula in arid irrigation district with shallow groundwater *J. Hydrol.* **582** 124496

Cheng L, Xu Z, Wang D and Cai X 2011 Assessing interannual variability of evapotranspiration at the catchment scale using satellite-based evapotranspiration data sets *Water Resour. Res.* **47** W09509

Cheng S, Cheng L, Qin S, Zhang L, Liu P, Liu L, Xu Z and Wang Q 2022 Improved understanding of how catchment properties control hydrological partitioning through machine learning *Water Resour. Res.* **58** e2021WR031412

Chiew F H S 2006 Estimation of rainfall elasticity of streamflow in Australia *Hydrol. Sci. J.* **51** 613

Chiew F H S, Peel M C, McMahon T A and Siriwardena L W 2006 Precipitation elasticity of streamflow in catchments across the world *Climate Variability and Change-Hydrological Impacts* ed S Demuth, A Gustard, E Planos, F Scatena and E Servat (IAHS International Commission on Water Resources Systems) p 256

Chiew F H S, Wang Q J, McConachy R, James R, Wright W and deHoedt G 2002 Evapotranspiration maps for Australia *Hydrology and Water Resources Symp. (Institution of Engineers, Melbourne, 20–23 May 2002)*

Choudhury B J 1999 Evaluation of an empirical equation for annual evaporation using field observations and results from a bio-physical model *J. Hydrol.* **216** 99

Chung M, Dufour A, Pluche R and Thompson S 2017 How much does dry-season fog matter? Quantifying fog contributions to water balance in a coastal California watershed *Hydrol. Process.* **31** 3948

Churkina G, Running S W and Schloss A L 1999 Comparing global models of terrestrial net primary productivity (NPP): the importance of water availability *Glob. Chang. Biol.* **5** 46

Cramer W, Bondeau A, Schaphoff S, Lucht W, Smith B and Sitch S 2004 Tropical forests and the global carbon cycle: impacts of atmospheric carbon dioxide, climate change and rate of deforestation *Philos. Trans. R. Soc. Lond.* B **359** 331

Cramer W, Kicklighter D W, Bondeau A, Iii B M, Churkina G, Nemry B *et al* 1999 Comparing global models of terrestrial net primary productivity 247 (NPP): overview and key results *Glob. Chang. Biol.* **5** 1

Dai A, Qian T T, Trenberth K E and Milliman J D 2009 Changes in continental freshwater discharge from 1948 to 2004 *J. Clim.* **22** 2773

Dai A and Trenberth K E 2002 Estimates of freshwater discharge from continents: latitudinal and seasonal variations *J. Hydrometeorol.* **3** 660

Daly C, Neilson R P and Philips D L 1994 A statistical-topographic model for mapping climatological precipitation over mountainous terrain *J. Appl. Meteorol.* **33** 140

Dannowski M and Block A 2005 Fractal geometry and root system structures of heterogeneous plant communities *Plants Soil* **272** 61

Deusdará-Leal K *et al* 2022 Trends and climate elasticity of streamflow in south-eastern Brazil basins *Water* **14** 2245

de Wit C T 1965   Photosynthesis of leaf canopies, Agricultural Research Report No. 663. PUDOC

Donohue R J, Roderick M L and McVicar T R 2007 On the importance of including vegetation dynamics in Budyko's hydrological model *Hydrol. Earth Syst. Sci.* **11** 983

Dooge J C I 1988 Hydrology in perspective *Hydrol. Sci. J.* **33** 61–85

Dooge J C I 1992 Sensitivity of runoff to climate change: A Hortonian approach *Bull. Am. Meteorol. Soc.* **73** 2013

Duan Q, Schaake J, Andréassian V, Franks S, Goteti G, Gupta H V *et al* 2006 Model Parameter Estimation Experiment (MOPEX): an overview of science strategy and major results from the second and third workshops *J. Hydrol.* **320** 3

Eagleson P S 1978a Climate, soil, and vegetation: 1. Introduction to water balance dynamics *Water Resour. Res.* **14** 705

Eagleson P S 1978b Climate, soil, and vegetation: 2. The distribution of annual precipitation derived from observed storm sequences *Water Resour. Res.* **14** 713

Eagleson P S and Tellers T 1982 Ecological optimality in water-limited natural soil-vegetation systems: 2. Tests and applications *Water Resour. Res.* **18** 341

Egli M, Favilli F, Krebs R, Pichler B and Dahms D 2012 Soil organic carbon and nitrogen accumulation rates in cold and alpine environments over 1 ma *Geoderma* **183–84** 109

Egli M, Hunt A G, Dahms D, Raab G, Derungs C, Raimondi S and Yu F 2018 Prediction of soil formation as a function of age using the percolation theory approach *Front. Environ. Sci.* **6**

Esser G 1991 Osnabrück biosphere model: Structure, construction, results *Modern Ecology: Basic and Applied Aspects* ed G Esser and D Overdieck (Amsterdam: Elsevier) p 679

Esser G, Lieth H F H, Scurlock J M O and Olson R J 1997 *Worldwide Estimates and Bibliography of Net Primary Productivity Derived from pre-1982 Publications* p 122 ORNL Technical Memorandum TM-13485 Oak Ridge National Laboratory

Fahey B, Davie T and Stewart M 2011 The application of a water balance model to assess the role of fog in water yields from catchments in the east Otago uplands, South Island, New Zealand *J. Hydrol.* **50** 279 (New Zealand)

Fan Y, Miguez-Macho G, Jobbágy E G, Jackson R B and Otero-Casal C 2017 Hydrologic regulation of plant rooting depth *Proc. Natl Acad. Sci. USA* **114** 10572

Field J P, Belnap J, Breshears D D, Neff J C, Okin G S and Whicker J J 2010 The ecology of dust *Front. Ecol. Environ.* **8** 423

Fischer D T, Stillz C J and Williams A P 2009 Significance of summer fog and overcast for drought stress and ecological functioning of coastal California endemic plant species *J. Biogeogr.* **3** 783

Flury M and Flühler H 1994 Susceptibility of soils to preferential flow of water: A field study *Water Resour. Res.* **30** 1945–54

Fu B 1981 On the calculation of the evaporation from land surface *Sci. Atmos. Sin.* **5** 23 (in Chinese)

Gan G, Liu Y and Sun G 2021 Understanding interactions among climate, water, and vegetation with the Budyko framework *Earth Sci. Rev.* **212** 103451

Garcia E S and Tague C L 2015 Subsurface storage capacity influences climate-evapotranspiration interactions in three western United States catchments *Hydrol. Earth Syst. Sci.* **19** 4845

Garrels R M and Mackenzie F T 1972 A quantitative model for the sedimentary rock cycle *Mar. Chem.* **1** 27

Gentine P, D'Dodorico P, Linter B R, Sivandran G and Salvucci G 2012 Interdependence of climate, soil, and vegetation as constrained by the Budyko curve *Geophys. Res. Lett.* **39** L19404

Greve P, Gudmundsson L, Orlowsky B and Seneviratne S I 2015 Introducing a probabilistic Budyko framework *Geophys. Res. Lett.* **42** 2261

Guswa A 2008 The influence of climate on root depth: A carbon cost-benefit analysis *Water Resour. Res.* **44** W02427

Guswa A 2010 Effect of plant uptake strategy on the water–optimal root depth *Water Resour. Res.* **46** W09601

Hanisch S, Lohrey C and Buerkert A 2015 Dewfall and its ecological significance in semi-arid coastal south-western Madagascar *J. Arid Environ.* **121** 24

Harman C J, Troch P A and Sivapalan M 2011 Functional model of water balance variability at the catchment scale 2. Elasticity of fast and slow run-off components to precipitation change in the continental United States *Water Resour. Res.* **47** 02523

Henley B J, Thyer M A, Kuczera G and Franks S W 2011 Climate-informed stochastic hydrological modeling: Incorporating decadal-scale variability using paleo data *Water Resour. Res.* **47** W11509

Horton R 1931 The field, scope, and status of the science of hydrology *Eos. Trans. Am. Geophys. Union* **12** 189

Hunt A G 2017 Spatio-temporal scaling of vegetation growth and soil formation: explicit predictions *Vadose Zone J.* **16** 1

Hunt A G 2021 Soil formation, vegetation growth, and water balance: A theory for Budyko *Hydrogeology, Chemical Weathering, and Soil Formation* ed A G Hunt, M Egli and B A Faybishenko (Washington, DC: American Geophysical Union) Geophysical Monographs

Hunt A G, Faybishenko B A, Egli M, Ghanbarian B and Yu F 2020a Predicting water cycle characteristics from percolation theory and observational data *Int. J. Environ. Res. Public Health* **17** 734

Hunt A G, Faybishenko B A and Ghanbarian B 2021 Predicting characteristics of the water cycle from scaling relationships *WAter Resour. Res.* **57** e2021WR030808

Hunt A G, Faybishenko B A and Powell T L 2020b A new phenomenological model to describe root-soil interactions based on percolation theory *Ecol. Model.* **433** 109205

Hunt A G, Faybishenko B A and Powell T L 2022 Test of prediction of equivalence of tree height growth and transpiration rates in percolation-based phenomenology for root-soil interaction *Ecol. Model.* **465** 109853

Hunt A G and Ghanbarian B 2016 Percolation theory for solute transport in porous media: geochemistry, geomorphology, and carbon cycling *Water Resour. Res.* **52** 7444

Hunt A G, Holtzman R and Ghanbarian B 2017 A percolation-based approach to scaling infiltration and evapotranspiration *Water* **9** 104

Hunt A G and Manzoni S 2016 *Networks on Networks: The Physics of Geobiology and Geochemistry* (Bristol: IOP Publishing)

Hunt A G and Sahimi M 2017 Flow, transport, and reaction in porous media: Percolation scaling, critical-path analysis, and effective medium approximation *Rev. Geophys.* **55** 993

Hunt A G, Sahimi M, Faybishenko B A, Egli M, Ghanbarian B and Yu F 2023a Interpreting water demands of forests and grasslands within a new Budyko formulation of evapotranspiration using percolation theory *Sci. Total Environ.* **877** 162905

Hunt A G, Sahimi M and Ghanbarian B 2023b Predicting streamflow elasticity based on percolation theory and ecological optimality *AGU Adv.* **4** e2022AV000867

Hunt A G, Sahimi M, Ghanbarian B and Poved G 2024 Predicting ecosystem net primary productivity by percolation theory and optimality principle *Water Resour. Res.* **60** e2023WR036340

Hunt A G and Wu J 2004 Climatic influences on Holocene variations of soil erosion rates on a small hill in the Mojave Desert *Geomorphology* **58** 263

Huxman T E, Wilcox B P, Breshears D B, Scott R L, Snyder K A, Small E E, Hultine K, Pockman W T and Jackson R B 2005 Ecohydrological implications of woody plant encroachment *Ecology* **86** 308

Ivancic T J and Shaw S B 2015 Examining why trends in very heavy precipitation should not be mistaken for trends in very high river discharge *Clim. Change* **133** 681

Jacobs A F G, Heusinkveld B G, Wichink Kruit R J and Berkowicz S M 2006 Contribution of dew to the water budget of a grassland area in the Netherlands *Water Resour. Res.* **42** W03415

Jasechko S 2019 Global isotope hydrogeology *Rev. Geophys.* **57** 835

Kalthoff N, Fiebig-Wittmaack M, Meissner C, Kohler M, Uriarte M, Bischoff-Gauss I and Gonzales E 2006 The energy balance, evapo-transpiration, and nocturnal dew deposition of an arid valley in the Andes *J. Arid Environ.* **65** 420

Ketipearachchi K W and Tatsumi J 2000 Local fractal dimensions and multifractal analysis of the root system of legumes *Plant Prot. Sci.* **3** 289

Khan Z, Khan F A, Khan A U, Hussain I, Khan A, Shah L A and Różkowski K 2022 Climate-streamflow relationship and consequences of its instability in large rivers of pakistan: an elasticity perspective *Water* **14** 2033

Khormali F, Ghergherechi S, Kehl M and Ayoubi S 2012 Soil formation in loess derived soils along a sub-humid to humid climate gradient, northeast Iran *Geoderma* **179–80** 113

Kidron G 1999 Altitude dependent dew and fog in the Negev Desert, Israel *Agric. For. Meteorol.* **96** 1

Klemes V 1988 A hydrological perspective *J. Hydrol.* **100** 3

Krause-Jensen D and Sand-Jensen K 1998 Light attenuation and photosynthesis of aquatic plant communities *Limnol. Oceanog.* **43** 396–407

Lan T, Lin K R, Liu Z Y, He Y H, Xu C Y, Zhang H B and Chen X H 2018 A clustering preprocessing framework for the subannual calibration of a hydrological model considering climate-land surface variations *Water Resour. Res.* **54** 10034

Lee Y, Andrade J S, Buldyrev S V, Dokholoyan N V, Havlin S, King P R *et al* 1999 Traveling time and traveling length in critical percolation clusters *Phys. Rev. E* **60** 3425

Levang-Brilz N and Biondini M E 2003 Growth rate, root development and nutrient uptake of 55 plant species from the Great Plains Grasslands, USA *Plant Ecol.* **165** 117

Lieth H 1975 Modeling the primary productivity of the world *Primary Productivity of the Biosphere* ed H Lieth and R H Whittaker (Berlin: Springer) p 237

Liu J, Zhang Q, Singh V P, Song C, Zhang Y, Sun P and Gu X 2018 Hydrological effects of climate variability and vegetation dynamics on annual fluvial water balance in global large river basins *Hydrol. Earth Syst. Sci.* **22** 4047

Liu Y, Wagener T, Beck H E and Hartman A 2020 What is the hydrologically effective area of a catchment *Environ. Res. Lett.* **15** 104024

Lvovitch M I 1973 The global water balance *Eos Trans. Am. Geophys. Union* **54** 28

Lynch J 1995 Root architecture and plant productivity *Plant Physiol.* **109** 7

Madany T S, Reichstein M, Carrara A, Martin M P, Gonzales-Cascon R and Penuelas J 2021 How nitrogen and phosphorus availability change water use efficiency in a Mediterranean savannah ecosystem *J. Geophys. Res.: Biogeosci.* **126** e2020JG006005

Maher K 2010 The dependence of chemical weathering rates on fluid residence time *Earth Planet. Sci. Lett.* **294** 101

Manabe S 1969 Climate and ocean circulation. I. Atmospheric circulation and hydrology of the Earth's surface *Mon. Weather Rev.* **97** 739

Martinez G F and Gupta H V 2010 Toward improved identification of hydrological models: A diagnostic evaluation of the "abcd" monthly water balance model for the conterminous United States *Water Resour. Res.* **46** W08507

Matimati I 2009 The relevance of fog and dew precipitation to succulent plant hydrology in an arid South African ecosystem *Master's Thesis* University of the Western Cape

Milly P C D 1993 An analytic solution of the stochastic storage problem applicable to soil water *Water Resour. Res.* **29** 3755–8

Milly P 1994 Climate, soil-water storage, and the average annual water balance *Water Resour. Res.* **30** 214

Milly P C D, Dunne K A and Vecchia A V 2005 Global pattern of trends in streamflow and water availability in a changing climate *Nature* **438** 347–50

Milne B and Gupta V 2017 Horton ratios link self-similarity with maximum entropy of eco-geomorphological properties in stream networks *Entropy* **19** 249

Mianabadi A, Davary K, Pourreza-Bilondi M and Coenders-Gerrits A M J 2020 Budyko framework; towards non-steady state conditions *J. Hydrol.* **588** 125089

Mirtl M, Borer E T, Djukic I, Forsius M, Haubold H, Hugo W and Haase P 2018 Genesis, goals and achievements of long-term ecological research at the global scale: a critical review of ILTER and future directions *Sci. Total Environ.* **626** 1439

Moratiel R, Martínez-Cob A, Tarquis A M and Snyder R L 2016 Water balance correction due to light rainfall, dew, and fog in Ebro river basin (Spain) *Agric. Water Manage.* **170** 61

Moretti G and Montanari A 2007 A spatially distributed grid based rainfall–runoff model for continuous time simulations of river discharge *Environ.l Model.Softw.* **22** 823–36

Moro M J, Were A, Villagarcia L, Canton Y and Domingo F 2007 Dew measurement by Eddy covariance and wetness sensor in a semiarid ecosystem of SE Spain *J. Hydrol.* **335** 295

NAS 1991 Opportunities in the hydrologic sciences https://nap.nationalacademies.org/catalog/1543/opportunities-in-the-hydrologicsciences

National Research Council 1996 *Rock Fractures and Fluid Flow* (Washington, DC: National Academy Press)

Němec J and Schaake J 1982 Sensitivity of water resource systems to climate variation *Hydrol. Sci. J.* **27** 327–43

Nijzink R C and Schymanski S J 2022 Vegetation optimality explains the convergence of catchments on the Budyko curve *Hydrol. Earth Syst. Sci.* **26** 6289–309

Ning T, Li Z and Liu W 2017 Vegetation dynamics and climate seasonality jointly control the interannual catchment water balance in the loess plateau under the budyko framework *Hydrol. Earth Syst. Sci.* **21** 1515

Niu J, Shen C, Li S-G and Phanikumar M S 2014 Quantifying storage changes in regional Great Lakes watersheds using a coupled subsurface-land surface process model and GRACE, MODIS products *Water Resour. Res.* **50** 7359

Nordt L and Driese S G 2013 Application of the critical zone concept to the deep-time sedimentary record *Sediment. Rec.* **11** 4–9

NSF 2020 NSF 20-560: Hydrologic Sciences (HS) https://new.nsf.gov/funding/opportunities/hydrologic-sciences/nsf20-560/solicitation

Odum E 1959 *Fundamentals of Ecology* (Philadelphia, PA: W.B. Saunders)

Oki T and Kanae S 2006 Global hydrological cycles and world water resources *Science* **313** 1068

Oldekop E M 1911 On evaporation from the surface of river basins *Trans. Meteorol. Obs., Univ. Tartu* **4** 200

Oppelt A L, Kurth W, Dzierzon H, Jentschke G and Godbold D L 2000 Structure and fractal dimensions of root systems of four co-occurring fruit tree species from Botswana *Ann. For. Sci.* **57** 463

Palmroth S, Katul G G, Hui D, McCarthy R, Jackson R B and Oren R 2010 Estimation of long-term basin scale evapotranspiration fromstreamflow time series *Water Resour. Res.* **46** W10512

Peel M C, Chiew F H S, Western A W and McMahon T A 2000 Extension of unimpaired monthly streamflow data and regionalisation of parameter values to estimate streamflow in ungauged catchments, Report prepared for the Australian National Land and Water Resources Audit (2000) http://audit.ea.gov.au/anra/water/docs/national/streamflow/stream-flow.pdf

Pike J G 1964 The estimation of annual run-off from meteorological data in a tropical climate *J. Hydrol.* **2** 116

Planchon O and Mouche E 2010 A physical model for the action of raindrop erosion on soil microtopography *Soil Sci. Soc. Am. J.* **74** 1092

Porporato A, Daly E and Rodriguez-Iturbe I 2004 Soil water balance and ecosystem response to climate change *Am. Nat.* **164** 625

Porto M, Havlin S, Schwarzer S and Bunde A 1997 Optimal path in strong disorder and shortest path in invasion percolation with trapping *Phys. Rev. Lett.* **79** 4060

Potter N J, Zhang L, Milly P C D, McMahon T A and Jakeman A J 2005 Effects of rainfall seasonality and soil moisture capacity on mean annual water balance for Australian catchments *Water Resour. Res.* **41** W06007

Pregitzer K S and Euskirchen E S 2004 Carbon cycling and storage in world forests: biome patterns related to forest age *Glob. Chang. Biol.* **10** 2052

Raich J W *et al* 1991 Potential net primary productivity in South America: Application of a global model *Ecol. Applic.* **1** 399–429

Reaver N G F, Kaplan D A, Klammler H and Jawitz J W 2022 Theoretical and empirical evidence against the Budyko catchment trajectory conjecture *Hydrol. Earth Syst. Sci.* **26** 1507

Roberts S, Vertessy R and Grayson R 2001 Transpiration from *Eucalyptus sieberi* (L. Johnson) forests of different age *For. Ecol. Manag.* **143** 153

Rodríguez-Iturbe I and Porporato A 2007 *Ecohydrology of Water-controlled Ecosystems: Soil Moisture and Plant Dynamics* (Cambridge: Cambridge University Press)

Rodriguez-Iturbe I, Porporato A, Ridolfi L, Isham V and Cox D R 1999 Probabilistic modelling of water balance at a point: The role of climate, soil and vegetation *Proc. R. Soc. Lond.* A **455** 3789

Rosenzweig M L 1968 Net primary productivity of terrestrial communities: prediction from climatological data *Am. Nat.* **102** 67

Sahimi M 1987 Hydrodynamic dispersion near the percolation threshold: Scaling and probability densities *J. Phys.* A **20** L1293

Sahimi M 2023 *Applications of Percolation Theory* 2nd edn (Berlin: Springer)

Sankarasubramanian A and Vogel R M 2003 Hydroclimatology of the continental United States *Geophys. Res. Lett.* **30** 1363

Sankarasubramanian A, Vogel R M and Limbrunner J F 2001 Climate elasticity of streamflow in the United States *Water Resour. Res.* **37** 1771

Schenk H J and Jackson R B 2002 Rooting depths, lateral root spreads and below-ground/above-ground allometries of plants in water-limited ecosystems *J. Ecol.* **90** 480

Schlesinger W H and Jasechko S 2014 Transpiration in the global water cycle *Agric. For. Meteorol.* **189** 115

Schreiber P 1904 Ueber die Beziehungen zwischen dem Niederschlag und der Wasserfuehrung der fluesse in Mitteleuropa *Meteorol. Z.* **21** 441

Scurlock J M O and Olson R J 2002 Terrestrial net primary productivity–a brief history and a new worldwide database *Environ. Rev.* **10** 91

Sheppard A P, Knackstedt M A, Pinczewski W V and Sahimi M 1999 Invasion percolation: New algorithms and universality classes *J. Phys. A: Math. Gen.* **32** L521

Slack J R, A M Lumb A M and Landwehr J M 1993 Hydroclimatic data network (HCDN): a U.S. Geological Survey streamflow data set for the United States for the study of climate variation, 1874–1988 *U.S. Geol. Surv. Water Resour. Invest. Rep. [CD-ROM]* **93**

Slater L J and Villarini G 2006 Recent trends in US flood risk *Geophys. Res. Lett.* **43** 12428

Smakhtin V U 2001 Low flow hydrology: A review *J. Hydrol.* **240** 147–86

Spieler D, Mai J, Craig J R, Tolson B A and Schütze N 2020 Automatic model structure identification for conceptual hydrologic models *Water Resour. Res.* **56** e2019WR027009

Sposito G 2017a Incorporating the vadose zone into the Budyko framework *Water* **9** 698

Sposito G 2017b Understanding the Budyko equation *Water* **9** 236

Stagnitti F, Parlange J-Y, Steenhuis T S, Boll J, Pivetz B and Barry D A 1995 Transport of moisture and solutes in the unsaturated zone by preferential flow *Environmental Hydrology* ed V P Singh (Dordrecht: Kluwer Academic) p 193

Stauffer D and Aharony A 1994 *Introduction to Percolation Theory* 2nd edn (London: Taylor and Francis)

Stothoff S A, Or D, Groeneveld D P and Jones S B 1999 The effect of vegetation on infiltration in shallow soils underlain by fissured bedrock *J. Hydrol.* **218** 169

Subramanian A and Rao A V R K 1983 Dewfall in sand dune areas of India *Int. J. Biom.* **27** 271

Sun P, Wu Y, Xiao J, Hui J, Hu J, Zhao F and Liu S 2019 Remote sensing and modeling fusion for investigating the ecosystem water-carbon coupling processes *Sci. Total Environ.* **697** 134064

Tang Q and Lettenmaier D P 2012 21st century runoff sensitivies of major global river basins *Geophys. Res. Lett.* **79** L06403

Thompson E 2021 A well-balanced ecosystem uses water most efficiently *Eos* **102**

Turc L 1954 Le bilan d'eau des sols: Relation entre les précipitations, l'évaporation et l'écoulement *Ann. Agron.* A **5** 491–595

Uclés O, Villagarcía L, Moro M J, Canton Y and Domingo F 2014 Role of dewfall in the water balance of a semiarid coastal steppeecosystem *Hydrol. Process.* **28** 2271

Vertessy R, Watson F, O'Sullivan S, Davis S, Campbell R, Benyon R and Haydon S 1998 Predicting water yield from mountain ash forest catchments Industry Report, Report 98/4, Cooperative Research Centre for Catchment Hydrology

Vertessy R, Watson F G R and O'Sullivan S K 2001 Factors determining relations between stand age and catchment water balance in mountain ash *For. Ecol. Manag.* **143** 153

Vörösmarty C J, Fekete B M, Meybeck M and Lammers R B 2000 Global system of rivers: Its role in organizing continental land mass and defining land-to-ocean linkages *Glob. Biogeochem. Cycl.* **14** 599–621

Vörösmarty C J *et al* 2010 Global threats to human water security and river biodiversity *Nature* **467** 555–61

Wada Y, Reagers-Benjamin J T, Chao F, Wang J, Lo M-H, Song C *et al* 2017 Recent changes in land water storage and its contribution to sea level variations *Surv. Geophys.* **38** 131

Wang D, Wang G and Anagnostou E N 2007 Evaluation of canopy interception schemes in land surface models *J. Hydrol.* **347** 308

Wang T, Istanbulluoglu E, Lenters J and Scott D 2009 On the role of groundwater and soil texture in the regional water balance: an investigation of the Nebraska Sand Hills, USA *Water Resour. Res.* **45** W10413

Webb W L, Lauenroth W K, Szarek S R and Kinerson R S 1986 Primary production and abiotic controls in forests, grasslands, and desert ecosystems of the United States *Ecology* **64** 134

Wells S G, McFadden L D, McDonald E V, Eppes M C, Young M H and Wood Y A 2014 Desert pavement process and form: Modes and scales of landscape stability and instability in arid regions *Geophys. Res. Abstr.* **16** EGU2014-8060-1

Westbeld A, Klemm O, Grießbaum F, Sträter E, Larrain H, Osses P and Cereceda P 2009 Fog deposition to a Tillandsia carpet in the Atacama Desert *Ann. Geophys.* **27** 3571

Westra S, Thyer M, Leonard M, Kavetski D and Lambert M 2014 A strategy for diagnosing and interpreting hydrological model nonstationarity *Water Resour. Res.* **50** 5090–113

Williams C A, Reichstein R, Buchmann N, Baldocchi D, Beer C and Schwalm C 2012 Climate and vegetation controls on the surface water balance: Synthesis of evapotranspiration measured across a global network of flux towers *Water Resour. Res.* **48** W06523

Xiao H, Meissner R, Seeger J, Rupp H and Borg H 2009 Effect of vegetation type and growth stage on dewfall determined with high precision weighing lysimeters at a site in northern Germany *Hydrology* **377** 43

Xing W, Wang W, Zou S and Deng C 2018 Projection of future runoff change using climate elasticity method derived from Budyko framework in major basins across China *Glob. Planet. Change* **162** 120

Yamanaka N, Hou Q-C and Du S 2013 Vegetation of the loess plateau *Restoration and Development of the Degraded Loess Plateau, China* (Springer) p 49

Yang D, Shao W, Yeh P J-F, Yang H, Kanae S and Oki T 2009 Impact of vegetation coverage on regional water balance in the nonhumid regions of China *Water Resour. Res.* **45** W00A14

Yang F-L, Yue P, Yao T and Wang W-Y 2015 Characteristics of dew formation and distribution and its contribution to the surface water budget in a semi-arid region in China *Bound.-Layer Meteorol.* **154** 317

Yang H, Yang D, Lei Z and Sun F 2008 New analytical derivation of the mean annual water-energy balance equation *Water Resour. Res.* **44** W03410

Yang Q, Cheng W, Hao Z, Zhang Q, Yang D, Teng D, Zhang Y, Wang X, Shen H and Lei S 2022 Study on the fractal characteristics of the plant root system and its relationship with soil strength in tailing ponds *Wirel. Commun. Mobile Comput.* **2022** 9499465

Yu F, Faybishenko B, Hunt A G and Ghanbarian B 2017 A simple model of the variability of topsoil depths *Water* **9** 460

Yu F and Hunt A G 2017a An examination of the steady-state assumption in soil development models with application to landscape evolution *Earth Surf. Process. Landforms* **42** 2599

Yu F and Hunt A G 2017b Damköhler number input to transport-limited chemical weathering and soil production calculations *Earth Space Chem.* **1** 30

Yu F and Hunt A G 2017c Predicting soil formation on the basis of transport-limited chemical weathering *Geomorphology* **301** 21

Yu F, Hunt A G, Egli M and Raab G 2019 Comparison and contrast in soil depth evolution for steady-state and stochastic erosion processes: Possible implications for landslide prediction *Geochem. Geophys. Geosyst.* **20** 2886

Zhang D, Cong Z, Ni G, Yang D and Hu S 2015 Effects of snow ratio on annual run-off within the Budyko framework *Hydrol. Earth Syst. Sci.* **19** 1977

Zhang L, Dawes W R and Walker G R 2001 Response of mean annual evapotranspiration to vegetation changes at catchment scale *Water Resour. Res.* **37** 701

Zhang L, Hickel K, Dawes W R, Chiew F H S, Western A W and Briggs P R 2004 A rational function approach for estimating mean annual evapotranspiration *Water Resour. Res.* **40** W02502

Zhang L, Potter N, Hickel K, Zhang Y and Shao Q 2008 Water balance modeling over variable time scales based on the Budyko framework–model development and testing *J. Hydrol.* **360** 117

Zhang X, Li J, Dong Q and Woods R A 2022 An analytical generalization of Budyko framework with physical accounts of climate seasonality and water storage capacity *Hydrol. Earth Syst. Sci. Discuss.* 1–28

Zhang Y, Viglione A and Bloeschl G 2021 Temporal scaling of streamflow elasticity to precipitation: a global analysis *Water Resour. Res.* **58** e2021WR030601

Zheng J, He Y, Jiang X, Nie T and Lei Y 2021 Attribution analysis of runoff variation in Kuye River basin based on three Budyko methods *Land* **10** 1061

Zheng H, Zhang L, Zhu R, Liu C, Sato Y and Fukushima Y 2009 Responses of streamflow to climate and land surface change in the headwaters of the Yellow River Basin *Water Resour. Res.* **45** W00A19

Zhou S, Yu B, Huang Y and Wang G 2015a The complementary relationship and generation of the Budyko functions *Geophys. Res. Lett.* **42** 1781

Zhou X, Zhang Y and Yang Y 2015b Comparison of two approaches for estimating precipitation elasticity of streamflow in China'a2s main river basins *Adv. Meteorol.* **7** 1–8

IOP Publishing

Networks on Networks (Second Edition)
Role of connectivity in physics of geobiology and geochemistry
Allen G Hunt and Muhammad Sahimi

Chapter 12

Hazards to plants and vegetation: disease propagation, deforestation, and forest fires

12.1 Background

Several types of vegetation play key roles in sustaining humanity and Earth. They can be edible, medicinal, and ornamental. For example, many view plants as Nature's 'pharmacy,' with plant-based medicines continuing to be discovered. From the anthropocentric standpoint, vegetation plays a key role in the economy, because it produces wood, fibre, and chemicals. Non-edible vegetation is still highly important, because it is essential for crop production, as it helps preventing soil erosion, provides wind-breaks or shade against high temperatures, and aides in conditioning the soil. Most importantly, vegetation provides oxygen for the planet by absorbing CO_2, without which organic life would be impossible to sustain. Thus, vegetation plays a fundamental role in all aspects of Nature on Earth, more important than the roles that animals play.

But, similar to every living being on Earth, vegetation and plants are also exposed to a number of natural dangers. Examples include various diseases, adverse weather, and insect depredations, as well as such man-made disasters as pollution, excessive rapid development that causes deforestation, urbanization, and fires. Vegetation is also exposed to hazards that are common among human beings, such as getting old and genetic 'accidents.' Remarkably, another hazard that is not talked about much, but it is very real, is *plant wars*. In fact, similar to the smallest organisms, the largest and most intelligent plants are involved in the struggle for life among themselves, as well as being the natural prey of animals; there are even plant predators.

Description and modeling of most of such hazards are well beyond of the scope of this book, and the expertise of the authors. In this chapter, however, we describe three of such hazards that have been shown to be influenced strongly by connectivity, the main theme of this book.

doi:10.1088/978-0-7503-5698-5ch12

12.2 Improving production of plants with plague susceptibility

The first problem that we describe and study in this chapter is propagation of a disease in plants and vegetables, an important problem, not only from a scientific perspective, but also for its social and economical consequences. Plagues of insects or gastropods, on one hand, and the spread of diseases caused by bacteria, fungi, and oomycetes, on the other hand, pose a great threat to the production of fruits and vegetables, because they can reduce their production and even destroy them completely. They can also transmit the disease to other plantations that share the irrigation system. Several models of disease propagation in plants have been proposed (see, e.g., Shaw 1994, Bailey *et al* 2000, Mundt and Sackett 2012). Our interest in this chapter is to explore how percolation theory can help us gain a better understanding of the problems, and factors that contribute to disease propagation in plants and vegetables.

A percolation model was proposed by Ramírez *et al* (2018) in which the sites in a lattice represent two different types of plants, say A and B, which are growing on a given type of soil. Each tree type has a particular pathogen susceptibility, defined as the probability of being infected by a specific pathogen, which we denote by χ_A and χ_B for the two types. Next, consider a percolation lattice in which a fraction p of the sites are occupied by the two types of tree. Suppose that p_A is the probability that a site is occupied by tree A, so that $p_B = 1 - p_A$ is the probability that an available site is occupied by a plant of variety B.

The average number of sites available for spreading a disease is, $N_d = \langle N_A \rangle + \langle N_B \rangle$, with, for example, $\langle N_A \rangle$ being the average number of susceptible sites of type A. Thus, $\langle N_A \rangle = N p_A \chi_A p$, and $\langle N_B \rangle = N(1 - p_A)\chi_B p$, where N is the total number of sites in the lattice. One must also take into account the fraction of sites that are inoculated with the pathogen, which are located in the occupied sites or in resistant plants. Suppose that p_I is the probability that a site in the lattice is inoculated. Then, the average number $\mathcal{N}$ of sites through which a disease can propagate is given by

$$\mathcal{N} = N p_e = N_d + (N - N_d)p_I, \tag{12.1}$$

where the second term on the right side accounts for the inoculated sites that conform to the situation described above. Equation (12.1) implies that one has a percolation process on the lattice with an effective probability p_e given by

$$p_e = p_I + (1 - p_I)[p_A \chi_A + (1 - p_A)\chi_B]p. \tag{12.2}$$

The effective percolation threshold p_c^e of the process is obtained from equation (12.2) by setting $p_e = p_c$ and $p = p_c^e$, which yields

$$p_c^e = \frac{p_c - p_I}{(1 - p_I)[p_A \chi_A + (1 - p_A)\chi_B]}, \tag{12.3}$$

where p_c is the percolation threshold of the original lattice.

As discussed by Ramírez *et al* (2018), one must also consider the extent of disease incidence on the sown plants, since the pathogen can spread on plants that are susceptible and belong to the same cluster. Some pathogens may exhibit latency stages when they are in an adverse environment and, thus, any point in the lattice can be the source of infection, including a site with no plant, or one with a plant that resists the pathogen. Therefore, if the initial point of infection is an empty site, or one occupied by a plant resistant to the pathogen, then the disease may be transmitted to more than one adjacent cluster, even if they are disjoint. Thus, the average number of sites at which the pathogen causes damage is slightly larger than the average cluster size for $p_c^e \sim p_c$, as a result of spreading to disjoint clusters by the initial infection point. For $p_e < p_c$, on the other hand, the contribution of the initial infection point is through finite clusters, including isolated sites, whereas for $p_e > p_c$ the initial infection site belongs to the sample-spanning cluster. Therefore, if there is more than one initial point of infection in the system, one expects the appearance of sites that connect two adjacent disjoint clusters to be amplified.

Ramírez *et al* (2018) carried out simulations on a square lattice, since crops are planted in parallel rows on the soil, implying that the seeds are sown in a square lattice arrangement. The distance between the neighboring site was selected according to the maximum length that the pathogen can travel. Since one does not know *a priori* whether a seed was sown at a given site, one assigns 1 to each entry in the matrix based on the occupation probability p. Given a value of p_A, one assigns randomly to each occupied site — to each entry with a value of 1 — a plant of type A or B by generating for each site a random number $0 < R < 1$ such that if $R < p_A$, a plant of type A is sown there Otherwise, a plant of type B is sown.

The inoculated cells in the initial configuration were taken randomly and uniformly in the lattice with probability p_I, which represents the fraction of inoculated sites in the lattice, which may or may not propagate the pathogen to neighboring plants, depending on the pathogen susceptibility of each type of plant. Thus, a plant of type A (B) is infected and develops the disease if a random number R, selected uniformly, is less χ_A (χ_B); otherwise, the plant remains healthy. If the plant is infected and develops the disease, then its site value is changed from 1 to 0.

Remírez *et al* (2018) also carried out an experimental study of the same problem in a system that mimicked their simulated model. They prepared the substrate by mixing peat moss and sieved soil (2-mm mesh) in a 1:2 volume-volume mixture, which was placed in plastic double bags of 6 kg of high density polyethylene, and was sterilized in an electric autoclave. Three types of chili seed were used, from which groups of 100 seeds were selected for carrying out the experiments. For each bioassay, aluminum trays were used, and sterile substrate to each was added, after which the seed was spread homogeneously on the trays and covered with more of the substrate and moistened with enough water. The trays were watered daily to maintain the humidity until the buds of seedlings emerged. It took eight days after sowing to observe the initial growth, after which they were fertilized every seven days with a fertilizer.

The microorganisms were taken from a phytopathogenic oomycete strain. Each oomycete was inoculated in the same sterile substrate used for the preparation of the trays. Segments of the growths were inoculated in plastic bags containing the substrate, and were mixed by shaking them for three weeks to ensure the growth of the oomycete throughout the substrate, after which they were incubated at room temperature. Each tray planted contained, on average, about 80 seedlings, and each of the oomycetes was inoculated into the three trays corresponding to the three types of chili. The fertilization of the plants was stopped after the microorganisms were inoculated, but the humidity was maintained at field capacity during the entire experiments. Thirty five days after sowing, live plants were counted in each tray and the survival percentage was calculated.

The survival rate was determined experimentally by exposing a number of plants to the pathogen and counting the number of alive plants after five months. If S (as a percentage) is the survival rate of a plant type exposed to a pathogen, then, the pathogen susceptibility is, $\chi = 1 - S/100$. Depending on the pathogen susceptibility of each plant, the probability p_A for which the pathogen will only spread on finite clusters was determined, even if all the sites were sown. This is shown in figure 12.1, where the combinations of pathogen susceptibility χ_B and mixture proportion (probability) p_A that prevent the formation of the spanning cluster of diseased trees for fixed $\chi_A = 1.0, 0.75, 0.50,$ and 0.25 in various regular lattices in the limit $p_I \to 0$, corresponding to a single initial inoculation point, is shown.

The simulations of Ramírez et al were in agreement with their experiments. Thus, using a percolation-based model, one can devise a strategy for preventing spread of a disease in plants, hence increasing their production of fruits. One strategy consists of sowing two types of plants with different susceptibilities to a specific pathogen, and organized as a percolating lattice in order to maximize the number of plants that survive an infestation. The lattice spacing coincides with the maximum distance that the pathogen can travel before entering a state of dormancy or dying due to starvation. Under particular conditions of pathogen susceptibility, there are values of p_A for which the disease will only propagate on finite clusters, even if the entire soil is sown.

12.3 Spread of fungi in soil

Fungi exist naturally in soil, and most of them are actually beneficial for plants and vegetation. But there are also thousands of fungi that appear to have no purpose other than damaging plants and vegetation, because they spread a variety of sicknesses, such as root rot that infects plant roots, disabling them from drawing water and nutrients into the plant, which eventually kills them. An important example is saprotrophic fungi. When most of them grow in plantation and farms, they are actually useful and, therefore, do not need to be controlled, as they play an important role in the long-term health of plants. But a small number of such species possess poisonous fruiting bodies that can harm plants and vegetation, as well as humans and pets, if they consume them. The presence and persistence of harmful fungal parasites and saprotrophs in soil, and how to protect soil against them, have been long-standing problems. One highly important factor is the ability of the fungi to spread by mycelial growth and expansion of fungal colonies. The lateral spread of

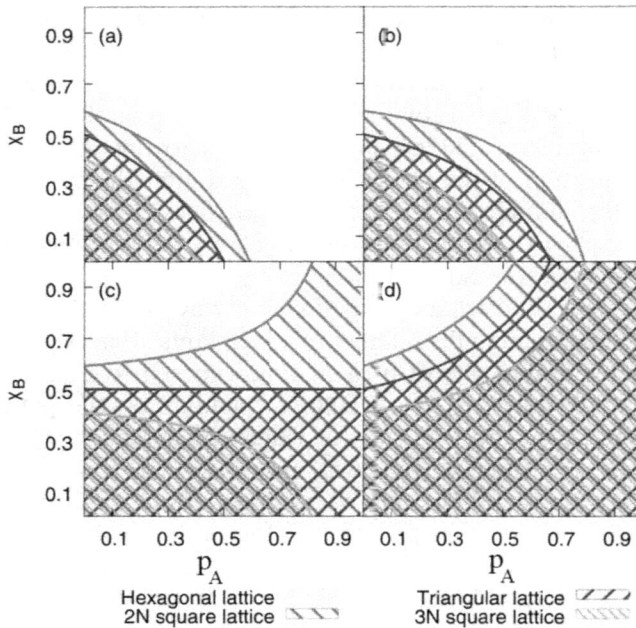

Figure 12.1. The conditions for no percolation in regular lattices of trees when the density of inoculated cells p_I is small, $p_I \to 0$ for susceptibilibies (a) $\chi_A = 1.0$; (b) 0.75; (c) 0.5, and (d) 0.25 (reproduced from Sahimi 2023; originally from Ramírez *et al* 2018).

parasites depends on the endogenous supply and movement of nutrients within the fungal colony, the growth-habit of the colony, and the distances between susceptible host roots or other organs. It is clear that if susceptible hosts are too far apart, local invasion ceases as the fungi exhaust their nutrient supply before infecting a new host. This points to the importance of connectivity and, therefore, the relevance of percolation theory.

Bailey *et al* (2000) developed a percolation model to study invasion and persistence of fungal parasites and saprotrophs in soil. They visualized spreading of harmful fungi as occurring through a population of discrete sites on a lattice which represent discrete nutrient sources, and consist of susceptible roots or discrete fragments of organic matter, with the entire lattice representing the oil in which the disease spread occurs. The lattice can be three dimensional (3D) for nutrient fragments, such as comminuted straw, leaf or root tissue. For lateral spread of disease, however, it can be restricted to a horizontal plane that passes through a population of seeds, roots, hypocotyls, or other organs. Therefore, as Bailey *et al* (2000) discussed, local spread and, hence, progressive invasion by colony expansion depend on the spatial distribution of uncolonized sites. The fungi will spread as long as they can make contact with these sites, in which case they create an expanding patch whose size is limited only by that of the system in which the fungi grow. The spreading will stop, however, if the fungi cannot make contact with new sources of uncolonized hosts or organic fragments, which results in a patch (cluster) of finite size. As our discussions throughout this book should have made clear by now, the crucial role of 'making contacts'—making connected path—is described by percolation theory. In other words, there is a critical distance r_c such that for distancees $r < r_c$ invasive spread of the pathogen amongst a population of nutrient sites can occur, whereas for distances $r > r_c$ the spread of the pathogen is finite. Clearly, the critical distance r_c corresponds to a percolation threshold.

Thus, Bailey *et al* (2000) developed a percolation model for describing saprotrophic growth of a soil-borne fungal plant pathogen, *Rhizoctonia (R.) solani*, which spreads in a finite system in a population of nutrient sites organized on a triangular lattice. They could predict the existence of a threshold probability of colonization between sites. If the probability of colonizing a neighboring site is above the threshold, the fungus can spread invasively and creates large patches (clusters), but the growth below the threshold is finite and is restricted to small patches.

To produce experimental data, small spots of agar of 1 percent (low nutrient) or 10 percent (high nutrient) (weight/volume) potato dextrose agar (PDA) were used to provide reproducible substrates at each site on a structure corresponding to a triangular lattice, with no significant decay during the course of the experiment. Pairs of agar sites that consisted of a donor and recipient were positioned at 4, 6, 8, 10, 12, 14, 16, 18, 20, or 22 mm apart (from center to center) in Petri plates. Fifteen replica of each distance were generated for completely randomized design, and the experiment was repeated for low and high nutrient sites. The donor site of each agar pair was inoculated with a single hyphal strand, *c.*, 1 mm in length, removed from the growing edge of a four-day old colony of *R. solani* Kühn, grown on water agar. The plates were assessed daily for 20 days for colonization of the recipient sites.

Using their experimental data, Bailey *et al* (2000) constructed a probability profile, which describes the change in probability of colonization with distance between a donor (colonized) and a recipient (uncolonized) site, and used percolation threshold p_c to estimate the critical distance r_c. One example is shown in figure 12.2. To test the model, they also used the known r_c from their experiments to estimate p_c. As figure 12.2 indicates, the probability p of colonization decays sigmoidally as the distance r between donor and recipient sites increases. Threshold distances r_c for bond percolation on the triangular lattice (for which, $p_c \simeq 0.35$) were obtained by inverse prediction by fitting the data to a functional form, which was done by maximum likelihood with the assumption that the errors were binomial. As figure 12.2 also indicates, increasing the nutrient status of the agar resulted in a steeper profile, shifted to the right and an increase in the threshold distance between sites from $r_c = 8.1 \pm 0.49$ mm for low nutrient sites, to $r_c = 11.8 \pm 0.93$ mm for sites with a high nutrient status.

Perhaps, the most important insight that percolation theory provided in this case was the fact that one can distinguish between invasive (sample-spanning) and noninvasive (finite) saprotrophic spread of *R. solani*. Moreover, percolation theory provided the tool to predict how changing the distance between uncolonized sites affects the colonization of organic matter. Colonization profiles that summarize the probability of transmission of the fungus with distance between donor and recipient sites identified a threshold distance r_c for either invasive or noninvasive spread of the

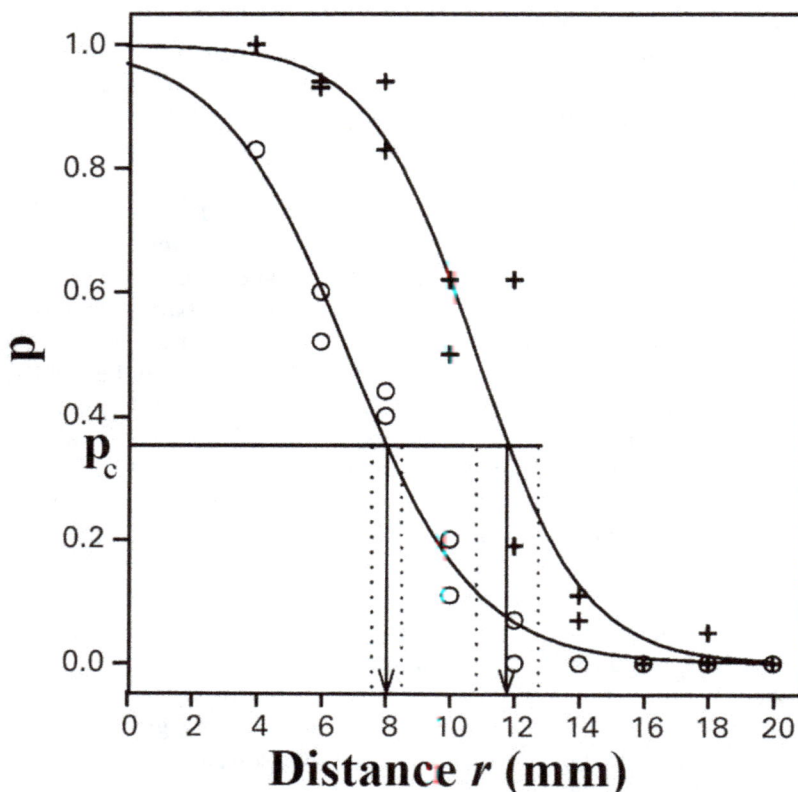

Figure 12.2. Dependence of the probability p of colonization of *Rhizoctonia solani* on the distance r between donor and recipient sites with low [1% potato dextrose agar (PDA), open circles] and high (10% PDA, crosses) nutrient agar. The experimental data (symbols) were fitted with the functions, $p = 1/\{\exp[0.51(r - 6.85)]\}$ and $p = 1/\{\exp[0.61(r - 10.83)]\}$, respectively, in order to provide profiles for the probability p of colonization with distance. Dotted vertical lines represent 95% confidence intervals about the estimated threshold distance r_c for invasion (reproduced from Sahimi 2023; originally from Bailey *et al* 2000).

fungus in a population of agar sites. Invasive spread did not depend on the furthest extent of mycelial growth, evident in the tails of colonization profiles from the placement experiments, but, instead, it was associated with a threshold distance r_c that was different for low and high nutrient sites, but corresponded to a common percolation threshold, $p_c \simeq 0.35$ for the triangular lattice.

12.4 Pattern of tropical deforestation

The significance of tropical forests to the global carbon cycle cannot be overstated. More than half of the known species worldwide reside in these forests, and have been studied intensively for decades. The rapid increase in agriculture, logging, and urban growth, as well as climate change that interrupts the water cycle brought about by severe draught, have, however, led to unprecedented losses of tropical forest, with the average annual deforestation rates since the 1990s being around 0.5 percent. Deforestation rates are not the same around the globe, with Asia and particularly Brazil being currently the hot spots. As the extent of forests decreases, it gives rise to fragmentation in which patches of forest are split into several smaller ones. The sizes of the fragments are not obviously the same and, therefore, there is a distribution of fragment sizes.

Taubert *et al* (2018) used percolation theory to analyze the current forest fragmentation structures in tropical and subtropical regions of the Americas, Africa, Asia, and Australia (denoted here as Asia–Australia) on a high-resolution forest cover map with approximately 21 billion pixels, each 30×30 m^2. For each continent, they used a clustering algorithm to count and analyze the size and perimeter distribution of all detectable forest fragments separately, Taubert *et al* (2018) identified more than 130 million forest fragments across all continents, with their sizes ranging over eleven orders of magnitude and reaching 427 million hectares (Mha). The largest forest fragment in south America, located in the Amazon, spans around 45 percent of its total forest area, whereas the largest fragment on Borneo in Asia covers only 18 percent of the forest.

Analysis of the data by Taubert *et al* (2018) indicated that the fragment-size distribution, i.e., the number n_s of fragments of size s, follows the power law of percolation for cluster-size distribution, i.e.,

$$n_s \sim s^{-\tau}. \tag{12.4}$$

Figure 12.3 presents their results for three regions around the globe. The exact value of τ in 2D is, $\tau = 187/91 \simeq 2.05$. Taubert *et al*'s analysis yielded $1.9 \leqslant \tau \leqslant 1.98$, very close to the exact value. In addition, the fractal dimension D_f of the 2D

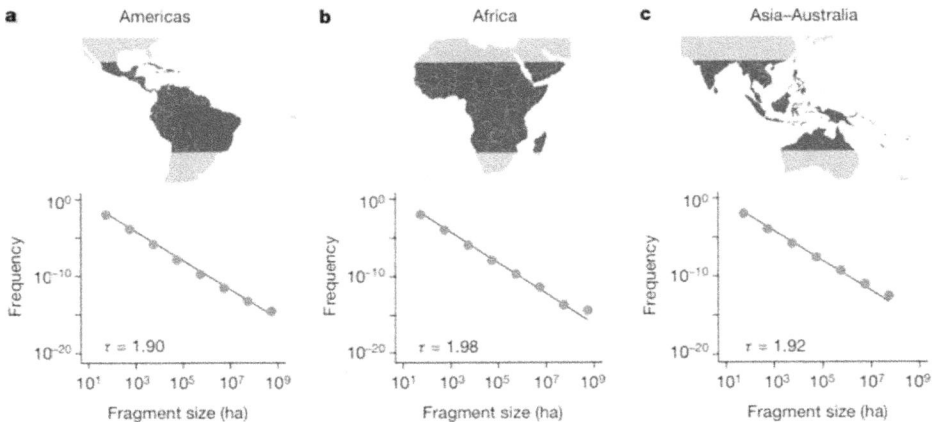

Figure 12.3. Continental–scale fragment size distribution of tropical and subtropical forests, showing the observed forest fragment size distribution with green dots indicating fragment sizes $\geqslant 10$ ha for the Americas (a, with 55.5 million fragments), Africa (b, with 44.8 million fragments) and Asia–Australia (c, with 30.5 million fragments). Solid lines represent the fits of power-law distributions with exponent τ. The maps show the entire land area of the tropical belt, and indicate the selected tropical regions in green (reproduced from Sahimi 2023; originally from Taubert *et al* 2018).

percolation clusters (fragments in the present context) is given by, $D_f = 2/(\tau - 1) = 91/48$, while the analysis by Taubert $et\ al$ (2018) yielded, $D_f \simeq 1.87$, for Asia–Australia, and $D_f \simeq 1.92$ for the Americas, again in excellent agreement with 2D percolation. Finally, the exponent τ_p (the analog of τ) for the perimeter of the fragments (Ziff 1986) was also estimated. The results were $\tau_p \simeq 2.16$, 2.21, and 2.23 for, respectively, the Americas, Asia–Australia, and Africa, to be compared with the exact value in 2D, $\tau_p = 15/7 \simeq 2.14$.

These results suggest that the tropical forest fragmentation is a critical phenomenon, occurring at or near a percolation threshold. In order to explain why this should be, Taubert $et\ al$ (2018) developed a forest fragmentation model, dubbed FRAG. A landscape with C_{max} cells of size s (30×30 m^2) was used. The cells were either forested or deforested. The simulation would begin with a fully forested area, i.e., all the cells were in the 'forest' state, with the total forest area (landscape area) being, $A_{max} = C_{max}s$. In each time step, taken to be one year, some forest area was cleared by a constant deforestation rate d_F in units of percentage per year, so that a certain number of forest cells, $d_F C_{max}$ were selected randomly and converted to a deforested area.

For simplicity, cells could not grow trees and go back to a forest state, so that the forest area was successively reduced over time, until it was cleared in its entirety. Note that deforestation rates used in the model could also be interpreted as net deforestation rates that result from reforestation and gross deforestation that occur at random sites. The results were then analyzed using the same methods that were used for the analysis of the high-resolution forest cover map in terms of remaining forest area, fragment numbers, and their mean size, as well as the fragment size distribution. Dynamics of fragment numbers scaled with landscape size C_{max}. Thus, normalized fragment numbers were computed by dividing the absolute number of fragments by C_{max}. The results reproduced those of the empirical data.

12.5 Forest fires

Forest fires are a worldwide phenomenon. In California, for example, forest fires occur for a variety of reasons, ranging from sparks by defective electric transmission lines, to natural causes such as lightning, and man-made fires (see https://www.fire.ca.gov/stats-events/). Spread of fire in a foreest belongs to a general class of problems in which a front propagates in a heterogeneous medium, similar to some of the phenomena associated with various diseases that were described above. Due to the destructive power of forest fires, their properties have been carefully monitored and recorded and, therefore, there are ample data in the literature.

The idea that if a tree in a forest is set on fire, it could spread to its nearest-neighbor tress and, then, depending on a variety of factors, spread throughout the forest, hints at the possibility of modeling forest fires as a critical phenomenon, and in particular percolation. Such factors include supply of adequate oxygen, wind velocity, land's topography, age and type of trees, and, of course, rainfall. The first step toward such goal was taken, to our knowledge, by MacKay and Jan (1984). They represented a forest by a densely packed triangular lattice, i.e., one in which all the sites were occupied by identical trees, and the spread of fire was a localized surface phenomenon, implying that a burning tree was able to ignite only its nearest neighbours.

In the model the fire begins with the central site ablaze at 'time' $t = 0$. The propagation of the fire to neighbouring 'warm' trees would occur with probability p, leading to four types of sites at any given time t, namely, (a) sites with burnt trees; (b) sites with burning trees; (c) warm trees (unburnt trees that are nearest neighbours to burning trees), and (d) the remaining trees (sites). At each time step of the simulation the propagation of the fire from ignited or burning trees to warm trees would occur with probability p. At the end of the time step all the previously burning trees are considered burnt (dead), and the warm trees that were ignited would be the new burning centres that create a new set of warm trees, which may include unignited warm trees from the previous time steps. The process stops when the system is 'exhausted,' meaning that there are no further ignited trees, or the fire reaches the edge of the forest (lattice). Note that there is a trapping effect in the model in that, an unburnt region surrounded by burnt or vacant sites cannot be ignited.

Others refined the model proposed by MacKay and Jan (1984); see for example, von Niessen and Blumen (1986) and Ohtsuki and Keyes (1986) who, in the presence of a wind during fire, related the problem to directed percolation. In particular, Albinet et al (1986) studied a model in which a fraction p of the sites of a lattice, either square or triangular, were combustible trees, while the rest were not. The combustible trees were characterized by two parameters, τ_i and τ_b, and by the type of interaction between the sites. τ_b was the number of time steps that an ignited site remained in the burning state and was, thus, a measure of the heat released by a site. τ_i, on the other hand, was a measure of the heat required to ignite a site and corresponded to the number of time steps required to ignite a site, if one site in interaction with it was burning. Every site could interact with up to its fifth neighbor.

At time $t = 0$ all the combustible sites in the first two rows were ignited. At each successive time step the heat transferred from each burning site to its neighbors was calculated from the type of interaction (i.e., in terms of the number of neighbors interacting with a given neighbor). A search was then carried out to see if new sites received enough heat to ignite, and whether burning sites have burned long enough to become extinguished. The process was continued until either (a) a site on the last row was ignited, or (b) all sites were extinguished, and the system was exhausted. Therefore, the model was essentially a site percolation problem with three additional features, namely, the two parameters τ_i and τ_b, and the interaction of any site with other sites that are farther than nearest neighbors.

In the simulation the lattice was scanned from left to right and row by row, and every time a burning site was identified, the heat transferred was computed and the affected sites were ignited, or extinguished, as necessary. One time step was equivalent to sweeping the lattice once, and in the limiting case of a densely populated lattice with $\tau_i = \tau_b = 1$, the fire front traversed the lattice in one time step. The critical densities, or percolation thresholds, were determined by using the algorithm. They depend, of course, on the range of interactions between a site and its neighbors.

As always, one must define an order parameter $P(p)$. Albinet *et al* (1986) defined the order parameter by

$$P(n,\, t,\, p) = \lim_{n,t \gg 1, n \leqslant L} \frac{\text{number of burnt trees in the } n\text{th row}}{pL}, \qquad (12.5)$$

for a lattice of size $L \times L$. This definition is essentially the same as the order parameter in classic percolation theory, which is defined by the fraction of sites belonging to the sample-spanning percolation cluster. If one defines a characteristic time, $\theta \propto [(p - p_c)/p_c]^{-\tau}$, then, it is straightforward to show that near the percolation threshold, the order parameter follows the following scaling law,

$$P(n,\, t,\, p) = n^{-\beta/\nu} f(n/\xi,\, t/\theta), \qquad (12.6)$$

where β and ν are the usual exponents for the order parameter and the correlation length ξ, and $f(x,\, y)$ is the scaling function. At p_c, one must have

$$P(n,\, t,\, p_c) = n^{-\beta/\nu} f_1(tn^{-\tau/\nu}), \qquad (12.7)$$

with $f_1(z)$ being another scaling function related to $f(x,\, y)$. This implies that the mean propagation speed n—the number of rows to which the fire has spread—must scale as

$$n \propto t^{\nu/\tau}. \qquad (12.8)$$

Note that the exponent ν/τ is a measure of the acceleration, or deceleration, of the fire front. The total number $N_b(t,\, p)$ of burnt trees is given by

$$N_b(t,\, p) = pL \int_n P(n,\, t,\, p) dn \propto t^{(\nu-\beta)/\tau}. \qquad (12.9)$$

The calculations by Albinet *et al* yielded $1.2 < \nu < 1.4$; recall that in 2D, $\nu = 4/3$.

Study of the mean speed $n(t)$ of fire propagation at $p = p_c$ indicated that one can discern three regions. In region I, the system is in a transitory state where the ignition of the entire first two rows ignites quick advance of the fire front, as the system 'remembers' the initial conditions. In region II one has a steady propagation in which the above scaling laws are valid. In region III the finite size of the system is the prevailing factor, so that fire propagation is restricted by the border of the forest, decreasing the speed of the mean position of the front.

Calculations of Albinet *et al* (1986), yielded, $\tau/\nu \approx 1.15$, so that the mean propagation speed scales as, $n(t) \sim t^{0.87}$. The sample-spanning percolation cluster of the burnt trees also had a fractal structure with a fractal dimension $D_f \approx 1.89$, in agreement with the exact value, $D_f = 91/48$. The length of the fire front—the number F of burned sites that are in contact with unburnt sites when the fire has just penetrated the forest—was also found to scale with L as, $F \propto L^{D_p}$, with $D_p \approx 1.8$.

Beer and Enting (1990) carried out actual laboratory experiments in order to test the predictions of the model by Albinet *et al* (1986). Their estimates for the exponents ν/τ and $(\nu - \beta)/\tau$ did not agree with what Albinet *et al* (1986) had reported. In particular, they noted that the estimate of ν/τ increases as the size of the burnt area does. The reason, as discussed by Beer and Enting (1990), is that the geometry of flame radiation is such that ν/τ increases as the curvature increases. Thus, as the size of the forest increases, large clusters can exist, hence larger scales and larger curvatures come into play. The conclusion was that, while percolation theory is certainly relevant, a simple site or bond percolation model may not be adequate to explain all aspects of forest fires. Perhaps, given that even a quiescent fire generates its own winds, its presence affects the outcome of the model and, in particular, universality class of the exponents. The presence of wind implies that directionality is an important factor and, thus, requires the use of directed percolation, as attempted by Ohtsuki and Keyes (1986).

Rabinovich *et al* (2002) suggested the conditions under which percolation theory is relevant to combustion of a heterogeneous medium, including forests. If the characteristic scale of combustion wave is smaller than the sizes of structural elements of the reacting system, one will have a percolation combustion regime in which the process has a form of successive ignition and burning of connected combustible elements that constitute a cluster spanning the opposite boundaries of the system. Their study indicated that the percolation regimes are possible if, (a) each element burns independently, meaning that the ratio of the combustion zone scale and the size of the burning elements should be less than unity, and the heat exchange between particles should be hampered; (b) the heat losses should be close to the extinction limit, and (c) the percentage of combustible elements should be in the vicinity of the percolation threshold. See also Grinchuck and Rabinovich (2004) who studied combustion of a heterogeneous mixture, and investigated the limit in which percolation theory is relevant, as well as Lam *et al* (2020) who used numerical simulations to study flame propagation in discrete particulate clouds, and linked the phenomenon to percolation.

References

Albinet G, Searby G and Stauffer D 1986 Fire propagation in a 2-D random medium *J. Phys.* **47** 1

Bailey D J, Otten W and Gilligan C A 2000 Saprotrophic invasion by the soil-borne fungal plant pathogen *Rhizoctonia solani* and percolation thresholds *New Phytol.* **146** 535

Beer T and Enting I G 1990 Fire spread and percolation modeling *Math. Comput. Model.* **13** 77

Grinchuk P and Rabinovich O 2004 Percolation phase transition in combustion of heterogeneous mixtures *Shock Waves* **40** 408

Lam F Y K, Mi X C and Higgins A J 2020 Dimensional scaling of flame propagation in discrete particulate clouds *Combust. Theor. Model.* **24** 486

MacKay G and Jan N 1984 Forest fires as critical phenomena *J. Phys.* A **17** L757

Mundt C C and Sackett K E 2012 Spatial scaling relationships for spread of disease caused by a wind-dispersed plant pathogen *Ecosphere* **3** art24

Ohtsuki T and Keyes T 1986 Biased percolation: forest fires with wind *J. Phys.* A **19** L281

Rabinovich S, Grinchuk P, Khina B B and Belyaev A V 2002 Percolation combustion: is it possible in SHS? *Int. J. Self-Propag. High-Temp. Synth.* **11** 257

Ramírez J E, Molina-Gayosso E, Lozada-Lechuga J, Flores-Rojas L M, Martínez M I and Fernández Téllez A 2018 Percolation strategy to improve the production of plants with high pathogene susceptibility *Phys. Rev.* E **98** 062409

Sahimi M 2023 *Applications of Percolation Theory* 2nd edn (Berlin: Springer)

Shaw M W 1994 Modeling stochastic processes in plant pathology *Annu. Rev. Phytopathol.* **32** 523

Taubert F, Fischer R, Groeneveld J, Lehmann S, Müller M S, Rödig E, Wiegand T and Huth A 2018 Global patterns of tropical forest fragmentation *Nature* **554** 519

von Niessen W and Blumen A 1986 Dynamics of forest fires as a directed percolation model *J. Phys.* A **19** L289

Ziff R M 1986 Test of scaling exponents for percolation-cluster perimeters *Phys. Rev. Lett.* **56** 545

IOP Publishing

Networks on Networks (Second Edition)
Role of connectivity in physics of geobiology and geochemistry
Allen G Hunt and Muhammad Sahimi

Chapter 13

Edaphic constraints: revisiting the gaia hypothesis

13.1 Background

In his book, *Gaia: A New Look at Life on Earth*, James Lovelock (1972) proposed that Earth's biosphere is a global-scale self-regulating organism. Evidence for Earth being a life-supporting system ('Homeostasis by and for the biosphere') is provided by the concentrations of CH_4, O_2, and CO_2 in Earth's atmosphere, which are more than 30 orders of magnitude out of their equilibrium values, in contrast to the atmospheric chemistry of Earth's nearest-neighbor planets that are in chemical equilibrium (Lovelock and Margulis 1974). Kleidon (2002) and Lenton (2002) refer to bounded fluctuations in Earth's mean temperature over the past ca. half billion years, and to models indicating that atmospheric and soil moisture conditions produced by life increase plant productivity by 250%. The original Gaia hypothesis has, however, been mostly abandoned (with reasons summarized by Schneider 1986, 2002), even though the concept that the biosphere is composed of interacting complex systems exhibiting emergent behavior (Margulis 1999), is rather generally accepted. Indeed, the history of Earth's climate system is taught within the discipline of Earth system science.

The need to cast Gaia as a planetary-scale organism traced originally to a result from biology in which organisms profit only from regulating their internal environment, whereas the external environment predominantly influences the organism (Kirchner 2002). In this way, however, the concept of homeostasis as a biosphere characteristic was also questioned for smaller scales associated with elements of the biosphere. Kirchner (1989) also pointed out the lack of testable predictions, as well as the difficulty to reconcile a directional regulator for reducing temperatures and atmospheric CO_2, from an early hot Earth that can also stabilize CO_2 at an optimal value for life (Kirchner 2002). Despite such shortcomings, the 'Gaia hypothesis' has contributed to progress in thinking about Earth (Gillon 2000, Grinspoon 2017).

doi:10.1088/978-0-7503-5698-5ch13
13-1

Moreover, we suggest in this chapter that it may be possible to verify the success of a previously made holistic prediction. From Lovelock's perspective (Ball 2014), the biosphere's ability to regulate Earth's climate to its advantage is not a coincidence (Waltham 2014), as is also reflected by a steadily increasing body of evidence connecting the biosphere and the physical planet Earth (see a bibliography compiled by Brig Klyce (2011)).

Similar to the previous chapters, the theory described in this chapter is based on applying statistical mechanics of heterogeneous media, and in particular percolation theory (Sahimi 2023) to predict a time scale for the formation of a continental-scale 'organism' (Hunt and Manzoni 2016). Although already derived in the previous chapters, the potential relevance of our power laws that were described in the previous chapters to Paleozoic ice ages is only now proposed. The subsequent discussion resolves some criticisms of the 'Gaia hypothesis.'

In the perspective presented in this chapter, biological innovations based on photosynthesis (bacteria and plants) that allow better exploitation of the atmosphere's carbon pool are viewed as (cooling) shocks to the system, leading to storage of energy derived from the Sun (Kleidon 2002), while the overall soil ecosystem, including bacteria, animals, and fungi can, by consuming some of that energy, respond so as to promote homeostasis (Bromhall 2019). Global temperature is then regulated not solely by the rate of injection of oxygen into the atmosphere through photosynthesis, but in tandem with the rate at which it is removed through metabolism. Since soil-based bacterial and fungal adaptation to new plant strategies, if given by the fundamental scaling functions we described in the previous chapters and used here, are slower than the relatively rapid spread of plants across the land through release of spores through the atmosphere, achievement of homeostasis is delayed.

The analysis presented in this chapter is based on the scaling relationship for plant growth (and fungal hyphae), which was described in the previous chapters, and was tested on pines (*eucalypts*) along a precipitation gradient of factor 4 (20) (Hunt *et al* 2020) and over time scales (Hunt and Manzoni 2016) from minutes to 100 000 years, while being used to 'extrapolate to a time scale when an "organism" with an optimal, hierarchical, structure would reach continental size (about 5000 km, if growing from the center). That time scale is less than 100 Ma (about 80 Ma).' The application to Earth's history was not appropriate: 'We will find that the time scales are adequate [short enough] if a symbiotic combination of plants and bacteria are envisioned. However, utilizing land plants in such a symbiosis would postpone the time of the development of a global "organism" to a date billions of years after the far-from-equilibrium composition of the atmosphere was obtained.' We show in this chapter that 80 Ma is very nearly an appropriate time span in a distinct context. We consider restoration of the Earth to higher temperatures subsequent to Paleozoic ice ages, triggered by colonization of the land by plants.

.2 A percolation model

ı any integrative sense, the existence of a global- or at least continental-scale organism'—think interacting ecosystems (Margulis 1999)—requires some means of communication over such a scale. If the communication is tied to the growth of plants and fungi, the assessment of controlled homeostasis should relate to the prediction for increase in root lateral spread (RLS) ℓ of plants/fungi as a function of time and water flow rates in the subsurface. As discussed in the previous chapters, roots within soil were proposed to follow paths of minimal cumulative resistance (Hunt and Manzoni 2016, Hunt 2017, Hunt and Sahimi 2017a). In a disordered pore network, such as soil, where local (pore-scale) resistances to movement or flow are broadly distributed, such a path becomes long and tortuous and may be described using fractal geometry (Porto *et al* 1998, Hunt and Sahimi 2017a). Then, the RLS relates to root length L through a power law, $L \approx \ell^\delta$. The exponent δ is the same as the fractal dimension D_{op} of the optimal paths in heterogeneous media described in chapter 4, which reflects confinement of plant root ecosystems globally to a very thin soil layer, with a universal value of 1.21 in two dimensions.

To describe the theory, its essential elements are summarized that were described in the previous chapters, which we use to make predictions for the RLS ('biological transport') of plants. We begin with

$$x = x_0 \left(\frac{t}{t_0} \right)^{1/D_{op}},$$
(13.1)

where $D_{op} \simeq 1.21$ is provided by percolation theory, valid when the paths are confined to a two-dimensional (2D) layer, such as the soil layer. To predict soil depth ('physical transport'), however, one may use the following equation (Hunt and Manzoni 2016, Hunt *et al* 2021a, Yu and Hunt 2017a, Yu *et al* 2019), which was also derived in the previous chapters

$$x = x_0 \left(\frac{t}{t_0} \right)^{1/D_{bb}},$$
(13.2)

with $D_{bb} \simeq 1.87$ bringing the fractal dimension of the percolation backbone (the flow-carrying part of the network) in 3D. On the other hand, although the flow, which supports soil development, is mainly vertical, the connectivity of the dominant flow paths is 3D. As discussed in chapter 5, solute transport limits chemical weathering and soil depths under nearly all natural conditions. Thus, to predict flow and crop heights, one should use the following equation that was already described in chapter 9:

$$x = x_0 \left(\frac{t}{t_0} \right),$$
(13.3)

which represents a scale-independent flow velocity. Here, water and nutrients are brought to the plants, eliminating constraints on root growth from the need to search for nutrients that are heterogeneously distributed within the soil. Note that, as discussed in the previous chapters, to make the proportionality useful for predicting plant growth as a function of time t, the scaling relationships require a fundamental length scale x_0 and a time scale t_0, which were given in chapter 9:

$$t = t_0 \left(\frac{x}{x_0} \right)^{D_{op}},$$

(13.4)

where the 2D value of D_{op} should be used, while identifying the RLS ℓ with the length x, and the ratio t/t_0 as the number of pores the root has grown through, which corresponds to root length L in multiples of x_0.

13.3 The data

The theory is tested by comparing its predictions with experimental data, the individual sources of which were mentioned in the previous chapters. The data have been reported for three categories, namely, crop height, plant height (or root lateral spread), and soil depth. Since the characteristics of natural vegetation growth are postulated to be governed by the need to find nutrients and/or water within heterogeneous soil, plants that are heavily fertilized and watered are considered separately and referred to as 'crops.' Their heights apply mostly to annual crops, such as beans, peas, corn, hemp, tomatoes, sunflowers, tobacco, wheat, and amaranth. When trees in plantations are heavily nourished, however, such as in *Eucalyptus* plantations, these are also treated as crops. In the case of short-term laboratory measurements of root tip extension rates, distinctions between crops and natural vegetation are made on the basis of the descriptions of the environment. When the environment had either too little water or too much salt, for example, such experiments are included within natural vegetation. Plants grown under conditions described as ideal are considered again as crops. The biomass allometry database (BAAD) (Falster *et al* 2015) with over 6000 entries for plant height was also divided in this fashion, with the relatively small database for fertilized *Eucalyptus* plantations added to the crops category, extending that dataset to more than two years. Climate, rather than temporal, effects on growth rates were inferred from, for example, dominant tree heights of *Eucalyptus regnans* along a climate gradient in Australia (Hunt *et al* 2020), which varied from 4 m to 88 m. The depth to the bottom of the B (commonly called the subsoil) or Bw horizon (horizons that have been changed by weathering), is considered to be the bottom of the soil (Hunt and Manzoni 2016, Yu and Hunt 2017b). Egli *et al* (2018) incorporated also the BC-horizon (representing the transition from the B to the C horizon, with the latter being the weathered, unconsolidated geologic material below the A or B horizon) by adding a percentage of its thickness to the total soil thickness. Data for soil depth from deep time (10 Ma and up) were described as 'deep tropical weathering,' with specific labels, such as 'laterite' and 'saprolite,' although climate regimes for some sources were humid temperate continental.

For the maximum crop height, natural vegetation RLS, and soil depth, the same values of x_0 and t_0 are used. Thus, the same network structure and flow properties are considered to govern all the above three scaling relationships, with the exponents used being universal and provided by percolation theory. The assumption throughout is that the association of local pores into dominant transport paths is described by percolation theory for identifying the paths of least resistance. The predictions are made without using any unknown/adjustable parameters.

13.4 Quantitative test of Gaia hypothesis using networks

Originally, x_0, taken to be, $x_0 \approx 1$ μm was estimated roughly as a fundamental pore scale, while the time scale t_0 was taken to be the time for water to flow across a 1 μm pore at a typical subsurface flow rate (Blöschl and Sivapalan 1995), 1 μm s^{-1} ($t_0 = 1$ μ m/1 μm s$^{-1} = 1$ s). This prediction is given by the dashed red line in figure 13.1 and compared with actual woody plant data. For lengths on the order of centimeters or shorter, the data reflect laboratory root tip or fungal hyphae extension rates. On length scales exceeding 120 m, data reflect the RLS of single organisms (clones) with multiple subaerial stems, both plants and fungi, while at intermediate length scales, the data are for vegetation height, which is known to be nearly equivalent to RLS on length scales between about 0.5 m and 40 m. For $x > 10$ km, neither the fundamental character of what could be called an 'organism,' nor its form of communication, are specified.

Figure 13.1 indicates near conformance of equation (13.4) with the upper bounds of the RLS data on time scales ranging from minutes to 100 ka, and length scales from 100 μm to 10 km. Using the same line, the time for a continental-scale (5000 km) 'organism' to develop is estimated to be 80 Ma. Although the slope of the line is universal, the value of the time scale required for an 'organism' of 5000 km changes, if the values of the parameters in equation (13.4) are different. As discussed in the previous chapters, x_0 was later on defined explicitly as a median particle size or plant xylem diameter. Then, if x_0 is not known explicitly, a better fundamental length scale is 10–30 μm, because this range is both near the middle of the silt particle size range and a geometric mean plant xylem diameter (Watt *et al* 2006). Since most variability in tree heights at a given time is linked with their actual transpiration rates, more slowly growing vegetation requires much less water. The earliest establishment of an 'organism' of continental size would then be associated with the fastest growth rates along the solid red line in figure 13.1.

Hunt and Manzoni (2016) proposed that a continental-scale 'organism' could consist of a synergistically operating system of plant roots and associated bacteria. Given that fungi follow the same growth model as plants and act to decompose wood, they may be included as well. Bacteria have recently been shown to act collectively by, for example, deferring consumption until better food sources are located, communicating at relatively short length scales through chemical 'quorum sensing,' and at longer length scales through electrical signaling along ion channels (Lyon *et al* 2021) [note that microbial communications and quorum sensing have been modeled as percolation processes (Larkin *et al* 2018); see Sahimi 2023]. Perhaps horizontal gene transfer serves at yet larger scales. We propose the possibility that a soil ecosystem that can produce a negative climatic feedback of continental size can

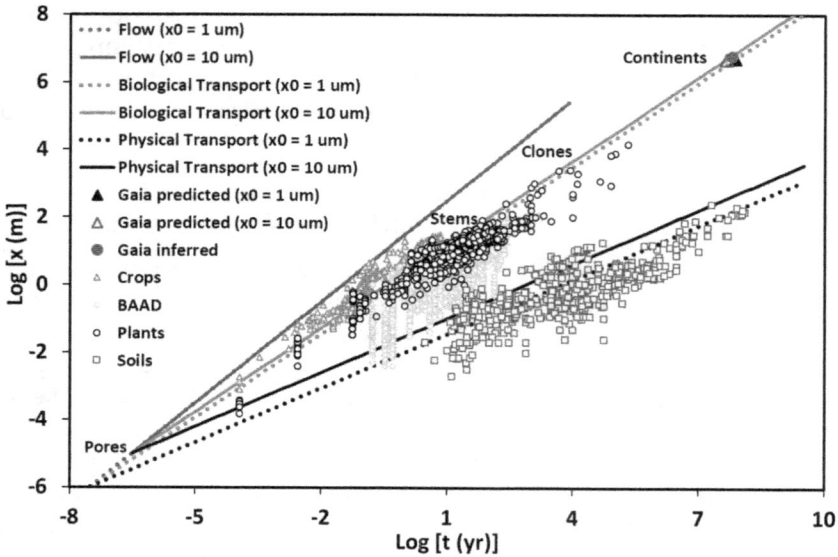

Figure 13.1. Application of the scaling relationships to biological and physical transport processes, assuming a flow speed independent of scale (blue) with its value taken from Blöschl and Sivapalan (1995). Root growth rates (red) decline over time according to equation (13.4) and the fractal dimension D_{op} of 2D optimal paths that percolation theory predicts, soil formation rates (black) decline over time according to the scaling of solute transport using the fractal dimension of the percolation backbone. BAAD refers to the biometric and allometric database for plant heights, 'plants' are from many sources, selected for faster growth, and 'soils' mean are soil depths (data compilation). Dashed lines reflect predicted times for biological organisms to reach a given RLS, and soil to reach a given depth using the originally suggested length scale of 1 μm, while solid lines indicate the changes resulting from choice of 10 μm as the fundamental length scale. Gaia predicted uses the time scale required for equation (13.4) to yield 5000 km, in agreement with the original prediction. 'Gaia inferred' pairs a physical extent of ca. 6200 km, obtained either as half the square root of the area of all the continents today, or the area of Pangaea, with the time for emergence from an ice age (60 Ma).

indeed be predicted through equation (13.4) that we tested at smaller length scales. Note that the span 10 μm to 10 km tests explicitly nine of the necessary 11–12 decades of length scale. In equation (13.4) the time t required for length scale x to reach continental size is required.

Since the initial test of equation (13.4) addressed the time required for drawdown of sufficient CO_2 to produce the great oxygenation crisis (Kopp *et al* 2005), equation (13.4) was judged irrelevant, as land plants arrived over a billion years too late. Producing an ice age through biological innovation, however, is better viewed as a climate crisis generated by a component of the ecosystem, namely, the plants, which robustly overproduce, whereas achievement of homeostasis afterwards would be consistent with establishment of a global-scale, adaptive 'organism,' or a nested system of ecosystems comprising a 'symbiotic planet'(Margulis 1999).

Thus, equation (13.4) is tested by applying it to the lifting of the climate shocks (ice ages) brought on by colonization of the land by plants. It has now become clear that the early colonization of land was by rooted plants (Morris *et al* 2018,

Puttick *et al* 2018), though fossil evidence suggests quite short root systems. Initiated near 500 Ma, the effects of the dramatic increase in photosynthesis on atmospheric CO_2 content appear to have led to large-scale glaciation by 488 Ma (Lenton *et al* 2012). The end of the glacial episode is estimated to be at 440 Ma (Lenton *et al* 2012), 60 million years after initial colonization. Given such a climate shock, return to equilibrium could roughly coincide with development of a 'symbiotic planet.' The prediction that it requires 80 Ma for adaptation to new conditions according to scaling relationship, equation (13.4), is only 33% larger than the actual value. Later colonization of the land by significantly rooted vascular plants at 420 Ma was followed by cooling and a glacial episode lasting approximately from 372 Ma to 359 Ma (Streel *et al* 2000). The time required for re-establishment of ideal environmental temperatures for life was approximately 60 million years since initial colonization, only 25% less than the prediction of equation (13.4). Although we do not here assert that our test constitutes a proof that Earth's biosphere may be identified with a global 'organism,' the predicted and verified time scales were, nonetheless, developed specifically from a strong Gaia hypothesis. Thus, we posit that any Gaia perspective, together with scaling relationships from complexity and percolation theory, may provide added tools for analysis of the biosphere's past, or projected future.

That the scaling relationship, equation (13.4), for plant growth could be extended to planetary scales as a means for understanding Gaia-like adaptation to land colonization by plants would be in accord with Lovelock's comment in a relatively recent interview (Ball 2014) in the journal *Nature*: 'I'm very intrigued by the latest attempt to resuscitate the idea that all of climate regulation is done by rock weathering. The geologists keep on ignoring the bacteria.' The work of Hunt and Manzoni (2016) is entirely consistent with Lovelock's implication. First, the primary dependence of rock weathering rates is on moisture fluxes, rather than temperature, which restricts the efficacy of the rock weathering thermostat to periods of time when the atmospheric moisture and temperature are in phase. One epoch when high temperatures and atmospheric CO_2 content appear not to have been in phase with greater precipitation, was during the great Permian extinction, a likely result of unification of Earth's land masses in Pangaea under relatively arid conditions (Hunt and Sahimi 2017b). Another instance was the termination of the Marinoan Snowball Earth episode at 635 Ma. The disruption of the atmospheric water cycle and reduction in exposed silicates led to a hiatus in weathering and massive build-up of CO_2 (Gan *et al* 2024). Absent higher rates of rock weathering, removal of CO_2 from the atmosphere would have been slow, indeed. Second, although the volume of solid inorganic carbon reservoir is much larger than the organic carbon reservoir, figure 13.1 demonstrates clearly that, particularly changes in its rate of storage, are much slower. Specifically, at 100 ka, the predicted and observed soil/weathering depth is at most a few meters, whereas the predicted and observed laterally integrated biological dimension reaches 10 km. Any pore-scale horizontal advection driven by plants is of a similar magnitude to the vertical advection driven by gravity (Hunt *et al* 2021a). Thus, spread of, and communication between, bacteria in the subsurface can be considerably enhanced in the horizontal directions (relative to vertical) through coupling with plant processes. Moreover, consistent with the original suggestion that the required global 'organism' could be a plant/bacteria soil ecosystem in which plant matter breakdown and production are at steady state, homeostasis occurs when a global balance in CO_2 drawdown and emission is achieved at the right atmospheric composition that a moderation of temperature is achieved. Here, such homeostasis emphasizes the key roles of bacteria and fungi, but allows their spread and adaptation to be controlled by the underground spreading rates of plant roots and fungal hyphae. The scaling relationships, on which this chapter is based thus tend to emphasize the relevance of biota, including bacteria, in developing a large-scale, restorative response, to climate perturbations, in comparison to the abiotic response, which is not even necessarily identified as a negative feedback. However, the negative feedback develops on time scales of approximately 60 Ma, much longer than the time scales at which biosphere feedbacks have been found to be positive.

As described in the previous chapters, the same scaling relationships used in the present chapter also form the basis for an accurate (within 1.5%) prediction of the global fraction of precipitation returned to the atmosphere through terrestrial evapotranspiration. The result is based on maximization of plant productivity using a thin root layer defined by the soil depth, and a global-scale assumption of neither energy nor water limitations. Thus, although a soil ecosystem may promote homeostasis, plants by themselves fit a model of 'greed' with respect to CO_2, taking as much as they can, in order to exploit solar and soil resources as possible to most effectively cover Earth's surface with plant matter (Hunt *et al* 2021b). Taken together, our results imply that negative feedback for climate change is not generated through plant interaction with atmospheric chemistry, in accord with cited evidence (Kirchner 2002) that a modern 25% increase in atmospheric CO_2 content led to only a 2% increase in productivity. Productivity is more directly tied to (evapo-)transpiration. Therefore, an additional criticism of the homeostasis hypothesis is removed.

The question of whether life on our planet tends to promote climate and chemical homeostasis (Gaia), or to cause fluctuations that endanger life (Medea), is of fundamental importance to understanding Earth's history, as well as the future of our species. The question appears, however, to be, as of the time of writing this book, unanswerable. The innovation of photosynthesis was deadly to a great deal of life, but promoted an environment that led to the explosion of multi-cellular life in the Cambrian, together with further crises along the way, from snowball Earth to Paleozoic ice ages. We looked at two of these major climate swings, arising from the conquering of the land by plants, and showed that a theoretical narrative already begun that had produced an explicit prediction for achievement of homeostasis through the linking of ecosystems up to the continental scale, is remarkably accurate. As described in the previous chapters, the same theoretical framework has also recently been applied to generate an accurate, parameter-free, prediction of the global water cycle through application of an ecological hypothesis that plant ecosystems, which are most successful at reproducing themselves (in terms of volume of CO_2 drawn down from the atmosphere) will dominate. Thus, through a kind of conceptual division and synthesis of ecosystems into their parts and their reassembly, we can identify the component—'greedy' photosynthesizers—which has led to climate shocks, and the whole that has restored equilibrium. In particular, we believe that what we described in this chapter regarding recovery from the Paleozoic ice ages, can help point the way to a future in which the opposite sides of the Gaia hypothesis can coexist meaningfully and usefully. Thus, whether the biosphere as a whole is Gaia-like or Medea-like (or neutral) is not put to the test, but a potentially important result and associated perspective are introduced.

References

Ball P 2014 James Lovelock reflects on Gaia's legacy *Nature* https://doi.org/10.1038/nature.2014.15017

Blöschl G and Sivapalan M 1995 Scale issues in hydrological modelling: a review *Hydrol. Process.* **9** 251

Bromhall N 2019 Role of fungi in the world ecology *Encyl. Brit.* available at https://www.britannica.com/video/81498/fungi-matter-molecules-soil-atmosphere

Egli M, Hunt A G, Dahms D, Raab G, Derungs C, Raimondi S and Yu F 2018 Prediction of soil formation as a function of age using the percolation theory approach *Front. Environ. Sci.* **6** 108

Falster D S, Duursma R A, Ishihara M I *et al* 2015 BAAD: a biomass and allometry database for woody plants *Ecology* **96** 1445

Gan T *et al* 2024 Lithium isotope evidence for a plumeworld ocean in the aftermath of the Marinoan snowball Earth *Proc. Natl Acad. Sci.* **121** e2407419121

Gillon J 2000 Feedback on Gaia *Nature* **406** 685

Grinspoon D 2017 It's time to take the Gaia hypothesis seriously *Nautilus* https://nautil.us/its-time-to-take-the-gaia-hypothesis-seriously-5935/

Hunt A G 2017 Spatiotemporal scaling of vegetation growth and soil formation: explicit predictions *Vadose Zone J.* **16** 1

Hunt A G, Faybishenko B A and Ghanbarian B 2021a Non-linear hydrologic organization *Nonlinear Process. Geophys.* **28** 599

Hunt A G, Faybishenko B A and Ghanbarian B 2021b Predicting characteristics of the water cycle from scaling relationships *Water Resour. Res.* **57** e2021WR030808

Hunt A G, Faybishenko B A and Powell T 2020 A new phenomenological model to describe root-soil interactions based on percolation theory *Ecol. Modell.* **433** 109205

Hunt A G and Manzoni S 2016 *Networks on Networks: The Physics of Geobiology and Geochemistry* (San Mateo, CA: Morgan & Claypool)

Hunt A G and Sahimi M 2017a Flow, transport, and reaction in porous media: percolation scaling, critical-path analysis, and effective-medium approximation *Rev. Geophys.* **55** 993

Hunt A G and Sahimi M 2017b The spaces in between *Eos* **98**

Hunt A G, Sahimi M, Faybishenko B, Egli M, Kabala Z J, Ghanbarian B and Yu F 2024 Gaia: complex systems prediction for time to adapt to climate shocks, to be published

Kirchner J W 1989 The Gaia hypothesis: can it be tested? *Rev. Geophys.* **27** 223

Kirchner J W 2002 The Gaia hypothesis: fact, theory, and wishful thinking *Clim. Change* **52** 391

Kleidon A 2002 Testing the effect of life on Earth's functioning: how Gaian is the Earth system? *Clim. Change* **52** 383

Klyce B 2011 Gaia, What's new since 1998? Available at: https://panspermia.org/gaia.htm.

Kopp R E, Kirschvink J L, Hilburn I A and Nash C Z 2005 The Paleoproterozoic snowball Earth: a climate disaster triggered by the development of oxygenic photosynthesis *Proc. Natl Acad. Sci. USA* **102** 11131

Larkin J W *et al* 2018 Signal percolation within a bacterial community *Cell Syst.* **7** 137

Lenton T M 2002 Testing Gaia: the effect of life on Earth's habitability and regulation *Clim. Change* **52** 409

Lenton T M, Crouch M, Johnson M, Pires N and Dolan L 2012 First plants cooled the Ordovician *Nat. Geosci.* **5** 86

Lovelock J E 1972 *Gaia: ANew Look at Life on Earth* (Oxford: Oxford University Press)

Lovelock J E and Margulis L 1974 Atmospheric homeostasis by and for the biosphere: The Gaia hypothesis *Tellus* **26** 1

Lyon P, Keijzer F, Arendt D and Levin M 2021 Basal cognition: Multicellularity, neurons, and the cognitive lens *Philos. Trans. R. Soc.* B **376** 132

Margulis L 1999 *Symbiotic Planet: A New Look At Evolution* (New York: Basic Books)

Morris J L, Puttick M N, Clark J W and Donoghue P C J 2018 The time scale of early land plant evolution *Proc. Natl Acad. Sci. USA* **115** E2274

Pennisi E 2018 Land plants arose earlier than thought, and may have had a bigger impact on animals https://www.science.org/content/article/land-plants-arose-earlier-thought-and-may-have-had-bigger-impact-evolution-animals

Porto M, Havlin S, Roman H E and Bunde A 1998 Probability distribution of the shortest path on the percolation cluster, its backbone, and skeleton *Phys. Rev.* E **58** R5205

Puttick M N *et al* 2018 The interrelationship of land plants and the nature of the ancestral embryophyte *Curr. Biol.* **28** 733

Sahimi M 2023 *Applications of Percolation Theory* 2nd edn (New Yok: Springer)

Schneider S H 1986 A goddess of the earth?: the debate on the Gaia hypothesis—an editorial *Clim. Change* **8** 1–4

Schneider S H 2002 Special theme: the Gaia hypothesis *Clim. Change* **52** 3

Streel M, Caputo M V, Loboziak S and Melo J H G 2000 Late Frasnian-Famennian climates based on palynomorph analyses and the question of the Late Devonian glaciations *Earth Sci. Rev.* **52** 121

Waltham D 2014 *Lucky Planet: Why Earth Is Exceptional–And What That Means for Life in the Universe* (London: Icon Books)

Watt M, Silk W K and Passioura J B 2006 Rates of root and organism growth, soil conditions and temporal and spatial development of the rhizosphere *Ann. Bot.* **97** 839

Yu F and Hunt A G 2017a An examination of the steady-state assumption in soil development models with application to landscape evolution *Earth Surf. Process. Landforms* **42** 2599

Yu F and Hunt A G 2017b Predicting soil formation on the basis of transport-limited chemical weathering *Geomorphology* **301** 21

Yu F, Hunt A G, Egli M and Raab G 2019 Comparison and contrast in soil depth evolution for steady state and stochastic erosion processes: Possible implications for landslide prediction *Geochemistry, Geophys. Geosystems* **20** 2886

www.ingramcontent.com/pod-product-compliance
Lightning Source LLC
Chambersburg PA
CBHW082136210326
41599CB00031B/6000